X-RAY ASTRONOMY WITH THE EINSTEIN SATELLITE

ASTROPHYSICS AND SPACE SCIENCE LIBRARY

A SERIES OF BOOKS ON THE RECENT DEVELOPMENTS
OF SPACE SCIENCE AND OF GENERAL GEOPHYSICS AND ASTROPHYSICS
PUBLISHED IN CONNECTION WITH THE JOURNAL
SPACE SCIENCE REVIEWS

VOLUME 87
PROCEEDINGS

X-RAY ASTRONOMY WITH THE EINSTEIN SATELLITE

PROCEEDINGS OF THE HIGH ENERGY ASTROPHYSICS DIVISION
OF THE AMERICAN ASTRONOMICAL SOCIETY MEETING
ON X-RAY ASTRONOMY
HELD AT THE HARVARD/SMITHSONIAN CENTER FOR ASTROPHYSICS,
CAMBRIDGE, MASSACHUSETTS, U.S.A., JANUARY 28–30, 1980

Edited by

RICCARDO GIACCONI
Harvard/Smithsonian Center for Astrophysics,
Cambridge, Mass., U.S.A.

D. REIDEL PUBLISHING COMPANY
DORDRECHT : HOLLAND / BOSTON : U.S.A.
LONDON : ENGLAND

Library of Congress Cataloging in Publication Data

CIP

Main entry under title:

X-ray astronomy with the Einstein satellite.

(Astrophysics and space science library. Proceedings ; v. 87)

Includes index.

1. X-ray astronomy–Congresses. I. Giacconi, Riccardo. II. American Astronomical Society. High Energy Astrophysics Division. III. Title. Einstein satellite. IV. Series.
QB472.A1X23 522'.686 81-7359

ISBN-13: 978-94-009-8461-5 e-ISBN-13: 978-94-009-8459-2
DOI: 10.1007/978-94-009-8459-2

Published by D. Reidel Publishing Company,
P.O. Box 17, 3300 AA Dordrecht, Holland.

Sold and distributed in the U.S.A. and Canada
by Kluwer Boston Inc.,
190 Old Derby Street, Hingham, MA 02043, U.S.A.

In all other countries, sold and distributed
by Kluwer Academic Publishers Group,
P.O. Box 322, 3300 AH Dordrecht, Holland.

D. Reidel Publishing Company is a member of the Kluwer Group.

TABLE OF CONTENTS

PREFACE

Riccardo Giacconi
Harvard/Smithsonian Center for Astrophysics

The meeting of the High Energy Astrophysics Division of the American Astronomical Society, held in Cambridge, Massachusetts on January 28-30, 1980, marks the coming of age of X-ray astronomy. In the 18 years since the discovery of the first extrasolar X-ray source, Sco X-1, the field has experienced an extremely rapid instrumentation development culminating with the launch on November 13, 1978 of the Einstein Observatory (HEAO-2) which first introduced the use of high resolution imaging telescopes to the study of galactic and extragalactic X-ray sources. The Einstein Observatory instruments can detect sources as faint as 10^{-7} Sco X-1 or about 17 magnitudes fainter.

The technological developments in the field have been paralleled by a host of new discoveries: in the early 1960's the detection of "X-ray stars", objects 10^9 times more luminous in X-rays than the Sun and among the brightest stellar objects at all wavelengths; in the late 1960's and early 1970's the discovery of the nature of such systems which were identified as collapsed stars (neutron stars and black holes) in mass exchange binary systems, and the detection of the first few extragalactic sources. The year 1972 also marked the discovery of X-ray emission from hot, optically thin gas in clusters of galaxies, whose detection was only possible at X-ray wavelength and whose mass equalled the mass of the luminous galaxies forming the clusters; in the mid-1970's the discovery of "bursters", currently believed to represent yet another aspect of the X-ray emission from binary X-ray sources and the study of X-ray emission from the nuclei of Seyfert, quasars and other active galaxies further enriched the X-ray sky.

The exciting discoveries sketched above placed X-ray observations at the forefront of observational relativistic astrophysics. Yet they were curiously decoupled from many of the traditional problems of astronomy dealing with the evolution of stars and galaxies, and with cosmological questions.

With the advent of the Einstein Observatory, X-ray observations have been extended to stars of all types in our own galaxy, to all classes

R. Giacconi (ed.), X-Ray Astronomy with the Einstein Satellite, vii–viii.

of external galaxies and to the most distant objects, clusters and quasars known at any wavelength.

This HEAD/AAS meeting represents the first attempt at an assessment by observational and theoretical astrophysicists of the impact of the new X-ray findings on all of astronomical knowledge.

The discussions ranged from non-degenerate stellar X-ray sources and stellar coronae, to supernovae, "bursters", globular clusters, normal galaxies, active galaxies and QSO's, clusters of galaxies, and finally to cosmology. In each field an attempt was made to bring together the relevant information obtained in radio, optical and X-ray observations in short reviews followed by theoretical discussion. A large number of contributed papers, both observational and theoretical, were also presented within this intellectual framework.

This book is a collection of all invited reviews at that meeting and gives a vivid portrait of the scope and depth of the investigations currently under way. While it is clear that progress in many of these topics will soon make the detailed contributions obsolete, I am confident that the overview presented by these brilliant reviewers will retain its validity for years to come.

Appreciation should go to those members of the Organizing Committee who assisted me in carrying out my responsibilities as Chairman. They are: George W. Clark, Giovanni Fazio, Jonathan Grindlay and Saul Rappaport. Also, the contributions of Polly Sullivan, of my Office, in coordinating and arranging the many details involved in carrying out such a meeting must be acknowledged.

Thanks also must go to those who chaired the Invited sessions, who are: Gordon Garmire, William Kraushaar, Jocelyn Bell-Burnell, George W. Clark, Bruno Rossi, Robert Novick, John Bahcall and Philip Morrison.

THE EINSTEIN/CFA STELLAR SURVEY: OVERVIEW OF THE DATA AND INTERPRETATION OF RESULTS

G. S. Vaiana
Harvard-Smithsonian Center for Astrophysics and
Osservatorio Astronomico di Palermo, Italy

ABSTRACT

We present results from an extensive survey of stellar X-ray emission, using the EINSTEIN Observatory. Over 140 stars have been detected to date, throughout the H-R diagram, thus showing that soft X-ray emission is the norm rather than the exception for stars in general. This finding is strongly at odds with pre-EINSTEIN expectations based on standard acoustic theories of coronal heating. We present typical examples of stellar X-ray detections and provide an overview of the survey data. In combination with recent results from solar X-ray observations, the new EINSTEIN data argue for the general applicability of magnetic field-related coronal heating mechanisms.

1. INTRODUCTION

The new data on X-ray emission from stars which have become available[1] with the launch of the EINSTEIN Observatory provide a new perspective on low luminosity ($L_x < 10^{35}$ erg s^{-1}) galactic X-ray sources. Emission is not an exceptional occurrence within a restricted group of stars, but rather is the norm. The EINSTEIN Observatory stellar surveys, which allow a limiting sensitivity $\sim 10^3$ greater than previous surveys, have demonstrated conclusively the ubiquitous nature of stellar X-ray emission.

In the following, I will present a synopsis of the data based on $\sim$ 140 low-luminosity ($L_x \lesssim 10^{35}$ erg s^{-1}) X-ray sources identified with stars, as part of the Stellar Survey program at the Center for Astrophysics.

Anticipating the major conclusions, we find that X-ray emission from stars appears to be the norm, rather than the exception and that the rather high observed X-ray luminosities are in gross qualitative and quantitative disagreement with respect to canonical theories of coronal formation. The standard theory, following the pioneering work of Biermann (1946) and Schwarzchild (1948) and the observations reported by Wilson (1966), Kraft (1967) and Skumanich (1972) looks at the problem of forming coronae as distinct from the generation of magnetic fields.

R. Giacconi (ed.), X-Ray Astronomy with the Einstein Satellite, 1–18.

In particular, the conventional theory of coronal formation[2] requires vigorous surface convection since the heating occurs by shocked acoustic modes generated by convective turbulence. Such theories make several specific predictions:

(a) Coronal activity should vary with the available acoustic flux; in particular, since the energy going into coronal heating cannot exceed the total from acoustic modes, stellar X-ray emission should be limited to spectral types F to G and should thus be a rare phenomenon;

(b) Since stellar mass loss correlates with the level of coronal activity under the Parker and Hearn types of wind theory, mass loss should vary with the available acoustic flux; and

(c) The presence of coronal activity and magnetic fields are only incidentally related, both owing their existence to the presence of an outer convective zone (and stellar rotation) in the case of the magnetic dynamo). X-ray emission, being tied primarily to convective activity, would depend mainly on spectral type and luminosity class and be relatively independent of such parameters as stellar rotation.

I will argue that the new data clearly indicate that the level of coronal emission does not primarily correlate with the level of convection and that the coronal heating mechanism is not likely to be acoustic. We have good indications that rotation plays a significant role and we suggest that the observed modulations of this pervasive coronal phenomenon are likely to be a reflection of the modulation of emerging surface magnetic flux by the stellar magnetic dynamo, and the interaction between emerged flux and surface turbulence.

In the following, I will first describe briefly the pre-EINSTEIN status of stellar X-ray results; second, I will review the various components of the survey and the context of this research; next, I will present the data and focus on the discrepancies with existing theories. At the end, I will briefly attempt to sketch the requirements for and the suggested ingredients of a possible new coronal scenario.

2. BRIEF REVIEW OF PRE-EINSTEIN RESULTS

Before the advent of EINSTEIN, there were already indications from both solar and stellar observations that the standard coronal scenario had difficulties. A foreshadowing of difficulty was already found in Ca II emission line data from "active" stars. Thus, Blanco et al. (1974) had in detail compared Ca II h and k line fluxes of "active" stars with predictions of chromospheric acoustic heating by de Loore (1970), and had found strong disagreement (theory predicting far lower emission levels than actually encountered); and Mullan (1975) had proposed a possible resolution based upon enhanced Alfven wave heating above star "spots" in dMe stars.[3] This suggestion, that enhanced stellar chromospheric activity and surface magnetic activity were coupled, of course dates back to the earlier work of Wilson (1966) and Skumanich (1972) (Fig. 1), and finds its ultimate roots in much earlier, classical work relating solar plages to surface magnetic fields (cf. Jordan 1973 for

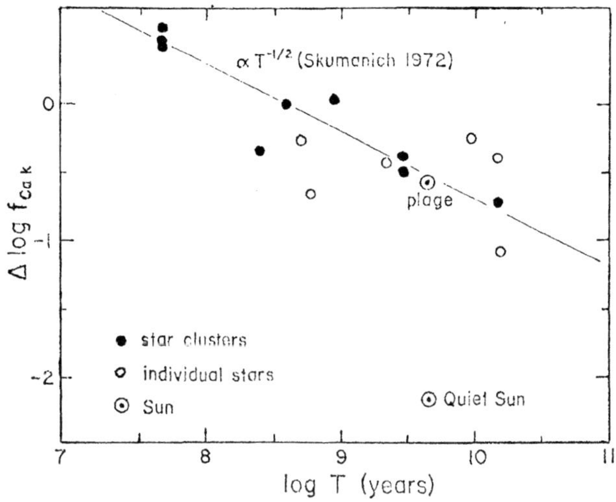

Figure 1. Ca II k emission line fluxes for stellar groups and individual stars as a function of age, normalized to the Hyades main sequence (after Blanco et al. 1974). Also shown are the corresponding fluxes for solar plages and the Quiet Sun, and the Ca II - age scaling law found by Skumanich (1972).

discussion and references; see also Leighton 1959).

Observationally, we had the finding that certain types of stars were found to be strong X-ray emitters, including Capella (Catura, Acton & Johnson 1975; Mewe et al. 1975) and the class of RS CVn stars. The latter have surface X-ray fluxes orders of magnitude larger than the Sun's (associated with plasma whose temperature is substantially higher than the Sun's corona), yet have effective surface gravities much lower than the Sun's; clearly, these coronae must be confined (Walter et al. 1979; Holt et al. 1979). For more details on observations of these RS CVn stars, see Walter et al. (1979) and references therein. Observations of the α Cen A/B system (G2 V and K0 V, respectively) by Nugent and Garmire (1978) confirmed that truly solar-like stars could indeed be observed (althrough identification of the dominant emitter, α Cen B, had to await the EINSTEIN observations).

A number of stars throughout the H-R diagram have been looked at with Copernicus, BUSS, and IUE; and UV line emission due to ionization states associated in the solar case with the transition region between the chromosphere and corona have been observed. Particularly striking have been: (i) observations of O VI and other transition region line emission from early-type stars (cf. Snow and Morton 1976), suggesting that these stars also may possess coronae; (ii) the possible correlation between UV emission and stellar rotation rate for late-type stars reported by Ayres and Linsky (private communication), suggesting that convection activity levels alone are not the sole determinant of coronal activity; and (iii) indications from the UV that the transition region structures of late-type dwarfs are relatively independent of the stellar effective temperature (Hartmann et al. 1979).

It should, however, be noted that the earlier discussions in general viewed magnetically enhanced atmospheric confinement and heating as supplementary processes, which would increase mechanical heating locally above quiescent levels determined by nonmagnetic heating processes (cf. Jordan 1973, Piepenhahn 1973). Prior to EINSTEIN discussions of quiescent stellar coronal activity were, with few exceptions (cf. Blanco et al. 1974 and Thomas 1975), phrased in terms of traditional acoustic heating theories (cf. Mewe 1979). Assuming that the literature reflects the scientific opinion of the day, one is forced to conclude that acoustic coronal heating represented the consensus view with magnetically-coupled heating processes viewed as relevant only to enhancements of coronal activity.

A similar situation has characterized solar studies (viz. Athay 1976) with only a minority insisting upon the dominance of magnetic field-related heating throughout the corona. It would seem that in the absence of detailed comparison between theory and observations, which only became feasible for solar coronal data after the flight of Skylab, alternatives to acoustic heating models were relatively speculative. The extrapolation of observations and interpretation of, for example, Ca II emission by "active stars," to a radically different coronal formation scenario involving magnetic fields as the essential element for "normal stars" in general, would not have been reasonable without the new results which are now available.

3. THE CFA STELLAR SURVEY

The EINSTEIN/CfA stellar survey (Vaiana et al. 1980) combines results from four separate programs: a pointed survey, which samples nearby optically well-characterized stars; the 8.5 magnitude survey (Topka, 1980), which systematically searches EINSTEIN/CfA fields for X-ray emission from stars down to visual magnitude 8.5, and the Deep Survey (Giacconi et al. 1979a) and Serendipitous Survey, which search for X-ray sources accidentally falling into the field of view which can be identified with stars. Each of these programs, depending upon the selection criteria used and on its degree of completeness, contributes unique information regarding the X-ray luminosity functions of stars.

Two of the EINSTEIN focal plane instruments have been used in the present survey: the Imaging Proportional Counter, which allows positional localization of $\sim$ one arcminute and the High Resolution Imager with a positional localization in the arcsecond range. Details of the EINSTEIN instrumentation may be found in Giacconi et al. (1979b). Most of our sources are well above the 3σ level and are subject to the additional constraint of optical positional coincidence; the quoted X-ray fluxes apply to the 0.3 to 3.5 keV passband.

Figure 2 shows an H-R diagram, on which are plotted the stars which have been observed as X-ray sources by EINSTEIN. We have sampled and seen the entire main sequence, as well as giants up to and including spectral type K, supergiants up to F, some subdwarfs, white dwarfs, pre-main sequence stars and exotic objects such as dwarf novae. We also have upper bounds on M giants and supergiants later than spectral type K.

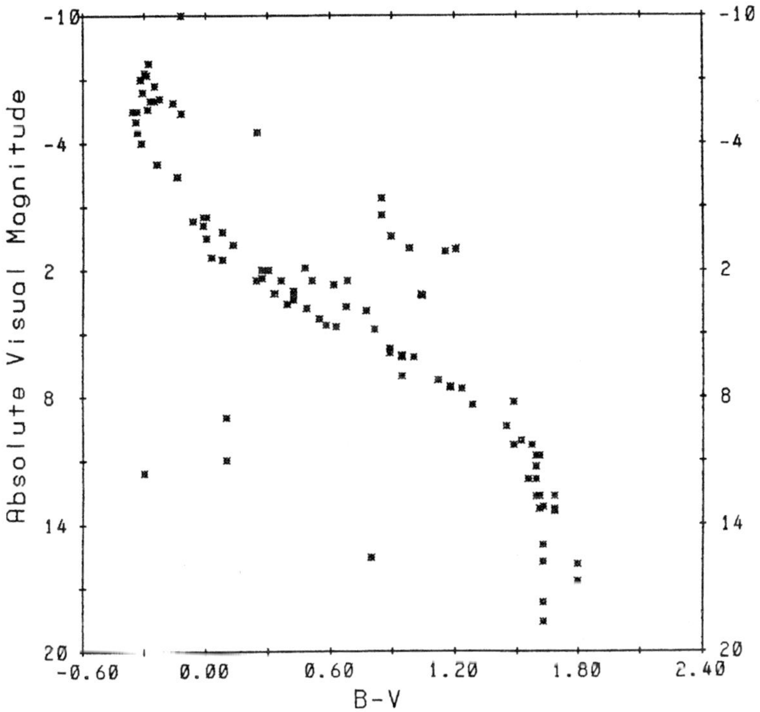

Figure 2. H-R diagram of the stars from which X-ray emission has been detected by the CfA stellar surveys (Vaiana et al. 1980)

3.1 Sample Stellar Detections

Before discussing the detailed quantitative analyses and interpretation which have been performed, we will provide a few examples of the stellar survey results to date. Figure 3 shows an HRI image of α Cen A/B (G2 V and K0 V, respectively), the nearest binary system and also the nearest solar-like star. The two components are clearly resolved, emitting at luminosities of 10^{27-28} erg s^{-1} (Golub et al. 1980) Note that the K dwarf is the dominant emitter, which is contrary to expectations (cf. Fig. 13 below and Nugent and Garmire 1978). Additionally, in a guest investigator observation Haisch and Linsky (Haisch et al. 1980) have detected Proxima Cen (α Cen C) emitting in steady state

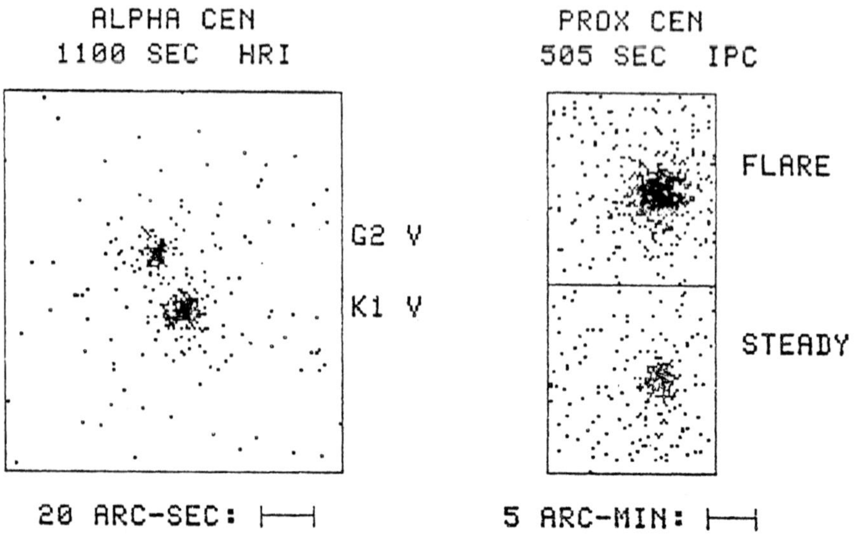

Figure 3. The three stars of the Alpha Cen system. Both the G- and the K-dwarf in α Cen A/B have been seen in soft X-rays by the CfA. The right-hand figure shows a flare observed on Proxima Cen in a collaborative program with guest observers Haisch and Linsky.

at comparable luminosity. During this observation Proxima Cen was caught flaring, as shown in the upper right portion of the fiture. The observed coronal temperature for Prox Cen is as low as 3×10^6 K during quiescent emission and rises to $\gtrsim 10^7$ K in a flare. The respective emission measures are 2×10^{49} cm^{-3} and 10^{50} cm^{-3}.

Going down the main sequence, Figure 4 shows EQ Vir, a K5e dwarf known as an active chromosphere star, emitting at a level of $\sim 10^{29}$ erg s^{-1}. Yet later spectral type is represented by YZ CMi (Fig. 5) which we observed twice and which shows clear variability. It is a dMe flare star and also emits at a level of 10^{29} erg s^{-1}. Both of these stars appear to be stronger sources than average for their spectral types, although they are within the 2 to 3 order of magnitude range in X-ray luminosity observed in K- and M-dwarfs.[2] However it is apparently not coincidental that these stars are both known to be optically active and emission-line stars. We believe that it is also significant that these stars are exceptionally rapid rotators for their spectral types.

Early-type stars have turned out to be relatively strong emitters as evidenced by the early observations of the Cygnus OB2 associations (Harnden et al. 1979a, b). This is dramatically seen in the IPC

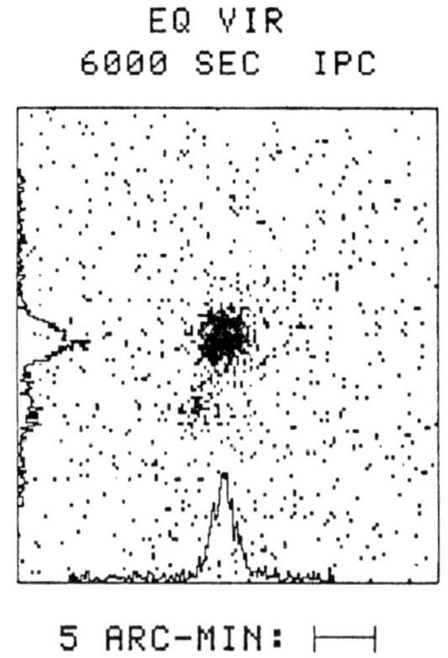

Figure 4. The dwarf K star EQ Vir appears to be an exceptionally strong X-ray emitter for its spectral type; it is also known to be a rapid rotator (see text).

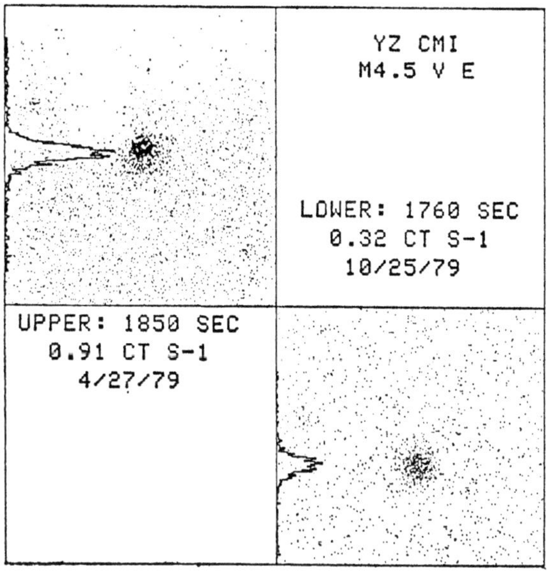

Figure 5. The flare star Yz CMi was observed at six-month intervals and shows substantial variation in X-ray emission level.

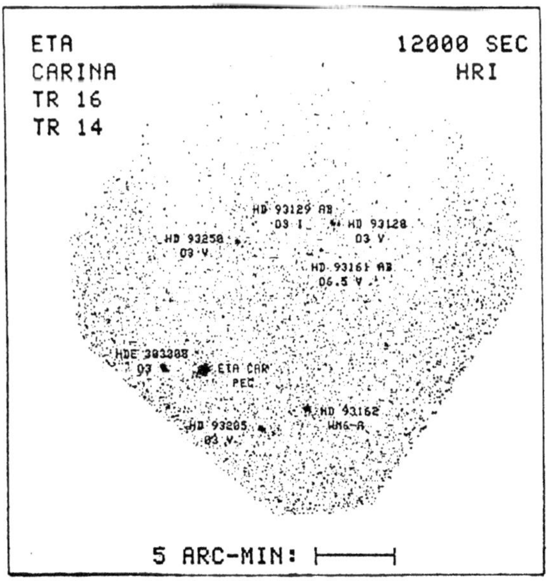

Figure 6. HRI image of the central portion of the η Car field, with individually resolved sources identified.

observations of η Carina (Seward et al. 1979), in which in addition to η Car itself, a WR star and many O main main sequence and giant stars are seen as sources. The added power of the HRI in constraining source location and extent is demonstrated in Figure 6, an image of the central portion of the η Car region. There is no room for doubt as to the optical identifications, and the strong constraint upon the size of the source region argues for a coronal interpretation of the X-ray emission. Contrary to the result for late-type stars, we find that for these early stars the ratio L_x/L_{opt} is approximately constant (Rosner et al. 1979; Harnden et al. 1979b).

3.2 Principal Survey Results

The quantitative analysis of the data focusses on those general aspects which allow us to draw conclusions about the luminosity functions of the overall sample (for a more detailed discussion, see Vaiana et al. 1980). Rather than discuss each spectral type individually, Figure 7 shows a synopsis of the results for main sequence stars (top) and for giants and supergiants (bottom). The data for these

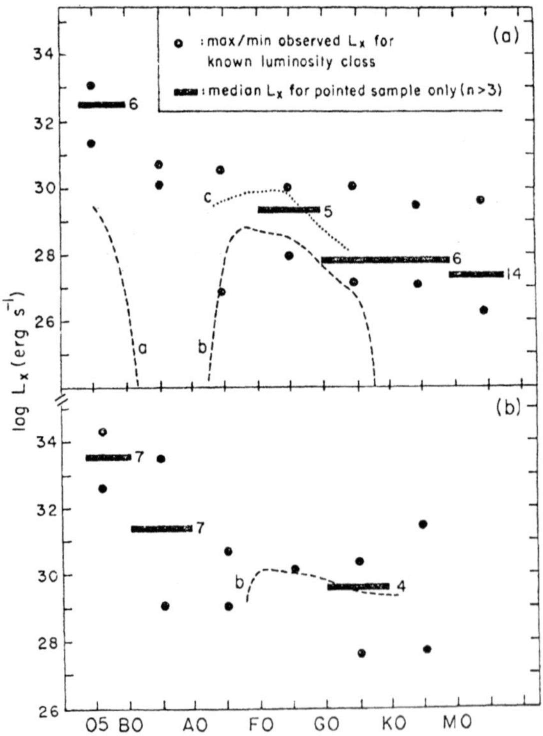

Figure 7. Variation in X-ray luminosity L_x vs. spectral type for the optically well-classified sample of stars: a) main sequence, b) giants and supergiants. Circles indicate the maximum and minimum value of L_x found in this optically well-classified sample and horizontal bars the median value of L_x for the subset of Pointed-survey stars.

We also plot several theoretical predictions of X-ray emission levels based on acoustic coronal heating (a and b: from Mewe 1979; c: from Landini & Monsignori-Fossi 1973).

(from Vaiana et al. 1980)

figures are drawn from a subset of all the stars in the survey (85 out of 140), namely those for which the luminosity classification is available.

The primary result of this survey is that all categories of stars have been detected as X-ray emitters, with the sole exception of very late-type supergiants. Except within a very narrow range of spectral types, both the levels of X-ray emission and the general behavior of the median luminosities disagree with theoretical predictions. The disagreement appears not only in the behavior of median X-ray luminosity as a function of spectral type, but also in the very wide range of observed X-ray luminosities at each spectral type (Fig. 7). This finding is quite general: within each spectral type and luminosity class there is observed a large range of X-ray luminosities (two to three orders of magnitude).

This range is also large when compared to the variation of bolometric luminosity within each spectral type and to the change in median X-ray luminosity. The question therefore is raised: what stellar characteristic determines the total level of X-ray emission? It clearly is not spectral type, luminosity class, or mass, those very parameters that locate main-sequence stars on the H-R diagram. What then is it that determines the level of X-ray emission?

We do have some hints about that in the data. For each spectral type (G to M) along the main sequence the brightest stars in X-rays appear to be rapid rotators for their spectral type, including some stars already well-known to be active, such as π^1 UMa and EQ Vir. The drop in the median X-ray luminosity in going from F to G along the main sequence is also consistent with rotation dependence of X-ray emission, because as a class, F stars are substantially more rapid rotators than G and later stars.

Some preliminary data are shown in Figure 8, in which we plot X-ray luminosity vs. equatorial rotation rate (v sin i) for main-sequence stars and subgiants. Rotation data are from a compilation of Uesugi and Fukuda (1970) and from a recent review paper by M. Smith (1979) on slow-rotating late-type main-sequence stars. Rotation rates for dMe stars are determined from the photometric period, attributed to spot rotation; those of RS CVn-like stars have been estimated from the orbital period, assuming synchronous rotation.

In spite of the large scatter, there is a tendency for rapid rotators to have higher X-ray luminosity. This tendency is stronger than one would expect from the variation of median X-ray luminosity and rotational velocity with spectral type and is made sharper, as indicated above, where only stars of the same spectral type are considered. Rotational data in Figure 8 were obtained with various dispersions and are of different qualities so only preliminary conclusions can be drawn at this stage. (Work is in progress to determine more precisely the relationship between rotation velocity and

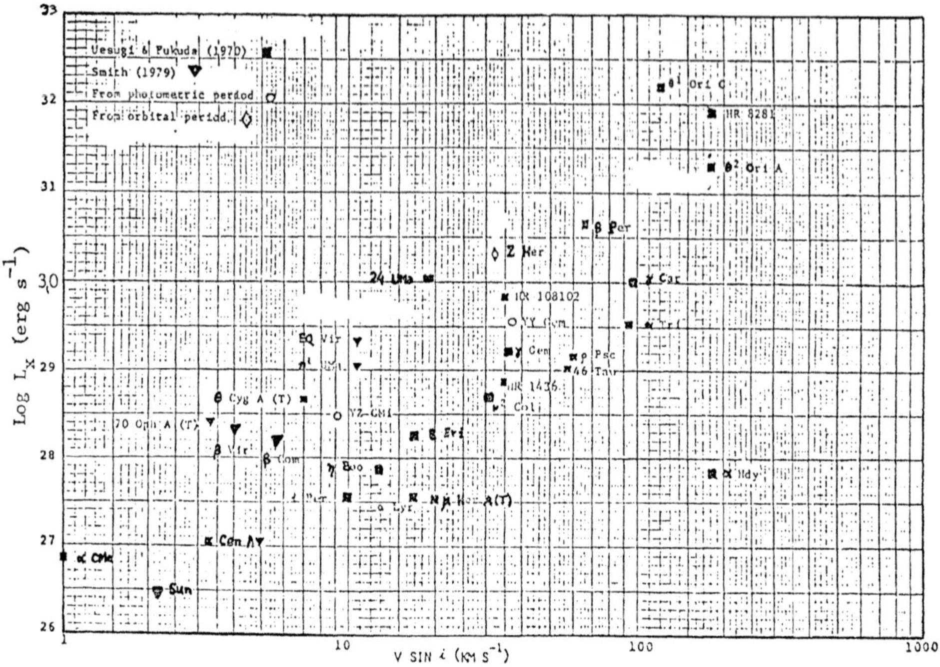

Figure 8. Preliminary plot of X-ray luminosity vs. v sin i for a sample of stars observed by EINSTEIN and for which rotation data are available. Note in particular that within each spectral type there are well known active chromosphere stars, such as EQ Vir, π'UMa and YZ CMi, which are known rapid rotators and are now found to be strong X-ray sources.

X-ray luminosity, including giants and supergiants). The preliminary findings, however, are already sufficient to indicate the importance of rotation in determining the X-ray luminosity levels of stellar coronae.

Some aspects of this discussion become clearer if we present the results in terms of stellar surface fluxes, i.e., taking into account the radii of the stars. Figure 9 shows the variation of the derived median X-ray surface brightness (f_X) for main-sequence stars as a function of spectral type and the range in f_X within each spectral type; this range is typically large in every spectral class. For comparison, the values of surface flux observed in individual features on the Sun are indicated at the right.

The fundamental point is that, although the median value of surface flux does show some variation along the main sequence, the ranges of surface fluxes encountered all overlap and in fact coincide

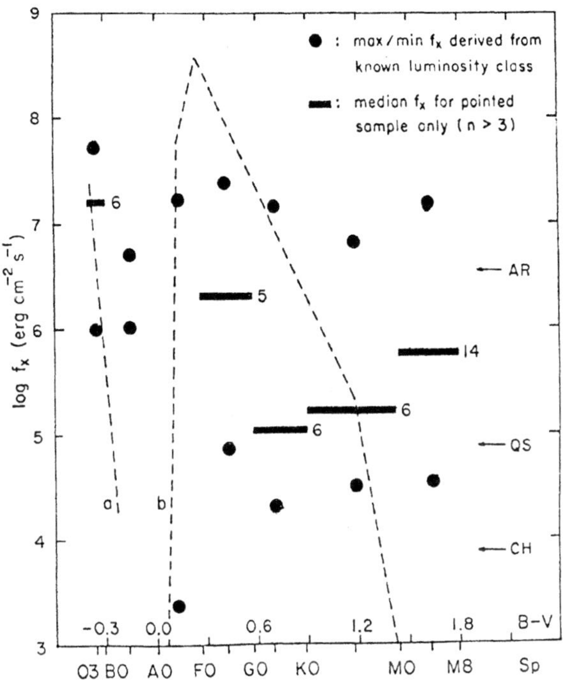

Figure 9. Variation of derived X-ray surface flux f_x vs. spectral type for main-sequence stars only. Circles indicate maximum and minimum observed f_x for main-sequence stars; bars indicate median f_x for the Pointed sample of main-sequence stars (calculated only if sample contained more than three stars).

(from Vaiana et al. 1980)

with the range of surface brightness observed for various discrete features on the Sun. This fact suggests that the Sun may serve as a useful guide or model in interpreting these observations. Note, however, that the values plotted are averages over the entire stellar surface. Thus, active region level fluxes occupy only a small portion of the solar surface even at solar maximum, whereas the brighter stars in our sample must be fairly well covered by such active regions if we insist on a solar comparison. Alternatively, lower area coverage factors are allowed if the "active region" pressures exceed solar values.

Figure 9 also compares the observations with theoretical predictions of the total acoustic flux available for coronal heating. The peak in the theoretical result finds a counterpart in the median observed X-ray surface flux; however, little else agrees. That is: (1) the X-ray surface fluxes for early and very late-type stars are underestimated by many orders of magnitude. (2) The very substantial range in emission levels encountered in every spectral class is completely unaccounted for, yet is the dominant feature of the data. The conclusion seems unavoidable that acoustic coronal heating, at least as proposed in the literature, is incompatible with the stellar experimental data taken *in toto*.

Before proceeding to discuss the implications of these new observations, it is instructive to review briefly what has been learned about coronal physics from recent solar studies. In particular, it is of the greatest importance to appreciate the fundamental role played by inhomogeneity in the solar case and to examine the extent to which this work applies in the stellar area.

Figure 10 is an image of the Sun obtained by us on SKYLAB in the same spectral band as EINSTEIN's. It is shown as a reminder that the solar corona is not a diffuse, homogeneous hot atmosphere, but rather

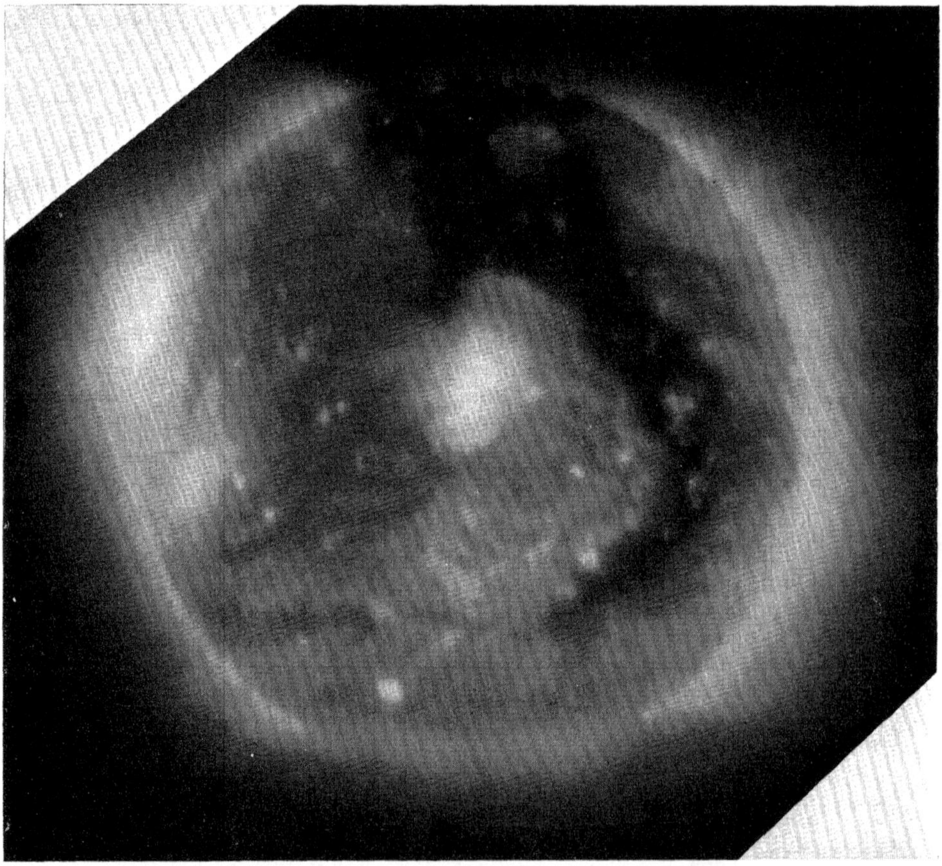

is strongly structured by the surface magnetic fields into topologically closed features dominated by magnetic confinement. The association of high speed solar wind streams with open structures, such as the large polar coronal hole in this image, shows that mass loss is predominantly associated with regions of open magnetic fields and low coronal activity (cf. Vaiana et al. 1973).

The principal results relevant to the present discussion are:

(a) The solar atmosphere is fundamentally inhomogeneous, from the chromosphere through the transition region and corona to the solar wind;

(b) Steady mass loss is spatially anticorrelated with the level of X-ray emission, since the recurrent high-speed streams in the solar wind originate in coronal holes, which are regions of very low X-ray emission (Timothy, Krieger and Roelof 1973; Vaiana, Krieger and Timothy 1973).

(c) Magnetic fields and hot coronal plasma on the Sun are intrinsically related, with both a strong qualitative correspondence between coronal pressures and coronal field topology (Maxson and Vaiana 1977; Rosner, Tucker and Vaiana 1978) and a quantitative association between the gas pressure and associated magnetic field in closed coronal structures (Golub et al. 1980).

4. TOWARD A NEW THEORY

Given the apparent failure of acoustic heating to account for stellar or solar data, the evident importance of magnetic fields in determining the activity level of the solar corona, and the suggestion from the stellar data that rotation is an important parameter in determining the level of X-ray emission, we have developed a scenario of coronal formation. This scenario is being used by us as a framework for defining further theoretical and experimental work (cf. Rosner and Vaiana, 1980). The basic assumptions underlying the scenario are first, that magnetic fields are a ubiquitous feature of stars and, second, that modest levels of surface turbulence exist on all stars. The basic process invoked is the interaction of surface flows with surface magnetic fields under high-β conditions, leading to field stresses which are free to propagate outwards and dissipate in the overlying low-β atmosphere, leading to hot coronal plasma.

A schematic of our approach is shown in Figure 11, together with a parallel listing of the principal theoretical problems which must be addressed. These are:

(i) What determines the total magnetic flux level at the stellar surface?

(ii) How are surface magnetic fields coupled to turbulent surface flows?

(iii) How do magnetic stresses (or perturbations) propagate from the photospheric "source surface" through the overlying atmosphere?

(iv) How do these perturbations thermalize in the chromosphere, transition region, and corona?

The classes of problems have all received extensive attention. I will focus on the first of them, primarily because the new solar and stellar data combined with recent theoretical developments may be

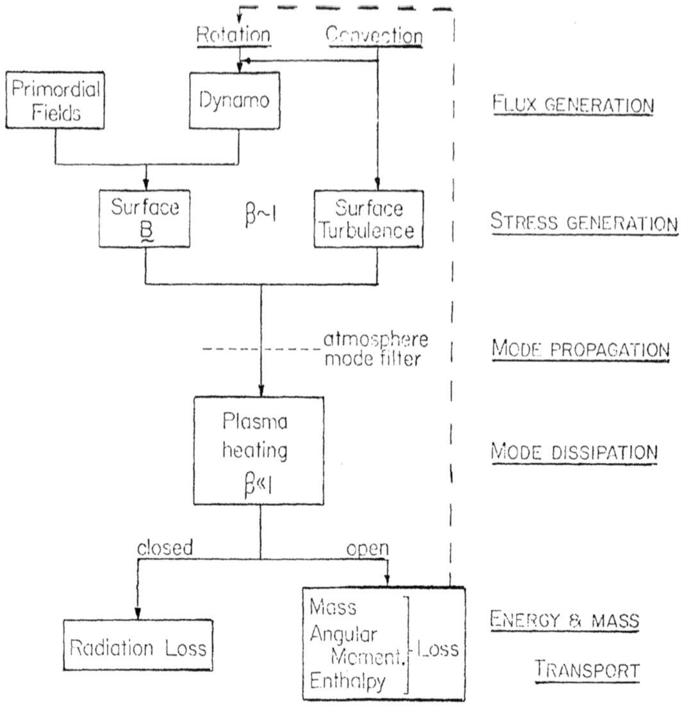

Figure 11. Possible schematic flow chart of processes leading to stellar coronae and winds. In contrast to previous scenarios (cf. Kippenhahn 1973), magnetic fields play a central role throughout.

pointing us toward a significant advance in our understanding of this problem.

Concerning the interaction with magnetic fields to produce coronal emission, there are two possible scenarios which can be envisaged. Depending upon the mechanism by which magnetic fields are provided, we may have an active dynamo process (very likely in late-type stars), or formation may take place through remnant primordial fields (as is more probable for early-type stars). In both cases, hot confined coronal plasma is produced because of the presence of magnetic fields; in the case of late-type stars, a causal connection is therefore established between coronal intensity and processes thought to determine magnetic activity, such as rotation and depth of convective zones (Figure 12).

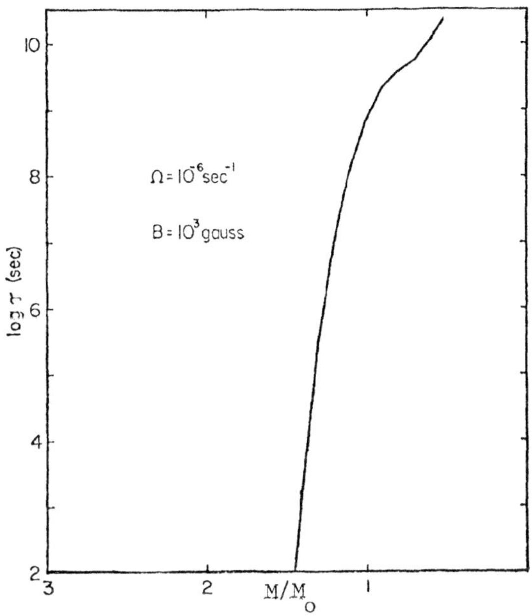

Figure 12. Time scale variation for magnetic flux eruption from "shell" dynamo layer vs. stellar mass (from Rosner & Vaiana 1980), for constant rotation rate and magnetic field strength.

Direct computation by Rosner and Vaiana (1980) shows that, for fixed rotation rate and field strength, the time scale for flux eruption progressively increases with decreasing stellar mass. Consequently, one expects relatively larger values of toroidal flux to be generated by the time $\tau^{-1} B_\phi$ becomes a significant loss effect as one goes to lower mass stars of equal rotation rate. We therefore expect larger values of erupting magnetic flux in M dwarfs than in G dwarfs (such as the Sun) having equal rotation rates; as there is substantial variation in observed values of ω for a given spectral type, similar variation in flux production within a given spectral type must also be expected; thus the above model makes the qualitative prediction that the mean magnetic flux generation follows the variation in mean rotation rate (viz. Fig. 8), with a scatter determined by the dispersion of rotation rates about the mean for the various types of stars. In light of the magnetic-field coupled coronal heating theory discussed above, this would lead directly to the observed behavior of coronal activity along the main sequence, from dG to dM, e.g.

(i) increasing strong emission-line behavior (associated with photospheric field activity) among late K and M stars;

(ii) appearance of "spot" stars among M dwarfs, indicating progressively larger magnetic fluxes at the stellar surface;

(iii) progressively larger mean X-ray surface fluxes, consistent with a systematically larger fraction of the stellar surface occupied by (in the solar analogy) confined coronal plasma.

Qualitatively similar behavior is expected in going from dF to dG stars (with the larger rotation rates associated with dF stars accounting for larger magnetic flux production and hence for their high activity levels).

It may still be argued that several distinct processes give rise to stellar coronae, and that the standard acoustic heating theories survive only for stars in the spectral range ∿ A5 to ∿ G5, with other (perhaps as yet unknown) processes responsible for coronae on other types of stars. This line of reasoning, which attempts to minimize the necessary adjustments, must, however, contend with the results of recent solar coronal research, which have shown that solar coronal X-ray emission is largely modulated by solar surface magnetic fields (see review by Vaiana and Rosner 1978); that is to say, straightforward acoustic heating theory is inadequate to explain observations of the corona of even the G2 V star for which it was originally developed.

5. CONCLUSIONS

The EINSTEIN stellar surveys have established that stars in general constitute a new class of low-luminosity galactic X-ray sources. X-ray emission appears to be characteristic of stars from youth, crossing virtually all spectral type and luminosity class boundaries, and is in all probability associated with the presence of stellar coronae. The preliminary EINSTEIN results thus settle the long controversy surrounding the pervasiveness of the coronal phenomenon in the affirmative, raise a multitude of new questions regarding the formation process responsible for stellar coronae, and clearly contradict standard theories of coronal X-ray emission.

High sensitivity stellar X-ray observations now join the more traditional tools of stellar astronomy in constraining theories of stellar structure and evolution. Problems of rotation, magnetic field generation and spindown, stellar activity from the very early stages of stellar formation to the very late stages, and throughout the mass spectrum, can now be attacked from a new perspective with renewed vigor. Also, a number of further new and unexpected areas of research have already opened up. For example, the relatively high X-ray luminosity of dMe stars means that at the limiting sensitivity of the Deep Survey, EINSTEIN can reach dM stars of 17 to 20 magnitude, corresponding to a distance of 50 to 100 pc. With sensitivities comparable to AXAF one would be able to reach and recognize 22nd magnitude M stars, enabling one to study the space distribution of low mass objects (which is important for determining the galactic mass) from a perspective complementary to that of the optical work currently being performed.

X-ray observations of pre-main sequence stars (Ku and Chanan, 1979; Vaiana et al. 1980) provide another example of new research possibilities. They allow us to probe stellar evolution in a new way, particularly exciting because of the possibility discussed above that magnetic fields are involved in the heating process which leads to X-ray emission. If this suggestion is confirmed, it means that we have found a direct way of probing the magnetic field activity in such stars, and studying in a new light the role magnetic fields may play in early stages of stellar evolution.

There are many other, similar examples one can come up with, and we are in fact pursuing a number of them. I feel quite safe in predicting that, given sufficient opportunities and effort, this new field of stellar X-ray astronomy opened up by EINSTEIN will stride a rich course at all levels of astrophysics, from the traditional to the frontier areas.

ACKNOWLEDGMENTS

This work is based upon the contributions of a large number of people in the EINSTEIN program. A list of some of those more directly involved in various aspects of the stellar program includes G. Fabbiano, R. Giacconi, L. Golub, P. Gorenstein, F.R. Harnden, Jr., E. Kwok, C.W. Maxson, R. Rosner, F. Seward, K. Topka; contributing guest investigators have included J. Cassinelli, B.M. Haisch, H. Johnson, J. Linsky, R. Mewe and C. Zwaan. Portions of this work have been sponsored by NAG 8302, NAS8-30751, the Langley Abbot Program of the Smithsonian Institution, NAS8-31374, NSG 7537, and the CNR (Italy).

REFERENCES

Athay, R.G. 1976, The Solar Chromosphere and Corona: Quiet Sun (Dordrecht: Reidel).

Biermann, L. 1946, Naturwiss. 33, 118.

Blanco, C., S. Catalano, E. Marilli and M. Rodono' 1974, Astron. Ap.33, 257.

Catura, R.C., L.W. Acton and H.M. Johnson 1975, Ap.J. (Lett.) 196, L47.

Durney, B.R. 1972, Proc. Asilomar Conf. on Solar Wind (NASA Sp. Rpt.)

Giacconi, R. et al. 1979a, Ap.J. (Lett.) 234, L1.

Giacconi, R. et al. 1979b, Ap.J. 230, 548.

Golub, L., F.R. Harnden, Jr., R. Rosner, K. Topka and G.S. Vaiana 1980a, submitted to Ap.J. (Lett.)

Golub, L., C.W. Maxson, R. Rosner, S. Serio, and G.S. Vaiana 1980b, Ap.J. (in press).

Gorenstein, P. and W.H. Tucker 1976, Ann. Rev. Astron. Ap. 14, 373.

Haisch, B.M., J.L. Linsky, F.R. Harnden, Jr., R. Rosner, F. Seward, and G.S. Vaiana 1980 (preprint).

Harnden, F.R., Jr., G. Branduardi, M. Elvis, P. Gorenstein, J. Grindlay, J.P. Pye, R. Rosner, K. Topka, and G.S. Vaiana 1979a, Ap.J. (Lett.) 234, L51.

Harnden, F.R., Jr., L. Golub, R. Rosner, F. Seward, K. Topka and G.S. Vaiana, 1979b, Bull. AAS, 11, 775.

Hartmann, L., R. Davis, A.K. Kupree, J. Raymond, P.C. Schmidtke and R.F. Winer 1979, Ap.J. (Lett.) (in press).

Holt, S.S., N.E. White, R.H. Becker, E.A. Boldt, R.F. Mushotzky, P.J. Serlemitsos and B.W. Smith 1979, Ap.J. (Lett.) 234, L65.

Kraft, R.P. 1967, Ap.J. 160, 551.

Ku, W. and G.A. Chanan 1979, Ap.J. (Lett.) 234, L59.

Linsky, J.L. 1977 in The Solar Output and Its Variation, ed. O.R. White (Boulder: Colo. Assoc. Univ. Press), 477.

Mewe, R. 1979, Space Sci. Rev. 24, 101.

Mewe, R., J. Heise, E.H.B.M. Gronenschild, A.C. Brinkman, J. Schrijver and A.J.F. den Boggende 1975, Ap.J. (Lett.) 202, L67.

Nugent, J. and G. Garmire 1978, Ap.J. (Lett.) 226, L83.

Rosner, R., J. Grindlay, R. Harnden, F. Seward and G.S. Vaiana 1979, Bull. AAS 11, 446.

Rosner, R., R. Giacconi, L. Golub, F.R. Harnden, Jr., K. Topka and G.S. Vaiana 1980 (in preparation).

Rosner, R., W.H. Tucker and G.S. Vaiana 1978, Ap.J. 220, 643.

Rosner, R. and G.S. Vaiana 1980, in X-Ray Astronomy, ed. R. Giacconi and G. Setti (D. Reidel, Dordrecht), p. 129.

Schwarzschild, M. 1948, Ap.J. 107, 1.

Seward, F.D., W.R. Forman, R. Giacconi, R.E. Griffiths, F.R. Harnden, Jr., C. Jones and J.P. Pye 1979, Ap.J. (Lett.) 234, L55.

Skumanich, A. 1972, Ap.J. 171, 565.

Smith, M.A. 1979, PASP 91, 737.

Snow, T. and D.C. Morton 1976, Ap.J. (Suppl.) 32, 429.

Thomas, R.N. 1975, Chem. Phys. 32, 259.

Topka, K. 1980, Ph.D. Thesis, Harvard University.

Topka, K., D. Fabricant, F.R. Harnden, Jr., P. Gorenstein and R. Rosner 1979a, Ap.J. 229, 661.

Topka, K., L. Golub, F.R. Harnden, Jr., P. Gorenstein, R. Rosner and G.S. Vaiana 1979b, Bull. AAS 11, 781.

Uesugi, A. and I. Fukuda 1970, Contr. Inst. Astrophys. Kwasan Obs. Univ. Kyoto, No. 189.

Vaiana, G.S. 1980, IAU Highlights of Astronomy 5 (D. Reidel, Dordrecht), p. 419.

Vaiana, G.S. and R. Rosner 1978, Ann. Rev. Astron. Ap. 16, 393.

Vaiana, G.S., W. Forman, R. Giacconi, P. Gorenstein, J. Pye, R. Rosner, F. Seward and K. Topka 1979, Bull. AAS 11, 446.

Vaiana, G.S., J.P. Cassinelli, G. Fabbiano, R. Giacconi, L. Golub, P. Gorenstein, B.M. Haisch, F.R. Harnden, H.M. Johnson, J.L. Linsky, C.W. Maxson, R. Mewe, R. Rosner, F. Seward, K. Topka, and C. Zwaan, 1980, Ap.J. (submitted).

Vaiana, G.S., A.S. Krieger and A.F. Timothy 1973, Sol. Phys. 32, 81.

Walter, F.M., W. Cash, P.A. Charles, and C.S. Bowyer 1979, Ap.J. (submitted).

Wilson, O.C. 1966, Science 151, 1487.

THEORY OF STELLAR CORONAE: AN INTERPRETATION OF X-RAY EMISSION FROM NON-DEGENERATE STELLAR SOURCES

Jeffrey L. Linsky[1]
Joint Institute for Laboratory Astrophysics
National Bureau of Standards and University of Colorado

ABSTRACT

I present arguments to demonstrate that the acoustic wave heating theory of stellar coronae is inadequate to explain *Einstein* observations of stellar coronae as well as previous ultraviolet and X-ray observations of the Sun and other stars. I outline the various lines of evidence that imply that magnetic fields, stellar rotation rates, and to some extent convection zone parameters are the important quantities in determining coronal heating and thus X-ray emission. These general results and the recent Einstein Observatory stellar observations suggest a speculative scenario of stellar coronae consisting of the following elements: Coronae in O-type stars are heated by the interaction of turbulent stellar winds with slowly decaying primordial magnetic fields or by radiative instabilities in the flow. The apparent absence of coronae in Ap stars, which have large but stable magnetic fields, is due to the absence of temporal changes in very stable atmospheres for which even weak convection is suppressed by the strong field. Weak X-ray emission in some normal A-type stars may be due to the beginnings of dynamo action, which is likely the dominant field regeneration process in cooler stars. In F- and later-type dwarfs, dynamo processes, which increase with stellar rotation velocity and convection zone depth, continually replenish surface magnetic fields. These in turn heat the corona through several possible mechanisms. The distinction between dMe and dM stars is probably a result of larger rotational velocities in the former, since these stars are very likely fully convective. For BY Draconis and RS CVn-type systems, which typically are close binaries, binarity is probably only important to the extent that tidal synchronism forces rapid rotation. K- and M-type giants and supergiants are very likely deficient in coronal material because in such stars strong chromospheric winds provide a cooling channel alternative to the conventional hot corona.

---

[1] Staff Member, Quantum Physics Division, National Bureau of Standards.

R. Giacconi (ed.), X-Ray Astronomy with the Einstein Satellite, 19–38.

This notion is reinforced by the "hybrid" character of early G supergiants, which provide an intermediate case. These stars show variable emission lines from hot plasma as well as strong chromospheric winds. The coexisting phenomena suggest the presence of two very distinct outer atmospheric structures, perhaps hot plasma entrained in closed magnetic loops and cool outflowing material in open field regions analogous to solar coronal holes.

1. INTRODUCTION

The study of stellar coronae is an excellent example of a field in which observations have generally led and dictated theoretical developments. In this review I will summarize how the recent Einstein Observatory observations, beautifully presented in the previous paper (Vaiana 1980), together with stellar ultraviolet (cf. review by Linsky 1980) and solar visual and X-ray observations have precipitated a major change in our understanding of stellar coronae. For example, until very recently acoustic wave dissipation was almost universally accepted as the dominant coronal heating mechanism, however in light of the new UV and soft X-ray measurements, this mechanism is now regarded as unimportant. In what follows, I will summarize what the important parameters in a new theory of coronal heating are likely to be, and then present a speculative scenario to qualitatively describe the different types of coronae that apparently exist in the HR diagram.

2. THE INITIAL PICTURE OF THE SOLAR CORONA

The discovery of visible emission lines of highly ionized iron and calcium, and high temperature radio emission from the Sun in the 1940's led to the first models of a spherically symmetric, million degree corona in hydrostatic equilibrium. Subsequently, Parker (1958) predicted the existence of a supersonic wind, and satellite observations led to estimates of mass loss in the range 10^{-13}–10^{-14} $M_\odot$ yr^{-1}.

Following the initial suggestions of Biermann (1946) and Schwarzschild (1948), a near consensus developed that the solar corona is heated by acoustic waves generated by turbulent motions in the hydrogen convection zone immediately below the visible photosphere. According to the theory, acoustic waves propagate upwards from the deep photosphere and steepen into shocks in the outer atmosphere as a consequence of the rapid decrease in density with height (cf. the reviews of Stein and Leibacher 1974, 1980). The role of the photosphere in this picture is to filter out the short period waves by radiative damping but to permit waves with periods larger than 300^s to shock and dissipate their mechanical energy as heat. The acoustic theory was generally accepted until very recently because of considerable corroborative empirical evidence. The existence of convective cells (granulation), photospheric motions, and the 300-second oscillatory pattern confirmed the existence of a turbulent medium in which acoustic waves can be generated. In addition, time series of solar Ca II K line and other line spectra presented by Liu (1974) and by Cram (1978) provided vivid evidence for the upward propagation of heating pulses into the low chromosphere.

Ulmschneider (1979) has reviewed the development of theoretical solar models computed on the basis of the acoustic wave theory. Kuperus (1965, 1969), Ulmschneider (1967), and de Loore (1970), among others, have extended the theory to the coronae of late-type stars. These authors have typically assumed the mixing length theory of convection, the Lighthill (1952) theory for acoustic wave generation, monochromatic acoustic waves, and grey opacity stellar atmospheres to compute coronal models as functions of stellar effective temperature and gravity. Such models (cf. Mewe 1979) generally predict bright X-ray emission in F dwarfs, weak emission in G-K dwarfs, essentially no emission in O-A and M dwarfs, and a rapid increase in X-ray surface flux with increasing stellar luminosity (decreasing stellar gravity).

3. PROBLEMS WITH THE ACOUSTIC WAVE HEATING THEORY FOR THE SOLAR CORONA

Gough (1976), among others, has emphasized the inherent inaccuracies of the acoustic wave heating theory that arise simply because the flux production is proportional to v_{conv}^8, where v_{conv} is the convective bulk velocity deep in the photosphere. Since v_{conv} cannot be measured directly and theoretical values computed on the basis of the mixing length theory are questionable, derived values of the initial acoustic flux at the base of the solar photosphere are highly uncertain. Furthermore, at least 90% of the acoustic energy flux generated in the convective zone is dissipated by radiative damping in the photosphere itself, and only long period waves escape without appreciable attenuation (Stein and Leibacher 1980). Consequently, the actual acoustic flux available to heat the corona is doubly uncertain. Finally, the acoustic wave spectrum is not known accurately.

An important goal of the OSO-8 ultraviolet spectrometer experiments was to measure the acoustic wave flux and heating rates in the solar chromosphere and transition region (the geometrically thin interface between the chromosphere and corona). Athay and White (1978,1979) have used their measurements of C IV line widths and power spectra to infer an acoustic flux $\lesssim 10^4$ ergs cm^{-2} s^{-1} passing through the transition region, a factor of 30 less than what is needed to balance measured coronal radiative, conductive, and wind energy losses. More recently, Bruner (1979) has revised the estimate of acoustic flux passing through the transition region to a value $<10^{-3}$ of that needed to heat the corona.

The preceding arguments provide strong evidence against the acoustic wave heating theory, but there are also several compelling arguments in favor of competing mechanisms -- namely, magnetic heating processes. For example, a major result of the high resolution S-054 X-ray imaging experiment on Skylab was the clear identification of strong X-ray emission with extended loop structures, presumably defined by the coronal magnetic field (cf. Vaiana and Rosner 1978), and the confirmation of the previous result (cf. Withbroe and Noyes 1977) that areas of weak, open magnetic field geometry are regions of low X-ray surface brightness (the so-called coronal holes). Significantly, the coronal holes are the origin of high speed solar wind streams. In fact, Pneuman (1973) and others have shown that in coronal holes the wind is the dominant coronal cooling mechanism, whereas in closed loops radiation and conduction are the dominant cooling mechanisms (cf. Withbroe and Noyes 1977). Thus the magnetic field geometry appears to dictate the plasma cooling mechanism in the solar corona.

Presumably the magnetic field configurations and strength also control the coronal heating rate. There is no proof for this conclusion yet, but there is strong circumstantial evidence, including: (1) the correlation of large scale photospheric magnetic fields with bright X-ray emission from overlying coronal active regions, (2) the loop-like structure of bright coronal X-ray emission just described, (3) the empirical correlation of magnetic field flux with X-ray emission (Golub et al. 1980a,b), and (4) the empirical correlation of photospheric magnetic field strength and chromospheric emission line flux in the supergranulation network (Skumanich et al. 1975). The nature of the magnetic heating mechanism is as yet unclear. It may involve the dissipation of MHD waves as originally proposed by Osterbrock (1961) and recently discussed by Ionson (1978), and others (cf. Wentzel 1980 for a review). Alternatively, heating may result from the dissipation of electrical currents which result from nonpotential magnetic field configurations in the corona which, in turn, are induced by photospheric velocity patterns that stress the loop footpoints. A third alternative is steady field annihilation also driven by the twisting of field lines that results from convective motions in the photosphere. Which of these processes is the dominant coronal heating mechanism and the detailed manner in which photospheric motions are converted to coronal heating are major questions for solar physicists in the 1980's.

4. PROBLEMS WITH THE ACOUSTIC WAVE HEATING THEORY FOR STELLAR CORONAE

If acoustic wave dissipation fails to explain the heating of the solar corona, where comparisons with observations are most detailed, then the theory must be regarded as highly questionable for stellar coronae in general. There are, however, at least two additional arguments against the acoustic theory in the stellar context.

First, the theory makes a clear prediction concerning the dependence of the acoustic wave flux on stellar surface gravity that can be tested against observations (cf. Linsky 1980). For stars of similar effective temperature, the effect of lowering the gravity is to decrease the densities deep in the photosphere. Since the transport of convective energy is proportional to ρv_{conv}^2, the decrease in ρ with decreasing gravity causes an increase in v_{conv} (for late-type stars most of the flux near the top of the convection zone is carried by convection rather than radiation, consequently ρv_{conv}^2 is roughly invariant among stars of similar effective temperature). As a result, the acoustic flux must increase with decreasing gravity. Stein and Leibacher (1980) show that the acoustic flux should be proportional to g^{-1}, and acoustic fluxes computed by Renzini et al. (1977) and those computed by Ulmschneider et al. (1977) including photospheric radiative damping losses show roughly the g^{-1} dependence. Linsky and Ayres (1978) and Basri and Linsky (1979) have searched for an increase in chromospheric radiative loss rates with increasing luminosity (decreasing gravity) by measuring Mg II resonance line surface fluxes in some 60 stars. They find no dependence on gravity, contrary to the $\geqslant 3$ order of magnitude increase from dwarfs to supergiants predicted by the theory. More recently, Stencel et al. (1980) have reexamined this question with a larger stellar sample, including additional high luminosity stars, and they find at most a factor of 2 increase in the Mg II surface flux between giants and supergiants. While the inference of chromospheric radiative loss rates (and thus heating rates) from Mg II surface fluxes is somewhat uncertain (cf. Ulmschneider 1979, Linsky 1980), it appears that the acoustic wave theory has difficulty heating stellar chromospheres, much less coronae.

A more direct test of the acoustic wave theory of stellar coronae might be to search for a luminosity dependence of the X-ray surface flux for stars of similar spectral type. This test is currently under way as a part of the Einstein/CFA stellar survey (Vaiana et al. 1980). The survey to date is very incomplete in late-type supergiants and giants, but preliminary results suggest a sharp decrease in the L_X/L_V ratio from dwarfs to supergiants at least among M and late K-type stars, consistent with the IUE results of Linsky and Haisch (1979) which show a substantial decrease in transition region line fluxes in the K giants compared with K dwarfs. This test is not conclusive, however, because chromospheric winds in the cool giants may supplant coronae as the dominant cooling mechanism for the outer atmosphere (e.g. Haisch et al. 1980).

A second argument against the acoustic wave theory is the observation that Mg II line fluxes range over an order of magnitude for stars of similar spectral type and luminosity class (Basri and Linsky 1979), while X-ray fluxes range over nearly three orders of magnitude (Vaiana 1980). These large variations in chromospheric and coronal radiative loss rates are found for stars with presumably very similar convective -- and hence also acoustic -- energy fluxes at the bases of their photospheres. It is difficult to reconcile the acoustic wave theory with such a wide range of heating rates, although purely acoustic heating may explain the lowest heating rates and magnetoacoustic heating may explain the larger heating rates.

Finally, the unexpected Einstein discovery that O and B type stars are prodigious X-ray emitters with fluxes up to 10^{33} ergs s^{-1} (e.g. Vaiana 1980, Cassinelli et al. 1979) cannot be explained by the acoustic theory, since OB stars are generally thought to be too hot to have any significant convective energy transport in their outer layers.

5. WHAT MIGHT THE IMPORTANT STELLAR PARAMETERS BE?

Before I propose a possible broad picture for stellar coronae, it is important to consider what solar and stellar observations suggest might be the important parameters controlling the structure and energy balance of coronae.

5.1. Magnetic Fields

In the past decade a major revolution has occurred in our understanding of solar magnetic field structure. Hale's early investigations presented a picture of the Sun with a weak general magnetic field of order 1 G and strong magnetic fields only in sunspots. However, high resolution magnetographs and spectral polarimetry have since revealed that the global weak field is composed of small-scale (sub arcsec) features (the flux tubes) having kilogauss field strengths. Consequently, virtually all solar fields occur in regions of high field strengths, but the fields have a small filling factor. Why the Sun has such a "curious" magnetic field structure is an important question for theoretical investigation. It is likely that a wide range of stars also have similar "curious" magnetic field structures.

Strong, albeit local, magnetic fields can alter in fundamental ways the energy balance in the overlying chromosphere and corona. For example, the spatial correspondence of strong photospheric magnetic fields (clumps of flux tubes) with overlying bright chromospheric emission and downflows in the supergranulation network (e.g. Skumanich et al. 1975) imply that the magnetic field plays important roles in supplying or channeling nonradiative heating, defining the plasma geometry, and controlling velocity fields in the chromosphere.

Krieger et al. (1971) first showed a general correlation of strong X-ray emission with magnetic flux in the underlying photosphere. Subsequently many investigators pointed out that the solar corona contains two basic structures: open field regions called coronal holes and closed flux tubes. Rosner et al. (1978) have discussed scaling relations between the temperature, pressure, and length of coronal flux tubes. Golub et al. (1980a,b) have shown further that the gas pressures, and hence also the total thermal energy content, in flux tubes are correlated with the total magnetic flux in the underlying photosphere, and that these correlations are consistent with semi-empirical scaling laws. Consequently, the magnetic field determines both the geometry and the energy balance of coronal structures.

Until recently there was no direct evidence for magnetic fields in late-type stars other than the Sun. Kelch et al. (1979) did show that chromospheric models of G-M dwarfs with enhanced Ca II K line surface fluxes for their spectral types are similar to solar plage models. Consequently, such stars, specifically ξ Boo A (G8 V), 70 Oph A (KO V), ϵ Eri (K2 V), and EQ Vir (K7 Ve), very likely are covered with structures comparable in chromospheric emission and perhaps also field strength to solar plages. Another important piece of indirect evidence is provided by star spots whose presence has been deduced photometrically in BY Draconis stars (cf. Kunkel 1975) and in RS CVn-type binary systems (cf. Eaton and Hall 1979). If the starspots are analogous to sunspots, they would indicate the presence of large-scale concentrations of strong magnetic fields.

Boesgaard (1974) presented marginal detections of magnetic fields in three stars using standard magnetograph techniques and photographic detection of the spectra. It is difficult to detect magnetic fields in stars using polarizing optics owing to global cancellation of opposite polarity fields for unresolved sources unless the field has a simple geometry (as in Ap stars). Anderson et al. (1976), however, presented evidence for a 40 kG longitudinal field in a starspot on BY Draconis based on Zeeman spectrograms taken when a large spot was on the stellar limb. A more promising technique, recently proposed by Robinson (1980), involves the direct measurement of Zeeman triplet patterns in unpolarized high resolution spectra by a comparison of the Fourier deconvolution of lines having different Lande g factors. This technique will yield meaningful results only if the magnetic fields have roughly similar strengths over the surface of a star as appears to be true for the Sun. Robinson et al. (1980) have used this technique to measure fields of 2.6±0.4 kG covering roughly 30% of ξ Boo A and fields of 1.9±0.4 kG covering perhaps 10% of the surface of 70 Oph A. The Sac Peak program of stellar magnetic field measurements is being continued actively.

These first direct and indirect studies of magnetic fields in late-type stars are consistent with the following picture: (1) stellar magnetic fields are spatially concentrated (cf. Zwaan 1977) and of roughly constant field strength, (2) stars with large chromospheric emission line surface fluxes have considerably larger magnetic field filling factors and perhaps a modest increase in field strength compared with the Sun, (3) starspots (or spot groups) can be much larger in geometrical extent and cover much greater fractions of the stellar surface than in the solar case, and (4) the magnetic field very likely controls both the geometry and energy balance of late-type stellar coronae. The direct correlation of magnetic field strength or flux with coronal X-ray emission has now been made for the Sun and will presumably be made for stars when the data become available. It is important to point out, however, that the star for which a large field filling factor has been inferred, namely ξ Boo A, has an X-ray luminosity $L_X \cong 2 \times 10^{29}$ ergs s^{-1} (Walter et al. 1978) some 30 times the quiet Sun, but comparable to the Sun entirely covered by plages (Vaiana and Rosner 1978).

5.2. Rotation

During the 1960's, a number of authors found a statistical correlation of decreasing Ca II K line strength with increasing age for solar-type stars. Significantly, such stars also spin down as they grow older owing to the loss of angular momentum by weak coronal winds. This work is summarized by Wilson (1966), and by Linsky (1977), among others. In particular, Skumanich (1972) proposed that both the stellar rotational velocity and Ca II emission strengths decrease with age proportional to $t^{-1/2}$, suggesting a causal relation between rotational velocity and chromospheric heating.

It is now possible to determine such correlations more quantitatively owing to the development of sensitive methods for measuring rotational velocities and quantitative means for deriving chromospheric radiative loss rates from emission line strengths. For example, the photometric periods of spotted stars such as the BY Draconis class yield accurate estimates of rotational velocities independent of uncertain inclinations (sin i). Using such data, Bopp (1980) and Carrasco et al. (1980) comment that as a class the dMe stars, which have larger chromospheric radiative loss rates and therefore larger heating rates than dM stars (Kelch et al. 1979), are also more rapidly rotating. Rapid rotation could be either a consequence of youth or tidally forced synchronism in short-period ($P \lesssim 20^d$) binary systems (Zahn 1977). Young and Koniges (1978) have shown that RS CVn-type binaries with periods less than 20 days have extremely bright Ca II emission lines and Basri and Linsky (1979) have shown that such systems are bright in Mg II as well. Young and Koniges propose that the enhanced chromospheric emission in the close binaries is due to amplification of the acoustic or mechanical wave energy transport driven by tidal coupling.

An alternative explanation is that rapid rotation itself is the underlying cause of enhanced chromospheric heating rather than tidal coupling per se or some other aspect of binarity. For example, Bopp (1980) argues that EQ Vir (dK7e) is a rapid rotator with all the photometric symptoms of spottedness and strong chromospheric emission line behavior of binary BY Draconis stars, but EQ Vir shows no evidence for a close companion. Ayres and Linsky (1980) make a similar argument by demonstrating that the rapidly rotating secondary star in the widely separated, nonsynchronous Capella system is the dominant source of chromospheric and transition region emission, while the slow rotating primary is a comparatively weak emission source.

Fourier deconvolution of high signal-to-noise line profiles has led to recent measurements of stellar rotation velocities with precision heretofore unattainable. Smith (1979) has reviewed recent developments in this field. The detailed comparison between rotational velocity, chromospheric emission and X-ray luminosity for late-type stars using a good statistical sample of stars has not yet been done, but Ayres and Linsky (1980) find a good correlation between X-ray luminosity and rotational velocity for a small sample of late-type stars observed by HEAO-1, and Vaiana (1980) finds a similar preliminary correlation using Einstein observations. It is probably not coincidental that the early G dwarf with the highest measured rotational velocity in Smith's (1979) list, π^1 UMa, has the brightest Ca II K line surface flux of a G dwarf in the Linsky et al. (1979) survey and the highest X-ray luminosity of a G star in the Vaiana et al. (1980) survey with Einstein.

While the connection between coronal X-ray emission and stellar rotation has not yet been established conclusively, there is ample direct and indirect evidence that it is very real. Furthermore, the likely mechanism for this connection in F and cooler stars is almost certainly dynamo action (Parker 1958), which continually regenerates magnetic fields deep in the stellar convection zone.

5.3. Convection Zone Parameters

Since O- and B-type stars, which lack appreciable convection zones, have coronae with X-ray luminosities up to 10^7 times larger than the quiet Sun, the existence of a convection zone must not be a prequisite for the existence of a corona. However, a striking result of the Einstein stellar survey is the rapid increase in typical values of L_x/L_v from G- to M-type stars (Vaiana 1980). The increase in L_x/L_v ratios is far faster than the decrease in L_v/L_{BOL} with decreasing effective temperature, consequently there must be real increase in L_x/L_{BOL} with decreasing effective temperature. Since there is no known statistical correlation of stellar rotational velocity with effective temperature for stars later than spectral type G, I conclude that some aspect of the stellar convection zone plays an important role in determining coronal properties of cool stars.

5.4. Coronal Base Pressure and Gravity

Hearn (1975) proposed that the properties of a stellar corona can be computed by minimizing the total energy loss due to radiation, conduction, and the stellar wind. This so-called minimum flux corona concept was initially applied to a plane-parallel, hydrostatic corona with no magnetic fields. Hearn (1977) subsequently generalized the minimum flux concept to include the effect of inclined magnetic fields on conductive cooling of the corona. In the absence of magnetic fields the minimum flux corona hypothesis predicts that increased coronal temperature produces both increased X-ray emission and stellar wind losses. However, such a correlation between X-ray emission and stellar wind loss is the opposite of what is seen on the Sun (see §3).

The minimum flux corona hypothesis is highly controversial. Vaiana and Rosner (1978), Antiocus and Underwood (1978), among others, have criticized the theory on thermal stability and other grounds, and Hearn (1980) has confronted his critics. It is not the intent of this paper to join the controversy other than to point out that the minimum flux corona hypothesis does not address what I consider to be the most important question, namely, how coronae are heated and what stellar parameters determine the coronal heating rates. Instead, a coronal base pressure is assumed as a lower boundary condition, but the base pressure must be a direct result of the heating rate. Thus the minimum flux corona hypothesis provides little guidance at this time for developing a broad picture of stellar coronae.

6. A SPECULATIVE PICTURE OF STELLAR CORONAE

While systematic surveys of stellar coronae in different regions of the HR diagram and as functions of such parameters as rotation, gravity, age, and chromospheric properties are incomplete, enough preliminary data are available and have been presented (Vaiana 1980) to begin to piece together a speculative picture of stellar coronae. In doing so I am guided by the principle that magnetic fields define the basic geometry and control the energy balance of stellar coronae as they clearly do for the Sun (see §3). Magnetic fields can either be slowly decaying remnants from the protostellar collapse or be regenerated by dynamo processes. In either case the fields are probably stochastic, as they are for the most part in the solar case, consequently stellar coronae should be temporally variable and spatially inhomogeneous.

Many of the speculations that follow are simple extrapolations of trends seen in the solar and stellar data. Rosner and Vaiana (1980) have independently arrived at similar conclusions. Finally, I should clearly state that what follows is highly speculative and in some cases not corroborated in any way by existing data.

6.1. O- and B-Type Stars

The O-type Einstein sources have X-ray luminosities in the range log L_x = 32-34 and surface flux ratios in the range log L_x/L_v = -5 to -3. The B-type stars detected tend to have somewhat smaller values of L_x and L_x/L_v. Since these stars have short evolutionary time scales, they all must be young and it seems reasonable to assume that they possess remnant magnetic fields that have not yet decayed. I assume that dynamo processes cannot operate in such stars owing to the absence of appreciable surface convection zones.

The OB stars generally have strong winds, especially the most luminous O- and Of-type stars (cf. Cassinelli 1979). A critical aspect of these winds is that they are not steady. For example, Conti and Niemela (1976) have found long-term changes in the Hα profiles of ζ Pup (O4If) and York _et al._ (1977) have found changes in the O VI absorption components of three O stars on time scales of hours. These and more recent data suggest turbulent flows. Such flows can twist and stretch embedded magnetic fields into configurations that are far from potential. The stressed fields may then dissipate their stored energy by mechanisms similar to those responsible for heating the _solar_ corona. An alternative mechanism, not requiring magnetic fields, has been proposed by Lucy and White (1980). They note that winds driven by radiation pressure in lines are unstable (cf. references in their paper), and that nonradiative heating occurs in bow shocks between accelerated cool blobs that plough through the ambient material, which is not radiatively driven due to shadowing by the blobs. Castor (1980) is investigating the likely geometry of shock waves that are produced in this flow. If either of these scenarios is correct, then the brightest hot-star X-ray sources should be those with the most turbulent wind flows. Consequently, the most luminous O stars, which have the greatest mass loss rates and likely also the most turbulent flows, should be the strongest X-ray sources.

6.2. A and Ap Stars

Those A-type stars which have been detected by _Einstein_ exhibit a very wide range of log L_X (26-31) and log L_X/L_V (-8 _to_ -3). The most interesting A stars from the perspective of coronae are the Ap stars. These stars are thought to possess kilogauss fields on the basis of direct measurements or by inference. In particular, the Ap stars often exhibit abundance anomalies which are commonly interpreted as the result of diffusion in an extremely stably stratified atmosphere in which convective motions have been suppressed by strong fields with a simple dipole geometry.

In a search of 18 B5-A7 stars with HEAO-1, Cash _et al._ (1979) detected one normal B star (π Cet) and one classical Ap star (φ Her). The remaining stars were not detected even though many are closer and optically brighter than the stars detected. Snow and Cash (1980) have continued this program by studying five bright Ap and Am stars with _Einstein_. They determined three upper limits (log $L_X < 28$) and two detections, but both are binary systems and the unresolved companions may be the X-ray sources.

The interesting question is why stars with kilograuss fields are weak X-ray sources if magnetic field strength appears to be intimately associated with coronae and strong X-ray emission, at least for solar-type stars. I think the answer is that magnetic fields *per se* do not do the heating; rather the heating results from either MHD wave processes or the relaxation of stressed (nonpotential) fields perhaps by particle acceleration, current dissipation, or magnetic field annihilation. All of these processes require turbulent motions in addition to magnetic fields. The detection of kilogauss fields by Babcock-type magnetographs implies simple, stable field configurations in the Ap stars. It is possible that there are two distinct classes of field configurations for A-type stars: either strong dipolar fields that suppress turbulent motions and thereby suppress coronal emission, or weaker fields that are disordered and somewhat turbulent owing to weak convective motions. Sirius, Vega, and the *Einstein* A-star sources may be of the latter type.

6.3. F-Type Dwarfs

F-type dwarfs are characterized by log L_x = 28.5-30.5 and log L_x/L_v = -5 to -3. Thus the X-ray emission is 10^1 - 10^3 times that of the quiet Sun and comparable to or somewhat brighter than solar plages. I assume that the typically large values of L_x are a consequence of the onset of dynamo-regenerated magnetic fields near spectral type F0, while the wide range of L_x is due to the range in rotational velocity, that in turn depends on age as described above.

The theory of stellar dynamos is far from understood, and there is no consensus yet as to which type of dynamo process operates in the Sun. Despite this important uncertainty, a critical parameter determining the kind of differential rotation expected in a stellar convection zone, and thus the efficiency of dynamo field regeneration, is the ratio of the convective turnover time to the stellar rotation period (Gilman 1980). Since the convective turnover time should be roughly proportional to the convection zone depth (d_{conv}), the rate of creation of magnetic flux should depend functionally on the product $v_{rot}d_{conv}$.

In Figure 1, I schematically plot the increase in d_{conv} toward cooler stars and the dependence of rotational velocity on spectral type for young and old stars. If turbulent magnetic fields are responsible for coronal heating and the field regeneration rate depends functionally on $v_{rot}d_{conv}$, then L_x/L_v should depend on spectral type and stellar age as shown schematically in Figure 1. In particular, there should be a rapid increase in L_x/L_v in the early F stars and L_x/L_v should depend on v_{rot} and thus age for single stars of the same spectral type. The L_x/L_v versus age prediction can be tested by observing clusters of different ages. In fact, Stern *et al.* (1979) have reported that F stars in the young Hyades cluster have relatively large L_x/L_v values compared with F stars in the solar neighborhood. Finally Vaiana *et al.* (1980) find that the largest L_x/L_v ratio for F stars is about 10^{-3}, implying that 10^{-3} is the maximum efficiency for converting convective flux to magnetic fields by dynamo processes in such stars.

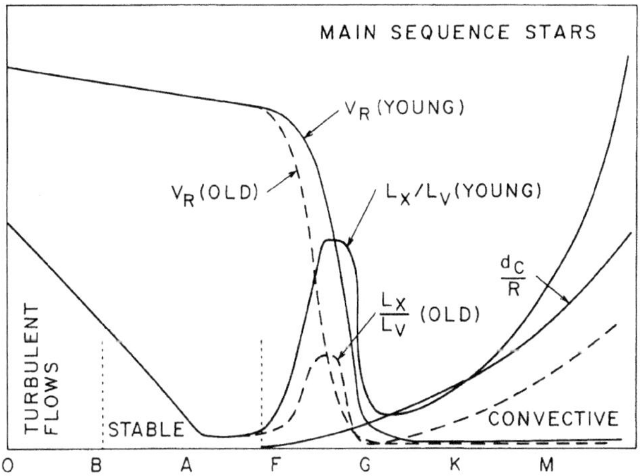

Fig. 1. A schematic representation of parameters relevant to stellar coronae for main sequence stars. Included are systematic trends of stellar rotational velocities (v_R) for young and old stars, the ratio of convection zone depth to stellar radius (d_c/R), and expected X-ray-to-visual luminosity ratios (L_X/L_V) for old and young stars. A rough division into regimes of turbulent flows, stable photospheres, and significant convective zones is indicated.

6.4. G and K Dwarfs

G- and K-type dwarfs have X-ray luminosities in the range 10^{27}–10^{30} ergs s^{-1} and log L_X/L_V in the range -6 to -2. The median X-ray luminosities are an order of magnitude smaller than for the F stars, presumably because the mean rotational velocities of field G- and K-type stars are smaller than field F-type stars. As previously noted, π^1 UMa is a very rapid rotator for its spectral type (G0 V) and is the G dwarf with the largest L_X so far detected.

6.5. dM and dMe Stars

The coolest dwarf stars also exhibit a wide range in L_X (log L_X = 27-30) and in L_X/L_V (log L_X/L_V = -3 to 0). Since M dwarfs are cool, only a small fraction of the bolometric luminosity appears in the visual band. Thus values of log L_X/L_{BOL} probably lie in the range -4 to -1, but these ratios are still far larger than for any other stellar types. One plausible reason is that these stars are the only dwarfs with convection zones that extend most or all of the way down to the stellar core. Thus if L_X/L_{BOL} depends functionally on the product $v_{rot}d_{conv}$, then EQ Vir (dK5e) and BY Dra (dM0e) should have far larger L_X/L_{BOL} ratios than π^1 UMa (G0 V), since all three stars have similar v_{rot} (cf. Smith 1979) but EQ Vir and BY Dra have much deeper convection zones. Also the type of dynamo and its efficiency in regenerating magnetic flux may be different for stars with deep convection zones compared to stars with shallow zones (Gilman 1980).

Although detailed calculations are not yet available, I presume that M dwarfs all have deep convection zones with comparable values of d_{conv}. Thus the wide range in L_X/L_{BOL} should be due mainly to differences in v_{rot}, which depends on stellar age (cf. Carrasco et al. 1980). M dwarfs are generally divided into two groups: dM stars with no emission lines in the visible except Ca II, and dMe stars which show emission in the Balmer and other lines. If, as I presume, rotational velocity is the single key parameter determining the X-ray emission of M dwarfs, then dM stars should be slower rotators than the dMe stars and the dM stars should occupy the low range of L_X/L_{BOL} in the M star distribution. Kelch et al. (1979) have found that the chromospheric radiative loss rate (and thus nonradiative heating rate) for 61 Cyg B (a dM0 star) is much smaller than for EQ Vir (dK7e), a result consistent with lower field regeneration rates in dM stars.

6.6. Close Binary Systems

Walter et al. (1980a) find that X-ray luminosities for RS CVn-type binaries typically lie in the range log L_X = 30.3-31.3 and log L_X/L_{BOL} = -5 to -2.5. These X-ray luminosities are large, but they are similar to the brightest of the G-K dwarf single stars (Walter et al. 1980b, Vaiana et al. 1980), and thus the mechanism for heating the coronae of RS CVn-type systems is probably similar to that operating in single stars. As previously noted, synchronism of rotation and orbital period is expected for RS CVn-type binaries with periods less than about 20 days. Thus the K0 IV active stars in such systems are probably strong X-ray emitters as a result of tidally-induced rapid rotation, whereas single stars with comparable X-ray luminosities are rapid rotators as a result of youth.

Simon and Linsky (1980) have computed chromospheric models for the K0 IV stars in two representative RS CVn systems, HR 1099 and UX Ari. Pressures at the top of the chromospheres in these stars lie in the range 0.2-1 dynes cm^{-2} and are consistent with density diagnostic line ratios for lines formed near 5×10^4 K. Rosner et al. (1978) have derived an empirical scaling law for closed flux tubes in the solar corona, $T_c = 1.4 \times 10^3 (P_c L)^{1/3}$, where T_c, P_c, and L are the coronal temperature, pressure, and flux tube length, respectively. Assuming $T_c = 10^7$ K (Walter et al. 1978) and coronal pressures equal to those derived for the top of the chromosphere, L ≈ 2-8 times the K0 IV star radius. Since the semimajor axis of the UX Ari system is 16 K star radii, these data suggest that coronal flux tubes from the K-star fill a significant portion of the volume of the system.

Short period RS CVn systems have been observed to flare at radio, optical, and ultraviolet wavelengths for periods of hours to almost two weeks. During one flare on UX Ari, Simon et al. (1980) observed with IUE a factor of six enhancement in ultraviolet emission line fluxes and asymmetries in the Mg II lines indicating flows of material from the K0 IV star to the G5 V star in the system. They interpret these data as resulting from the interaction of large flux tubes of both stars, field reconnection, and the temporary magnetic connection of the two stars via a single flux tube that permits flow of material from one star to the other. While this scenario is indeed speculative, it is based on the reasonable deduction that these stars have large closed field configurations which can have lengths comparable to the stellar separations.

6.7. C-M Giants and Supergiants

Except for G- and K-type giant components of long period RS CVn-type systems like Capella (e.g. Holt et al. 1979), G-M-type single giants and supergiants have not yet been detected as X-ray sources. I expect that G-type single giants and supergiants will eventually be detected, but the cooler luminous stars probably have no hot coronae at all or perhaps have only small regions in their atmospheres hot enough to emit X-rays. The upper limit on the X-ray surface flux of α Sco (M1Ib+B), for example, is 10^{-4} that of a coronal hole (Vaiana 1980).

The apparent absence of detectable coronae in such stars can perhaps be explained simply using Parker's wind theory. When the critical point lies at the base of a thermally-driven wind, the coronal temperature is maximized and is related to the stellar mass and radius by the equation

$$T_c = 5.5 \times 10^6 \text{ K} \left(\frac{M_*}{M_\odot}\right)\left(\frac{R_\odot}{R_*}\right) \quad .$$

For assumed masses of order unity, α Boo (K2 III) would have $T_C = 2 \times 10^5$ K and α Ori (M2 Iab) would have $T_C = 1 \times 10^4$ K. Thus one would not expect to see X-ray emission from such stars unless the masses are much larger than solar or closed flux tubes can survive disruption by the large stellar winds in some regions of these stars.

Stencel (1978) and Stencel and Mullan (1980) have shown that asymmetries in the Ca II and Mg II resonance line profiles are qualitatively different on either side of a roughly vertical line in the HR diagram near spectral type K2 III. They interpreted the change in asymmetry as due to the rapid onset of large winds as a star evolves toward cooler effective temperatures across the dividing line. On theoretical grounds, Mullan (1978) inferred that there should be a rapid increase in stellar winds on a line in the HR diagram that he called the supersonic transition locus (STL). This line is in roughly the same location on the HR diagram as the Ca II and Mg II asymmetry lines. Subsequently, Linsky and Haisch (1979) found from IUE observations that along a line in the HR diagram in a similar location to the STL and asymmetry lines, there is a rather abrupt change from stars with transition region emission lines ($T > 20,000$ K) to the left side of the line to stars with no or extremely weak transition emission lines on the right side of the dividing line. The sharpness of this dividing line may be less than originally thought (Dupree and Hartmann 1980). Thus there appears to be a good empirical correlation of bright transition region lines with weak winds and weak transition region emission with strong winds.

There are several probable consequences of this empirical correlation and these dividing lines in the HR diagram:

(1) The existence of a transition region strongly implies the presence of a hot corona. The apparent absence of coronae in single stars among giants and supergiants cooler than about spectral type K2 III is consistent with this picture.

(2) The apparent absence of coronae to the right of the dividing line implies that the winds are sufficiently strong to disrupt quickly any closed magnetic loop structures that are stable in stars to the left of the dividing line.

(3) The energy balance in the outer atmospheres of stars on either side of the dividing lines is probably very different, although the input of nonradiative energy could be similar. To the left of the dividing lines the nonradiative energy input to the corona is probably dissipated as X-ray emission and thermal conduction to the transition region. To the right, the energy input is probably converted into the expansion of the stellar wind.

(4) For stars to the right of the dividing lines the winds cannot be Parker-type flows because the material is so cool. Also, except for M-type supergiants, there is no measured infrared excess or other evidence for dust, so that radiation pressure on dust cannot accelerate the wind. Instead, Haisch et al. (1980) suggest that the winds are accelerated by the momentum deposition of Alfven waves and Lα radiation pressure. Hartmann and MacGregor (1980) have investigated winds driven by the momentum and energy deposition of acoustic and Alfven waves.

The early G supergiants, of which α Aqr (G2 Ib) and β Aqr (G0 Ib) are prototypes, are interesting in that they have bright transition region emission lines and strong winds (their Ca II and Mg II lines exhibit absorption components blue-shifted up to ~125 km s^{-1} [Dupree and Hartmann 1980, Hartmann et al. 1980]). Thus the outer atmospheres of these stars are hybrid in character. In addition, Dupree and Baliunas (1979) report that both the wind velocity and emission line fluxes in α Aqr are variable. One possible explanation for the hybrid character of the outer atmospheres of these stars is that they may consist of closed loops and open field structures similar to solar coronal holes.

The closed loops contain the hot, X-ray emitting material while the coronal holes produce the supersonic wind. The supergiant wind is much cooler and has far greater mass loss than the high-speed streams from solar coronal holes probably as a result of the three orders of magnitude smaller stellar gravity.

7. FINAL COMMENTS

During the next few years I would like to see major progress in addressing the following uncompleted tasks:

(1) First, we need a complete survey of X-ray emission from stars throughout the HR diagram. This work is fortunately far along and Vaiana et al. (1980) will soon publish the results of the first stellar survey with Einstein.

(2) Most of the observations to date have been single detections with no repeated observations to test for time variability. One anticipates variability owing to rotational modulation and evolution of loops, coronal holes, and active regions, as well as to flares and stellar cycles. Since a great deal of observing time is needed to search for variability on many time scales in a reasonable sample of stars, a dedicated X-ray and ultraviolet observatory is probably required.

(3) To date we have very little information on coronal temperatures, densities, and flares. To measure these critical quantities we need a high resolution X-ray spectroscopy experiment with sufficient sensitivity to observe a large number of stars.

(4) Finally we need a concerted effort to develop a theory of stellar coronae with high predictive power. In particular, we need further studies of dynamo processes, mechanisms for converting turbulent magnetic fields into heat, and acceleration mechanisms for stellar winds.

I wish to thank Drs. G. S. Vaiana and R. Rosner for communication of unpublished Einstein data and extensive discussions concerning all aspects of stellar coronae. This paper could not have been written without their important input. I would also like to thank Drs. T. R. Ayres and R. McCray for carefully reading the manuscript and suggesting changes. This work was supported in part through grants to the University of Colorado from the National Aeronautics and Space Administration.

REFERENCES

Anderson, C. M., Hartmann, L. W., and Bopp, B. W. 1976, Ap. J. (Letters), 204, L51.
Antiocus, S. K. and Underwood, J. H. 1978, Astr. Ap., 68, L19.
Athay, R. G. and White, O. R. 1978, Ap. J., 226, 1135.
_____. 1979, Ap. J. Suppl., 39, 333.
Ayres, T. R. and Linsky, J. L. 1980, Ap. J., submitted.
Basri, G. S. and Linsky, J. L. 1979, Ap. J., 234, 1023.
Biermann, L. 1946, Naturwiss, 33, 118.
Boesgaard, A. M. 1974, Ap. J., 188, 567.
Bopp, B. W. 1980, in Highlights of Astronomy, 5, p. 847.
Bruner, E. C., Jr. 1979, B.A.A.S., 11, 697.
Carrasco, L., Franco, J., and Roth, M. 1980, Astr. Ap., in press.
Cash, W., Snow, T. P., Jr., and Charles, P. 1979, Ap. J. (Letters), 232, L111.
Cassinelli, J. P. 1979, Ann. Rev. Astr. Ap., 17, 275.
Cassinelli, J. P. Waldron, W. L., Vaiana, G. S., and Rosner, R. 1979, B.A.A.S., 11, 775.
Castor, J. I. 1980, private communication.
Conti, P. S. and Niemala, V. 1976, Ap. J. (Letters), 209, L37.
Cram, L. E. 1978, Astr. Ap., 70, 345.
de Loore, C. 1970, Astr. Space Sci., 6, 60.
Dupree, A. K. and Baliunas, S. 1979, IAU Circular No. 3435.
Dupree, A. K. and Hartmann, L. 1980, in Stellar Turbulence, eds. D. F. Gray and J. L. Linsky (New York: Springer-Verlag), p. 279.
Eaton, J. A. and Hall, D. S. 1979, Ap. J., 227, 907.
Gilman, P. A. 1980, in Stellar Turbulence, eds. D. F. Gray and J. L. Linsky (New York: Springer-Verlag), p. 19.
Golub, L., Maxson, C., Rosner, R., Serio, S., and Vaiana, G. S. 1980a, preprint.
Golub, L., Rosner, R., Vaiana, G. S., and Weiss, N. O. 1980b, submitted to Ap. J.
Gough, D. O. 1976, Proc. IAU Colloq. No. 36, ed. R. M. Bonnet and P. Delache (Paris: Clermont-Ferand), p. 3.
Haisch, B. M., Linsky, J. L., and Basri, G. S. 1980, Ap. J., 235, 519.
Hartmann, L., Dupree, A. K., and Raymond, J. C. 1980, Ap. J. (Letters), in press.
Hartmann, L. and MacGregor, K. B. 1980, preprint.
Hearn, A. G. 1975, Astr. Ap., 40, 355.
_____. 1977, Solar Phys., 51, 159.
_____. 1980, Astr. Ap., in press.
Holt, S. S., White, N. E., Becker, R. H., Boldt, E. A., Mushotzky, R. F., Serlemitsos, P. J., and Smith, B. W. 1979, Ap. J. (Letters), 234, L65.
Ionson, J. A. 1978, Ap. J., 226, 650.
Kelch, W. L., Linsky, J. L., and Worden, S. P. 1979, Ap. J., 229, 700.

Krieger, A. S., Vaiana, G. S., and Van Speybroeck, L. P. 1971, in Solar Magnetic Fields, ed. R. Howard (Dordrecht: D. Reidel), p. 397.
Kunkel, W. E. 1975, in Variable Stars and Stellar Evolution, ed. V. E. Sherwood and L. Plant (Dordrecht: D. Reidel), p. 15.
Kuperus, M. 1965, Recherches Astr. Obs. Utrecht, 17, 1.
_____. 1969, Space Sci. Rev., 9, 713.
Lighthill, M. J. 1952, Proc. Roy. Soc. London, A211, 564.
Linsky, J. L. 1977, in The Solar Output and Its Variation, ed. O. R. White (Boulder: Colorado Associated University Press), p. 477.
_____. 1980, Ann. Rev. Astr. Ap., 18, in press.
Linsky, J. L. and Ayres, T. R. 1978, Ap. J., 220, 619.
Linsky, J. L. and Haisch, B. M. 1979, Ap. J. (Letters), 229, L27.
Linsky, J. L., Worden, S. P., McClintock, W., and Robertson, R. M. 1979, Ap. J. Suppl., 41, 47.
Liu, S. Y. 1974, Ap. J., 189, 359.
Lucy, L. B. and White, R. L. 1980, preprint.
Mewe, R. 1979, Space Sci. Rev., 24, 101.
Mullan, D. J. 1978, Ap. J., 226, 151.
Osterbrock, D. E. 1961, Ap. J., 134, 347.
Parker, E. N. 1958, Ap. J., 128, 664.
Pneuman, G. W. 1973, Solar Phys., 28, 247.
Renzini, A., Cacciari, C., Ulmschneider, P., and Schmitz, F. 1977, Astr. Ap., 61, 39.
Robinson, R. D. 1980, Ap. J., in press.
Robinson, R. D., Worden, S. P., and Harvey, J. W. 1980, Ap. J., in press.
Rosner, R., Tucker, W. H., and Vaiana, G. S. 1978, Ap. J., 220, 643.
Rosner, R. and Vaiana, G. S. 1980, in X-Ray Astronomy, eds. R. Giacconi and G. Setti (D. Reidel, Dordrecht), p. 129.
Schwarzschild, M. 1948, Ap. J., 107, 1.
Simon, T. and Linsky, J. L. 1980, Ap. J., submitted.
Simon, T., Linsky, J. L., and Schiffer, F. H. III 1980, Ap. J., in press.
Skumanich, A. 1972, Ap. J., 171, 565.
Skumanich, A., Smythe, C., and Frazier, E. N. 1975, Ap. J., 200, 747.
Smith, M. 1979, P.A.S.P., 91, 737.
Snow, T. P., Jr. and Cash, W. 1980, private communication.
Stein, R. F. and Leibacher, J. 1974, Ann. Rev. Astr. Ap., 12, 407.
_____. 1980, in Stellar Turbulence, eds. D. F. Gray and J. L. Linsky (New York: Springer-Verlag), p. 225.
Stencel, R. E. 1978, Ap. J. (Letters), 223, L37.
Stencel, R. E. and Mullan, D. J. 1980, Ap. J., in press.
Stencel, R. E., Mullan, D. J., Linsky, J. L., Basri, G. S., and Worden, S. P. 1980, Ap. J. Suppl., submitted.
Stern, R., Underwood, J. H., Zolcinski, M., and Antiocos, S. 1979, B.A.A.S., 11, 781.
Ulmschneider, P. 1967, Z. Ap., 67, 193.
_____. 1979, Space Sci. Rev., 24, 71.

Ulmschneider, P., Schmitz, F., Renzini, A., Cacciari, C., Kalkofen, W., and Kurucz, R. L. 1977, Astr. Ap., 61, 515.
Vaiana, G. S. 1980, paper presented at High Energy Astrophysics Meeting of the AAS, Cambridge, Mass., 28 January 1980.
Vaiana, G. S. and Rosner, R. 1978, Ann. Rev. Astr. Ap., 16, 393.
Vaiana, G. S. et al. 1980, in preparation.
Walter, F., Cash, W., Charles, P., and Bowyer, S. 1980a, Ap. J., 236, 212.
Walter, F., Charles, P., and Bowyer, S. 1978, Astr. Ap., 83, 1539.
Walter, F., Linsky, J. L., Bowyer, S., and Garmire, G. 1980b, Ap. J. (Letters), 236, in press.
Wentzel, D. G. 1980, in The Sun as a Star, Vol. VI A of the NASA-CNRS Monograph Series, preprint.
Wilson, O. C. 1966, Science, 151, 1487.
Withbroe, G. L. and Noyes, R. W. 1977, Ann. Rev. Astr. Ap., 15, 363.
York, D. G., Vidal-Madjar, A., Laurent, C., and Bonnet, R. 1977, Ap. J. (Letters), 213, L61.
Young, A. and Koniges, A. 1978, Ap. J., 211, 836.
Zahn, J.-P. 1977, Astr. Ap., 57, 383.
Zwaan, C. 1977, Mem. S. A. It., 48, 525.

X-RAY IMAGING: SUPERNOVA REMNANTS

David J. Helfand
Columbia Astrophysics Laboratory
Departments of Astronomy and Physics
Columbia University
New York, New York 10027

The Einstein Observatory has provided astronomers with the first opportunity to obtain optical quality images of extended X-ray sources. Such pictures of supernova remnants, for example, yield a wealth of information on the dynamics and evolution of the expanding blast wave, the possible collapsed remnants of the exploded star, and the structure of the interstellar medium. We herein review imaging observations of over 50 supernova remnants obtained during the first year of the Observatory's operation. Although analysis of these results is still in its infancy, it is already possible to draw important conclusions regarding models for stellar explosions, remnant evolution, neutron star formation, and the interstellar medium.

The Crab Nebula is object number one in the Messier Catalog of diffuse optical emission regions, testifying to the longstanding importance of optical imaging in the study of supernova remnants (SNR). Examples of recent progress in this area include the wide-field Hα surveys of Parker, Gull, and Kirschner (1979), which have revealed large shell-like features dotting the Milky Way, and the coronal Fe XIV line imaging data of several groups (Woodgate *et al.* 1974; Woodgate, Kirschner, and Balon 1977; Dopita and Mathewson 1979; Murdin, Clark, and Culhane 1978), which allow optical astronomers to trace out the hot gas associated directly with the high-velocity, shock-heated material. But optical emission from SNR (excluding the Crab) arises mostly from the relatively cool, dense regions associated with interstellar density enhancements or, as in the case of Cas A, with dense knots of stellar ejecta.

With the advent of interferometers, SNR were some of the first objects imaged by radio astronomers. The radio emission is thought to be more closely associated with the expanding shock, originating from the acceleration of high-energy particles in regions of turbulent gas and compressed magnetic fields. However, the total flux emitted in the radio regime is but a tiny fraction of the energy budget of the remnant.

R. Giacconi (ed.), X-Ray Astronomy with the Einstein Satellite, 39–49.

It is the X-ray emission that traces out the boundary of the expanding shock and that dominates the radiative energy losses of the remnant throughout much of its lifetime. It is the X-ray emission that can provide information on the possible collapsed remnants of the explosion — neutron stars and/or black holes. And it is the X-ray emission that allows us to use the supernova shock as a probe of the structure of the interstellar medium. As a result of this rich potential for SNR X-ray imaging, many rocket payloads have been dedicated over the past ten years to the imaging of nearby, large angular scale remnants such as Vela, Puppis A, the Cygnus Loop, and, through lunar occultation, the Crab Nebula. These observations, with a spatial resolution of a few arc minutes, showed that the X-ray emission from these objects followed the general outline of the radio contours although large variations in brightness over the face of the remnants were typical and the detailed correlation with optical and radio maps was poor. More importantly, perhaps, they gave us an indication of the wealth of information an X-ray imaging system with arc-second resolution could provide. For the sake of brevity, then, we dispense with a discussion of this early work and proceed directly to a review of the first year's results on supernova remnant imaging with the Einstein Observatory.

Over 30 known galactic remnants and a similar number of objects in other galaxies have been observed in the 0.15–4.5 keV band with the imaging instruments on board the Einstein satellite. The imaging proportional counter (IPC) provides a $1° \times 1°$ field of view with $1'$ spatial resolution and modest energy resolution. The high resolution imager (HRI) produces broad-band pictures with $4''$ resolution over a $20'$ field; further details of the instruments can be found in Giacconi *et al.* (1979). Nearly all of the sources observed have been detected, and many have been mapped in considerable detail. The only results to appear in print to date are those for the youngest galactic remnant, Cas A (Murray *et al.* 1979; Fabian *et al.* 1980), and for a dozen radio and optically selected remnants in the Large Magellanic Cloud (LMC) (Long and Helfand 1979). The overwhelming increase in both the quantity and quality of the new data has left the analysis somewhat behind the pace of the observations. Below, we briefly review the published work while concentrating on a presentation of preliminary results from a number of groups which give a glimpse of the major advances these data will provide.

Pictures for each of the seven historical remnants (Cas A, Tycho, Kepler, the Crab, SN1006, 3C58, and RCW86) are now available. In Cas A we see a diffuse circular shell of emission studded with bright X-ray knots (Murray *et al.* 1979). The brightest X-ray features are associated with optical knots exhibiting [S II] emission rather than those in which [O III] predominates. Indeed, as reported by Smith (1980), the X-ray spectrum of the remnant is dominated by strong Si and S lines. An X-ray jet in the northeast quadrant is associated with a string of high-velocity optical knots suggesting conductive heating of clumps of ejected material. However, at least one bright X-ray feature is coincident with one of the optical quasi-stationary flocci thought to represent circumstellar matter which resulted from an episode of mass loss prior to the supernova explosion. The general distribution of X-ray emission is consistent with a model consisting of two thin shells — the harder, outer emission from the high-velocity leading shock and the remainder from a reverse shock formed as a result of the deceleration of the initial blast wave and propagating inward, heating ambient material (Fabian *et al.* 1980). There is a hard X-ray plateau at a radius of approximately 140″ which corresponds to a similar emission region seen in radio maps, implying a possible synchrotron contribution to the hard X-ray flux from the remnant. Fabian *et al.* (1980) have calculated the mass of the X-ray emitting gas using the two-shell model and assuming pressure equilibrium. By requiring a fit to the X-ray counting-rate map and agreement between the computed spectrum and the spectrum observed by other instruments (Davison, Culhane, and Mitchell 1976; Becker *et al.* 1979), they arrive at a total of $\gtrsim 15\ M_{\odot}$. Coupled with the evidence supporting significant preexplosion mass loss, it appears certain that the Cas A progenitor was a massive star.

Although several of the other historical remnants are thought to have resulted from Type I supernova explosions and thus to have had lower mass progenitors, their X-ray morphology is rather similar to that of Cas A — circular, limb-brightened shells with strong variations around the shell rim which are often uncorrelated with bright spots in the radio and optical maps. In Tycho's remnant, the remarkable circular symmetry is broken only by a small protuberance in the southeast quadrant which corresponds to a similar feature in the radio image. The brightest X-ray region, however, is found in the northwest quadrant, whereas the brightest radio emission is found in the eastern limb. In Kepler, the circular symmetry is again striking although there is a ratio of 10:1 between the emissivity of the northwest and southeast regions of the shell (Pisarski and Long 1979). For the remnant of the nearest supernova of historical times, SN1006, we again see a circular limb-brightened shell (Pye *et al.* 1980), while for the oldest known recorded explosion, RCW86 (SN185 AD), the remnant shell is beginning to depart from circularity as it continues its expansion into a nonuniform medium.

The historical remnants, along with a few older, nearby sources, make up the first of three classes of X-ray morphology which characterize the SNR we have studied. The second type, exemplified by remnants such as W28, W44, and HB3, show no distinct shell, although the ragged outlines of the emission follow the radio map contours in a general way. These objects most likely represent older SNR in which the entire volume interior to the now slowly expanding blast wave has been heated to X-ray temperatures. A transition between remnants with strong shells and these "amorphous blob" sources is represented by the intermediate-aged remnants such as IC443 and Puppis A in which a still sharply defined shell encloses a more or less uniformly emitting volume of gas. Each of these morphological types is represented in the sample of over 20 SNR detected by the Columbia group (Long and Helfand 1979; Helfand and Long 1979) in the LMC. The low hydrogen column densities and well-determined distances for this set of objects make them a particularly valuable sample for the study of remnant evolution.

The third class of X-ray morphology found in our sample arises from a distinctly different physical process — the powering of a synchrotron emitting nebula by a central neutron star. The Crab Nebula is, of course, the prototype of this class. Its X-ray emission is strongly peaked toward the remnant center with a bright ridge of emission overlying the brightest optical filament and the pulsar standing out as a bright central point source (Harnden *et al.* 1979). The other historical remnant (SN1181) among the nine radio-selected Crab-like or plethoric remnants is 3C58. It has been detected as a point source in the IPC, and follow-up HRI observations are planned. Its spectrum, like that of the Crab, shows no line emission features and can be well represented by a power law of slope -2 (Smith 1980). Other detected members of this class include G21.5-0.9 and G74.9+1.2 (Wilson 1980), although the large X-ray extent (~20 pc) and relatively weak central condensation of this latter source may indicate that it is a more usual, old, amorphous remnant like W28 rather than a plerion. One or two of the LMC remnants may also belong to the Crab-like class of objects.

Observations of the historical remnants are used to derive estimates of the ejected and swept-up masses which imply progenitor types and explosion energies, while comparison with the large sample of older remnants provides important information on SNR evolution. For nearby objects such as the Cygnus Loop, however, detailed modeling of the supernova blast wave and its interaction with the interstellar medium becomes possible. Several X-ray imaging studies of this source have been carried out over the past five years. One of the more recent rocket flights (Tuohy, Nousek, and Garmire 1979) has shown that in the northeast quadrant, the X-ray emission is coming from up to 5′ outside of the bright optical filaments. The shock velocity derived from these observations and the optical coronal Fe XIV line studies of Woodgate, Kirschner, and Balon (1977) is ~300 km s^{-1}, in contrast to the $\lesssim$100 km s^{-1} which characterizes the excitation of the dominant optical filaments. In addition, Raymond *et al.* (1979) have recently studied a very faint, linear optical filament which is about 5′ outside of the western limb of the remnant. They find that this feature has a pure Balmer spectrum, similar to the spectra of optical filaments associated with the much younger Tycho remnant and in sharp contrast to the spectra of the bright optical filaments in Cygnus which are dominated by lines of [O III]. They argue that the Balmer lines cannot be due to recombination and hence must result from the collisional excitation of cold interstellar material swept up by a fast shock.

A 50-pointing IPC mosaic of the Cygnus Loop, supplemented by numerous HRI pictures with 1000 AU spatial resolution, is being assembled by the Columbia group. The first few pointings, primarily along the western limb, show a wealth of detail likely to keep theorists busy for years to come. In general, the brightest X-ray features overlie the bright optical filaments, arguing for a close association between the million degree ionized X-ray emitting gas and the 10^4 K material responsible for the optical line emission. Such cospatiality is expected in models where slow shocks, driven into dense interstellar clouds, excite the optical emission while the evaporating cloud–hot ISM boundary region contributes to the X-ray emission measure. An HRI pointing in the vicinity of the Hα filament, discussed by Raymond *et al.* (1979), reveals a coincident X-ray filament separated by a region devoid of any X-ray emission from the bright X-ray feature which overlies the optical western limb. This leading shock, quite possibly a piece of the original expanding blast wave, is running ahead through a low density portion of the ISM on its way to future encounters with dense clouds. Detailed images such as these, along with similar maps of Puppis A (Petre, Winkler, and Kriss 1980) and the Vela SNR (Harnden 1980) will greatly enhance our knowledge of SNR dynamics and the structure of the ISM.

Turning to the second major area germane to the X-ray imaging of SNR, we ask not what happened to the outer part of the exploding star as it expanded into the ISM, but rather, what became of the stellar core. Three scenarios are deemed possible by our current theoretical understanding of supernova explosions: the star disrupted completely leaving no remnant; the collapsing degenerate core bounced, leaving behind a neutron star; or the gravitational force overcame the neutrino and degeneracy pressures, and the core fell into a black hole. If complete disruption is the rule, nothing is left, and we should see nothing anomalous at the remnant center. If a black hole is formed, it would be very difficult to detect. A spherically accreting black hole of a few solar masses would be only marginally observable in SN1006 if it had a very low velocity with respect to the accreting material and if enough of that material were left following passage of the initial blast wave; black holes would not be observable in any other remnants. A neutron star with a mass of $M \lesssim 1.4\ M_\odot$ would be even harder to detect if its emission were powered by accretion. However, two other X-ray producing processes are expected to be operative. First, as in the case of the Crab, a rapidly spinning young star would be losing energy at a sufficient rate to produce a substantial synchrotron nebula in the surrounding medium, and second, the newly formed neutron star should be hot.

Formed in the imploding stellar core at a temperature of $T \gtrsim 10^{11}$ K, a young neutron star is essentially a very big, very hot atomic nucleus. Given a thorough understanding of the nuclear and particle physics involved in defining the equation of state for matter at supernuclear densities, the solid state physics of how the superfluid neutrons and superconducting protons of the core interact, the atomic physics of the crust where atoms are distorted into cylinders by the 10^{12} gauss magnetic field, and the radiative transfer and meteorology of the magnetosphere, it is a straightforward physics problem to calculate the cooling rate of this big hot nucleus; i.e., to predict the temperature of the star as a function of time. Alternatively, the observational determination of this cooling function may be viewed as

TABLE I

A Search for Thermal Emission from Young Neutron Stars in SNR

SNR	Age (yr)	Temperature (10^6 K)	Reference
Cas A	~300	<1.5	Murray *et al.* 1979
Tycho	376	<2.0	Helfand, Chanan, and Novick 1980a
Kepler	408	<2.0	Helfand, Chanan, and Novick 1980a
Crab	926	2.0	Harnden *et al.* 1979
SN1006	974	<0.8	Pye *et al.* 1979
RCW86	1,795	<1.8	Unpublished
RCW103	~2,000	2.2	Tuohy and Garmire 1980
W28	~3,500	<1.8	Helfand, Chanan, and Novick 1980a
G350.0-1.8	~8,000	<2.0	Helfand, Chanan, and Novick 1980a
G22.7-0.2	~10,000	<2.2	Helfand, Chanan, and Novick 1980a
Vela	~10,000	1.5	Harnden 1980

heady course in basic physics. Stimulated by the discovery of celestial X-ray point sources in the 1960's, Bahcall and Wolf (1965) were the first to address seriously the problem of neutron star cooling. They found that within a few hundred years, the stars should cool to $\lesssim 10^7$ K and that, consequently, the relatively hard-spectrum sources being cataloged in increasing numbers at the time were not young neutron stars. Over the next 15 years a number of people have worked on the cooling problem. The most extensive work, by Tsuruta (see Tsuruta 1979 for a recent review), led to predicted temperatures for a 300-yr old neutron star ranging between 5 and 15 $\times 10^6$ K, depending upon various parameters of the input physics including magnetic field strengths, stellar masses, equations of state, etc. At 1000 yr the temperature has fallen to $\sim 3 \times 10^6$ K and by 10^4 yr to below 10^6 K.

The earliest observational work on this problem resulted from several rocket experiments designed to observe lunar occultations of the Crab Nebula (Bowyer *et al.* 1964; Wolff *et al.* 1975; Toor and Seward 1977). The best limit obtained for the temperature of an unpulsed point source at the pulsar position was 3×10^6 K. The advent of SNR imaging with the Einstein Observatory has stimulated considerable new interest in this problem. The results for 11 remnants including six of the historical SNR are listed in Table 1. Surprisingly, of the five remnants known to be less than 10^3 yr old, only the Crab Nebula contains an X-ray point source at its center. Most of the X-rays from this source are pulsed at the radio pulsar period and have a spectrum which extrapolates smoothly into both the optical and γ-ray regimes: they are clearly part of the nonthermal pulsar emission. However, Harnden *et al.* (1979) have recently produced a set of phase-resolved images of the Crab which show that the point source never disappears, even during the interval between the interpulse and the main pulse where the pulsed emission in other spectral regimes is either extremely weak or nonexistent. If this residual flux is interpreted as blackbody radiation from a 10-km neutron star, the implied temperature is $\sim 2 \times 10^6$ K. Unfortunately, data from the instrument with sufficient spatial resolution to isolate the pulsar from the nebular background (the HRI) contain no energy information; thus, an examination of the spectrum of this off-pulse radiation must await the next generation of X-ray instruments. Since high signal-to-noise optical pulse profiles show emission over nearly the entire period, a nonthermal origin for this residual X-ray flux cannot be excluded. At worst, however, the current measurement sets a firm upper limit on the blackbody temperature of the known 926-yr old neutron star in the Crab Nebula: $T \lesssim 2 \times 10^6$ K.

No radio pulsars have been detected in any of the other historical SNR. A canonical explanation for this has been that the pulsars in these remnants are not pointed in our direction. Since this beaming effect leads us to expect that we will see about 20% of all radio pulsars, one (the Crab) out of five or six is reasonable. The thermal radiation from a hot young neutron star, however, is not expected to be highly beamed, although some modulation of the flux should occur owing to the nonuniform thermal conductivity of the surface induced by the star's strong magnetic field. The absence of X-ray point sources in these remnants, then, admits but one of two possibilities; either no neutron stars are present, or the available cooling calculations are incorrect. Considerable theoretical work has been stimulated by these results, and modifications to the predicted temperatures have recently been forthcoming. Sutherland and Glen (1979) and Lamb and van Riper (1979) have undertaken new calculations employing the appropriate general relativistic corrections to the thermodynamics of the problem and the latest nuclear, high-energy, and solid-state physics parameters. Their results, along with recent revised calculations of Tsuruta (1980), yield substantially lower predicted temperatures. In no case, however, can the limit of $<0.8 \times 10^6$ K for a neutron star in SN1006 be accommodated by the standard scenarios, and in a few other cases, low mass stars ($<1\ M_\odot$) must be invoked if the limits are to be satisfied. More rapid cooling is allowed if pion condensate cores form in the stars, although there is some doubt that densities sufficient to produce this eventuality exist for realistic masses and equations of state.

The alternative explanation, that no neutron stars exist in these remnants, holds serious implications concerning our understanding of neutron star formation and stellar evolution. The current best estimate for the pulsar birthrate in the Galaxy is 0.1 yr^{-1} (Taylor and Manchester 1977). The current best estimate for the galactic supernova rate is about one per 50 yr (Lerche 1980), although some authors (e.g., Tammann 1977) have argued for higher rates. This factor of five discrepancy is removed if the adopted beaming factor of 20% is in error; i.e., if we see essentially all of the pulsars that exist. However, if we are to interpret the X-ray results as implying that only one out of every five supernovae which produce a remnant creates a neutron star, the discrepancy reappears; with more realistic beaming factors, it is much worse than before. We are forced to conclude that the majority of radio pulsars are not formed in supernovae.

Results from two of the other remnants listed in Table 1 tend to support this suggestion. For both RCW103 and the Vela SNR, associated point sources have been detected. In the former, Tuohy and Garmire (1980) have discovered a faint point source coincident with the center of the circular X-ray shell of this young southern remnant. The source is too weak to search for pulsations. The implied blackbody temperature is $\sim 2 \times 10^6$ K. Along with the recently reported detection of nonpulsed X-rays from several older radio pulsars (Helfand, Chanan, and Novick 1980b), the data now strongly suggest that if a neutron star is hot enough, it will radiate thermal X-rays. Vela is the only other known case of an association between a young radio pulsar and a SNR. Harnden (1980) reported the detection of a strong IPC point source centered on the pulsar. A follow-up HRI image revealed that, in fact, this source consisted of an extended nebula about 1′ across, surrounding a central point source coincident with the pulsar. The IPC spectrum shows that this nebula is considerably harder than the surrounding remnant emission. It appears to be a small version of the Crab Nebula, scaled down by a factor ~ 5 in linear extent and a factor $\sim 10^3$ in X-ray luminosity. The point source is unpulsed and yields a temperature for a 10-km blackbody of 1.5×10^6 K; again, its spectrum is not separable from that of the surrounding nebula. While the point source may represent yet another detection of neutron star thermal emission, it is this second nonthermal nebula which provides the most important evidence regarding the birthplaces of pulsars. Not surprisingly, the X-ray luminosity of the two known such nebulae scales roughly as the rotational energy loss rate of the pulsars which power them. Our lack of detection of any such synchrotron nebulae in the other historical remnants, then, may be used to set a limit on the rotational energy loss rate of neutron stars they may contain. For Cas A, Tycho, Kepler, and SN1006, all of which are younger than or comparable to the age of the Crab, this rate must be $<10^{-2}$–10^{-3} that of the Crab pulsar. It is clear, then, that the standard scenario of neutron star formation, along with the canonical picture for the evolution of massive ($>6\ M_\odot$) stars, may have to be reexamined.

While studies of the evolution of the expanding blast wave and the fate of the presupernova core are both interesting and important areas of research, it may well be that X-ray imaging of SNR may make its most significant contribution in the study of the interstellar medium, with the supernova shock serving as the experimentalists' probe of the temperature, density, and pressure structure existing there. Our ideas concerning the ISM have changed drastically over the past decade. A picture with density and temperature variations of about two orders of magnitude between cloud and intercloud regions has given way to models which must incorporate conditions which range from

$T \sim 10$ K and $\rho \gtrsim 10^6$ in molecular clouds to $T \sim 10^6$–10^7 and $\rho \sim 10^{-2}$–10^{-3} in the ionized component of the ISM responsible for the diffuse soft X-ray background, O VI absorption lines, and pulsar dispersion measures.

Two detailed models of the ISM have been put forward in the last few years, and both not only rely on supernovae as a dominant energy input but require self-consistency between the stirring action of the evolving remnants and the ultimate structure of the medium. Cox and Smith (1974) envisioned cavities of hot ($T \sim 10^6$), low density ($\rho \sim 10^{-2.5}$) gas made up of intersecting remnants strung out into a series of interconnecting tunnels ($R \sim 10$ pc) through the action of magnetic stresses and differential galactic rotation. The material in the tunnels is continually reheated by new remnant blast waves impinging on them. The hot component has a filling factor of ~50%. In addressing the dual questions of SNR evolution and the structure of the ISM, McKee and Ostriker (1977) arrived at a picture that incorporates a hot diffuse medium, with a filling factor of ~90%, studded with cold dense clouds, each of which is surrounded by an evaporative halo of warm neutral and warm ionized matter. As a supernova blast wave propagates into this medium, its expansion is only minimally affected by the small clumps of neutral gas, and a smooth, uniform shock demarcation is seen. Slower shocks are driven into the dense cloudlets giving rise to the remnant's optical filaments. Initially, the hot gas filling the remnant cavity quickly evaporates these clouds so that, at early times, the mean density of the X-ray emitting gas is equal to the mean interstellar density of ~1 cm^{-3}. As the shock slows and the temperature declines, however, the cloud cores remain neutral, implying a decrease in the mean remnant density as calculated from the X-ray emissivity.

The preliminary evidence from the new X-ray imaging observations gives some support to both of the models discussed above. For the 14 LMC remnants reported by Long and Helfand (1979), a marked decrease was noted in the remnant density (inferred from application of the standard Sedov solution) with remnant diameter. However, the diameters (mostly from optical observations) for the remnants employed in that work need to be updated in light of our new HRI X-ray maps. In addition, Dopita and Mathewson (1979) and Murdin, Clark, and Culhane (1978) have recently obtained beautiful, spatially resolved optical spectroscopy of several LMC remnants, showing Hβ-emitting blobs (cloudlets heated by the slow shocks driven into them), surrounded by Fe XIV-emitting halos (the evaporating cloud envelopes); the HRI pictures show these same knots to be X-ray bright spots. On the other hand, the recurring pattern of horseshoe-shaped remnants (N63A and N132D in the LMC, SN1006 and Kepler among the historical remnants), which exhibit large-scale changes in emissivity around the rim of their shells, argues for some fairly large-scale anisotropies, such as the tunnel system of Cox and Smith (1974). RCW86 is an excellent example of this phenomenon. This remnant has reached an extent of nearly 40 pc in only 1,800 yr, suggesting expansion into a low-density medium. However, in the southwest quadrant where the brightest emission is seen, the arc of the remnant shell appears to have a radius of only a few parsecs, reminiscent of a blast wave propagating into a much denser region. There is an absence of emission along a northwest and southeast line, while the northeast quadrant shows a faint, large radius shell. The X-ray image is consistent with an explosion occurring near the edge of a low-density tunnel located toward the northeast. As the blast wave reaches the tunnel wall, the expansion velocity increases, and the radius of curvature of the shell expands. Along the axis of the tunnel (NW-SE) no emission is seen, whereas, where the far side of the tunnel is encountered (about 20 pc to the northeast), the blast wave again hits denser material, and the emission measure rises to detectable levels. The factor of ~10 change in radius of curvature (and, thus, tenfold change in shock velocity) is consistent with expected density ratio between tunnel and intertunnel regions ($v \propto \sqrt{1/\rho}$) and the tunnel diameter of ~20 pc is just that expected from the model. It is apparent, then, from even such qualitative descriptions of the SNR images, that both models have a role to play in developing a consistent picture of the ISM.

As of this writing, the data base continues to grow, and the analysis is still in its infancy. Over 100 remnants will have been imaged in soft X-rays by the end of the Einstein mission. From the wealth of detail already accumulated, it is clear that these data will have an important role to play for many years to come, in the study of remnant dynamics and evolution and in the quest for an understanding of the origin and physics of neutron stars and the structure of the interstellar medium.

This work was supported by the National Aeronautics and Space Administration under contract NAS8-30753. This paper is Columbia Astrophysics Laboratory Contribution No. 189.

REFERENCES

Bahcall, J. N., and Wolf, R. A.: 1965, Phys. Rev. 140, p. B1452.

Becker, R. H., Holt, S. S., Smith, B. W., White, N. F., Boldt, E. A., Mushotzky, R. F., and Serlemitsos, P. J.: 1979, Astrophys. J. (Letters) 234, p. L73.

Bowyer, S., Byram, E., Chubb, T., and Friedman, H.: 1964, Science 146, p. 912.

Cox, D., and Smith, B. E.: 1974, Astrophys. J. (Letters) 189, p. L105.

Davison, P. J. N., Culhane, J. L., and Mitchell, R. J.: 1976, Astrophys. J. (Letters) 206, p. L37.

Dopita, M., and Mathewson, D.: 1979, Astrophys. J. (Letters) 231, p. L147.

Fabian, A., Pye, J. P., Willingale, R., and Murray, S. S. 1980, preprint.

Giacconi, R., *et al.*: 1979, Astrophys. J. 230, p. 540.

Harnden, F. R.: May 1980, talk presented at Workshop on "Stellar Collapse, Supernovae, and Neutron Star Formation," Santa Barbara, California.
Harnden, F. R., Buehler, B., Giacconi, R., Grindlay, J., Hertz, P., Schreier, E., Seward, F., Tananbaum, H., and van Speybroeck, L.: 1979, B.A.A.S. 11, p. 789.
Helfand, D. J., Chanan, G. A., and Novick, R.: 1980a, Nature 283, p. 337.
Helfand, D. J., Chanan, G. A., and Novick, R.: 1980b, B.A.A.S. 12, p. 514.
Helfand, D. J., and Long, K. S.: 1979, B.A.A.S. 11, p. 796.
Lamb, D. Q., and van Riper, K. A.: 1979, B.A.A.S. 11, p. 779.
Lerche, I.: May 1980, talk presented at Workshop on "Stellar Collapse, Supernovae, and Neutron Star Formation," Santa Barbara, California.
Long, K. S., and Helfand, D. J.: 1979, Astrophys. J. (Letters) 234, p. L77.
McKee, C., and Ostriker, J. P.: 1977, Astrophys. J. 218, p. 148.
Murdin, P., Clark, D. H., and Culhane, J. L.: 1978, Monthly Notices Roy. Astron. Soc. 183, p. 79P.
Murray, S. S., Fabbiano, G., Fabian, A., Epstein, A., and Giacconi, R.: 1979, Astrophys. J. (Letters) 234, p. L69.
Parker, R. A. R., Gull, T. R., and Kirschner, R. P.: 1979, "An Emission-Line Survey of the Milky Way," NASA SP-434.
Petre, P., Winkler, P. F., and Kriss, G. A.: 1980, B.A.A.S. 12, p. 542.
Pisarski, R. L., and Long, K. S.: 1979, B.A.A.S. 11, p. 789.
Pye, J. P., Pounds, K. A., Rolf, D., Seward, F. D., Smith, A., and Willingale, R.: 1979, B.A.A.S. 11, p. 790.
Raymond, J. C., Davis, M., Gull, T. R., and Parker, R. A. R.: 1980, Astrophys. J. (Letters), in press.
Smith, B.: 1980, this volume.
Sutherland, P. G., and Glen, G.: 1979, B.A.A.S. 11, p. 779.
Tammann, G. A.: 1977, Ann. N.Y. Acad. Sci. 302, p. 61.
Taylor, J. H., and Manchester, R. N.: 1977, Astrophys. J. 215, p. 885.
Toor, A., and Seward, F. D.: 1977, Astrophys. J. 216, p. 560.
Tsuruta, S.: 1979, Phys. Reports 56, p. 237.
Tsuruta, S.: 1980, private communication.
Tuohy, I., and Garmire, G.: 1980, Astrophys. J. (Letters), in press.
Tuohy, I., Nousek, J. A., and Garmire, G. P. 1979, Astrophys. J. (Letters) 234, p. L101.
Wilson, A. S.: 1980, B.A.A.S. 12, p. 541.
Wolff, R. S., Kestenbaum, H. L., Ku, W., and Novick, R.: 1975, Astrophys. J. (Letters) 202, p. L77.
Woodgate, B. E., Kirschner, R. P., and Balon, R. J.: 1977, Astrophys. J. (Letters) 218, p. L129.
Woodgate, B. E., Stockman, H. S., Angel, J. R. P., and Kirschner, R. P.: 1974, Astrophys. J. (Letters) 188, p. L79.

X-RAY SPECTRA OF SUPERNOVA REMNANTS

Barham W. Smith
Los Alamos National Laboratory

Abstract

High quality X-ray spectra for the brightest remnants are now available from several satellites. The data for Tycho's remnant, Cas A, and Kepler's remnant have been used to derive elemental abundances which appear mostly larger than average interstellar values and increase with respect to those values as one progresses from Mg to Ca. Iron appears low. All the abundances appear particularly high in Tycho. Spectral observations above 20 keV for Cas A and Tycho show that electron-ion equilibration proceeds faster than the Coulomb rate. Ionization-recombination equilibration, however, is almost certainly not attained in these young remnants, and this may account for the two temperature components found to fit the spectra best by most X-ray observers. Spectral evidence, coupled with optical evidence that Tycho's remnant is in the Sedov phase of evolution, indicate that the ejecta were not primarily iron. The older remnant in Puppis has been well studied by a crystal spectrometer and shows many informative line features.

R. Giacconi (ed.), X-Ray Astronomy with the Einstein Satellite, 51–60.

Other articles in this volume contain examples of the beautiful X-ray maps of supernova remnants (SNR's) now available from the Einstein Observatory. This satellite has also provided X-ray spectra of SNR's with a quality only dreamed of ten years ago. It would be nice to have detailed spectra without giving up any spatial resolution, but this will have to await a generation of better X-ray detectors. Meanwhile, it behooves us to see what we can learn from the best spatially integrated X-ray spectra. It will be very difficult to do an adequate job of modeling a messy explosion like Cas A, so let us begin with the best understood SNR, Tycho.

Figure 1 shows the spatially averaged counting rate spectrum of Tycho as measured by the Goddard Space Flight Center Solid State Spectrometer (SSS), together with a model of the spectrum. This model attempts to fit the data with two components having temperatures of about 0.5 and 4.0 keV, representing the interstellar medium heated by the main shock and the ejecta heated by the reverse shock. Additional parameter values and experiment description may be found in Becker et al. (1980a). The two-temperature model clearly fits the qualitative character of the spectrum in this energy range, but the goodness-of-fit test shows that details of line blend shapes are not well represented. This situation is typical of all the young SNR spectra.

The model is constructed from collisional equilibrium spectral models for each abundant element (Raymond and Smith 1977, plus improvements too numerous to list). The model-data comparison leads directly to estimates of elemental abundances. For the composite model shown, some ratios of (abundance in Tycho) to (abundance averaged over the local interstellar medium) are Mg (0.1), Si (6.0), S (13.5), Ar (34.5), Ca (76), and Fe (0.15). Observations with broader band, lower resolution counters in the HEAO A-2 experiment (Pravdo et al. 1980) indicate somewhat lower abundance ratios for Si (3.0) and S (4.0) and higher ratios for Fe (1.2). The apparent overabundances are very sensitive to the model assumptions, especially to the use of collisional equilibrium component spectra and to the restriction of the elements responsible for the below-band continuum (He, C, N, O, Ne) to their cosmic abundances.

Tycho's remnant is thought to be from a Type I event and thus should have had about 1 $M_{\odot}$ of ejecta. Proper motion studies by Kamper and van den Bergh (1978) yield strong evidence that the bright filaments are expanding with the radius-time behavior characteristic of the Sedov blast wave phase of SNR evolution. This conclusion is independent of distance to the remnant. In order that the shock wave system be advanced enough to be in the Sedov phase, the main shock must have swept up an amount of interstellar matter at least equal to the original mass of ejecta, and probably several times more. This is the same epoch at which the reverse shock forms. We will return to these points later.

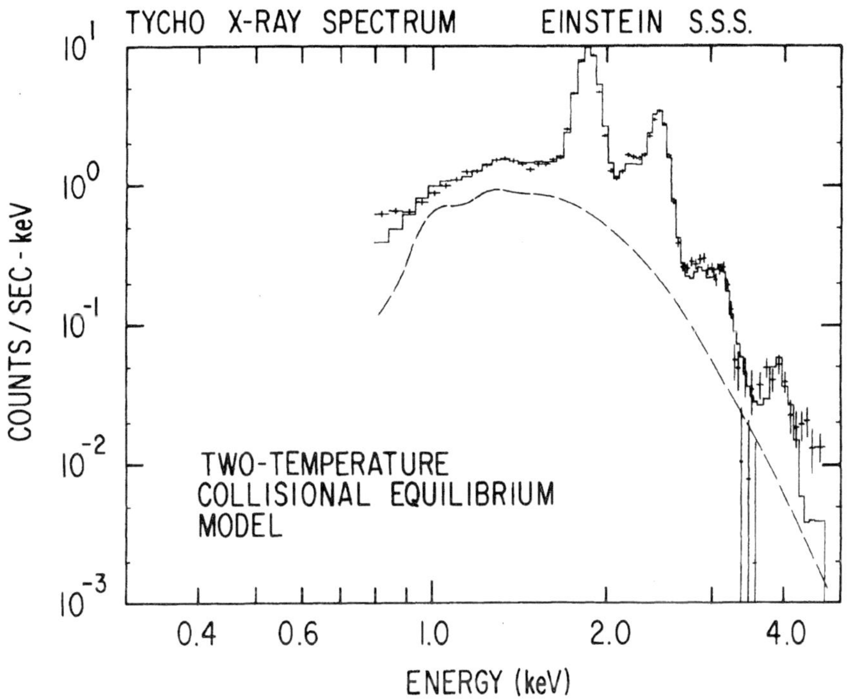

Fig. 1: The X-ray spectrum of Tycho's remnant as observed by the GSFC Solid State Spectrometer aboard the Einstein Observatory (Becker et al. 1980a). The crosses are the data, and the histogram is the sum of two collisional equilibrium, optically thin spectra (see text). The dashed line shows the contribution to the model of continuum processes, if two equilibrium spectra are appropriate. The remainder of the model is all due to narrow emission lines which are heavily blended.

Thus a model including main and reverse shocks fits our understanding of Tycho. To predict the spectrum for a model, we need to know what physical states the gas is in. Two kinds of equilibria must be determined, electron-ion equilibrium and ionization-recombination equilibrium. Pravdo and Smith (1979) show evidence from gas proportional counter observations that electrons in the Tycho main shock are being heated by ions at a faster rate than that due to long-range Coulomb collisions. Perhaps collective effects and plasma instabilities (McKee 1974) contribute. Certainly electron-ion equilibrium has already been much more closely reached than some authors had predicted. Ionization-recombination equilibrium pertains to the relative presence of different ionic charge states of an element. For example the quiet solar corona is thought to be in a state of statistical, collisional equilibrium. Behind a supernova shock, ions presumably heat electrons while the cooler electrons collide with the ions to ionize and excite them. Relaxation toward collisional equilibrium at the postshock temperature requires a long time in young supernova remnants. The timescale for equilibration depends on rates for the process of electron collisional ionization (known to within a factor of two for all elements of interest and all stages of ionization) and on the preshock density (known to a factor of three from optical observations such as Chevalier and Raymond (1978) which give the shock velocity, combined with the measured X-ray flux). This timescale is thousands of years for Tycho. Thus we expect the gas heated in the main shock of Tycho to be underionized, although the observational evidence is indirect at present. Such an underionized plasma has enhanced line emission, since the energies of available transitions are lower relative to the average free electron energy than they would be in equilibrium. The apparent increase in abundance ratios as one progresses from Mg to Ca are qualitatively in keeping with these enhancements, since the heavier the element the longer it requires to reach equilibrium. In contrast, the ejecta are dense enough that they may equilibrate after passing through the reverse shock. Thus only the lower temperature "component" in SSS spectra is likely to be unaffected by nonequilibrium.

The best observational evidence for nonequilibrium comes from the 0.5 to 20 keV spectrum obtained with gas proporonal counters on HEAO-1. Pravdo et al. (1980) were able to compare the energy centroids of emission of the 6.7 keV iron line blends in the Tycho and Cas A spectra by scheduling a pair of consecutive observations, so that gain variations were negligible. The line center in Tycho is distinctly lower in energy, indicating a greater contribution to the blend from $K\alpha$ satellites of the helium-like resonance line arising in iron ions with more than two electrons. Such ions would be nearly absent in collisional equilibrium at the main shock temperature of greater than 10 keV. The iron ionization state deduced from all observations is characteristic of equilibrium at less than 2×10^7K, far lower than the bremsstrahluug temperature from the main shock. Yet the dominant 6.7 keV line blend cannot arise in the much cooler reverse shocked ejecta.

Cas A is a very different object, as shown by the appearance of X-ray and optical maps. Interaction with the surrounding medium is greatly affected by the motion of fast knots. However, the ideas from one-dimensional analysis of SNRs are still valuable in relating the ionization state to hydrodynamic evolution. As for Tycho, Pravdo and Smith (1979) show that electron-ion equilibration is happening too fast for pure Coulomb heating. These authors also use the ratio of n=3 to n=1 ($K\beta$) and n=2 to n=1 ($K\alpha$) transitions in H- and He-like ions to conclude that ionization-recombination equilibrium does not hold in Cas A's main shock. In fact, the measured line blend ratio $K\beta/K\alpha$ implies an iron ionization temperature of less than 2×10^7 K, as for Tycho. Figure 2 shows the SSS spectrum of Cas A (Becker et al. 1979). Almost the entire remnant was within the six arcminute field of view.

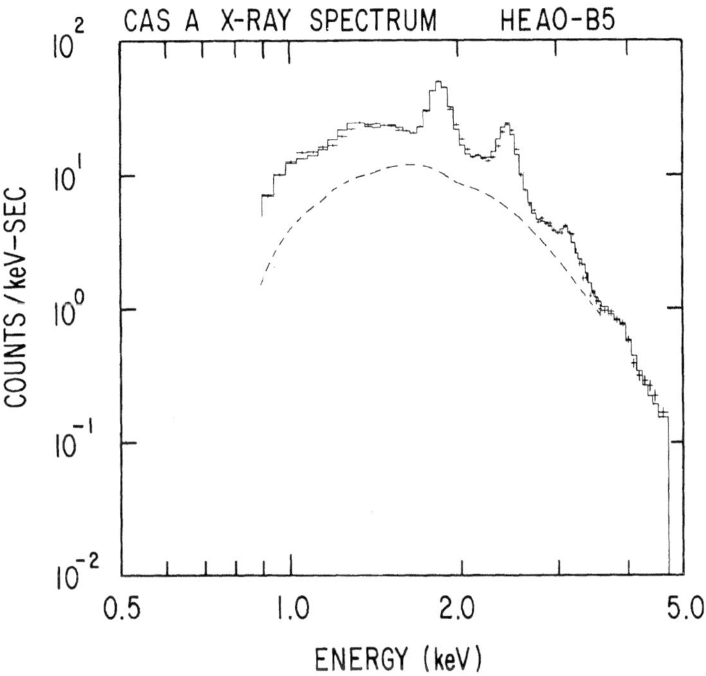

Fig. 2: The X-ray spectrum of Cas A as observed by the SSS (Becker et al. 1979). Model and data are shown as for Figure 1.

Although the ratios of Si and S lines to the continuum are lower than similar ratios in Tycho, both remnants show apparent increases in ratios of abundances to "cosmic" abundances as one progresses from Mg through Ca.

A different experiment on the Einstein Observatory, the MIT Focal Plane Crystal Spectrometer (FPCS), has observed the group of strong lines near 2.5 keV due to S XV (Winkler et al. 1980a). They find the triplet lines to be weaker than the singlet, indicating that the emission is not dominated by recombination cascades. This is not surprising, since we expect the gas to be in an underionized state. Inner shell ionization of Li-like ions can also strengthen the triplets relative to the singlet (Mewe and Schrijver 1975), but this is a weaker effect and one cannot conclude from the data that the gas is not underionized.

An X-ray spectrum for Kepler's SNR similar to those for Cas A and Tycho has been obtained by the Einstein SSS (Becker et al. 1980b). Kepler is also thought to be from a Type I event, more distant than

Tycho. Since it was only recently confirmed as an X-ray source, the quality of the SSS spectrum is very satisfying. Another young remnant, MSH 11-54, has a good SSS spectrum (Clark, Tuohy, and Becker 1980), and it is similar to the ones discussed above. It is tempting to say that if we can understand the X-ray spectral features of one young SNR, the similarity of other spectra allows us to generalize our qualitative interpretation to them all. However, the remnant of SN1006 has also been carefully observed, both with the SSS and with gas proportional counters on OSO-8, and its spectrum is apparently a featureless power law (Becker et al. 1980c). Neither Si, S, or Fe lines have been positively detected. This resemblance to the Crab Nebula is surprising, since studies of optical filaments in the remnant of SN1006 have led to the conclusion that it has a structure like Tycho (e.g. Chevalier and Raymond 1978).

Puppis A is an example of an older SNR. The SSS spectrum (Holt 1980) is softer than those of previously mentioned SNRs, but the dramatic Si and S line features are still present. Of eight pointing positions spaced over the X-ray remnant of Puppis, two seem to have significantly different abundance ratios than the average of the other six (Becker, private communication). One position show ratios three times the average, another twice. Could this be evidence for assymetric ejection by the explosion?

The FPCS group has spent considerable observation time on Puppis, taking advantage of its soft, line-rich spectrum and small interstellar absorption (Winkler et al. 1980b). Figure 3 shows the fruits of their effort, a line spectrum covering 600 eV eith resolution varying from 2 to 10 eV. The continuum does not appear because their narrow energy windows are not sensitive to it. A careful analysis of the O VII and O VIII line blends shows some indication that collisional equilibrium is satisfied in the emission region. This is of interest since it sets a lower limit on the preshock density. Three other old SNRs have been studied by the SSS, namely IC 443, RCW 103 and W44, and their spectra confirm the ideas we have formed in studying Puppis. Lastly, the Crab Nebula has been observed and shows a featureless spectrum as expected.

The goals of interpreting these observations include determining states of equilibrium in the emitting gas, estimating the preshock density and the mass of ejecta, and discovering how the emission is apportioned between the main and reverse shock waves. Furthermore, one would like to know if the apparent overabundance anomalies catalogued by the SSS group are real. Most of these goals have been reached to some extent in the papers referenced above. The question of abundance anomalies remains because of insufficient and inadequate theoretical modeling.

Optical line spectrophotometry shows abundance anomalies in SNRs, and we know that the ejecta are from processed outer layers of evolved stars. Fast knots in Cas A with exotic abundances are direct evidence that some such material has intermingled with interstellar gas (see

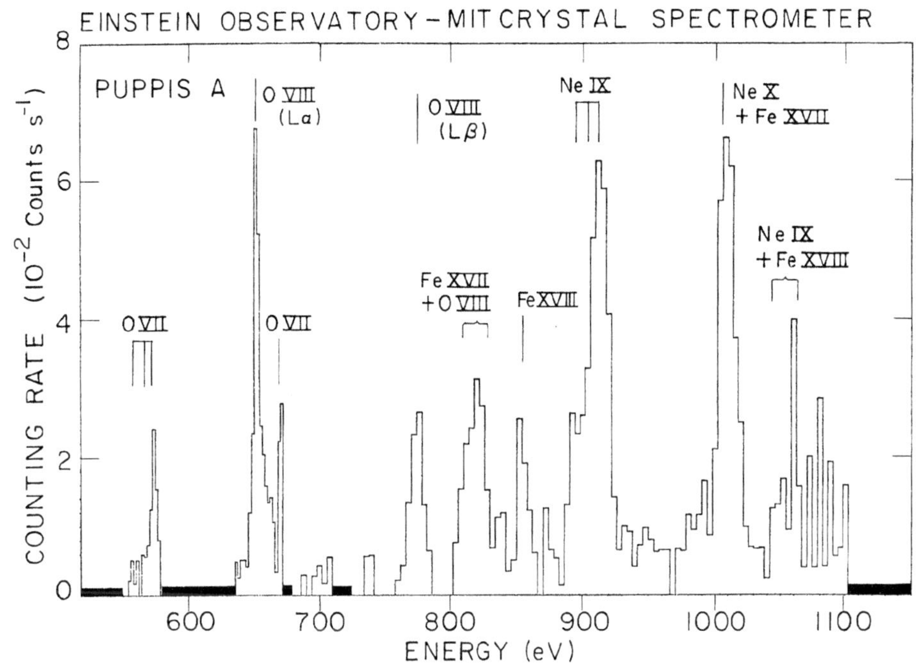

Fig. 3: X-ray emission spectrum from a portion of the bright eastern region of Puppis A, a composite of preliminary data from fourteen scans with the Einstein Observatory FPCS. Total exposure was approximately 1.2 x 10^5 sec. Heavy lines on energy axis indicate energy bands with no exposure. This raw counting rate spectrum has not been corrected for instrumental response.

Chevalier and Kirshner 1979). Therefore one might conclude that abundance anomalies are to be expected in X-ray spectra of SNRs.

However, the fraction of the apparent abundance anomalies which is real is still uncertain. None of the younger remnants (Tycho, Cas A, Kepler) has such a high estimated preshock density that collsional equilibrium can be enforced except in the densest shell of ejecta.

Nonequilibrium X-ray spectra of underionized interstellar gas are so distorted from spectra of single temperature equilibrium plasmas that it is amazing how well they can be fit with two equilibrium spectra (see below). Since two equilibrium component fits work fairly well for all the known thermal SNR spectra, it may be that none are in equilibrium. If such is the case, then all conclusions on abundances up to now are subject to modification. We are much in need of consistent nonequilibrium calculations of SNR spectra with sufficient detail for comparison with high quality SSS and FPCS spectra. At the present time, the greatest uncertainties in the interpretation of SNR X-ray spectra lie with theory, both atomic physics theory and hydrodynamic/ionization-evolution modeling. In the meantime, we can draw on the excellent papers by Itoh (1977, 1979).

Itoh has shown that the spectrum calculated for a Sedov blast wave, moving into a uniform interstellar medium of typical density, mimics almost perfectly the "two-component" spectra found by most observers. Refer in particular to Figure 10 in Itoh (1979). The coincidence of strong lines in both spectrum and (inconsistent) fit is remarkable. If one then concludes that the results of the SSS observations of Tycho, Cas A, Kepler, Puppis, etc. should be interpreted this way, then the "soft component" found for each source is mostly heavily blended iron lines. The shape of this iron emission blend is qualitatively different from collsional equilibrium iron blends, and furthermore, sits on strong 2-photon continuum emission from O, C, N, and Ne. The spectra are masquerading skillfully as two-component models with lower-than-cosmic iron abundances. Unfortunately, Itoh's published spectra are not suitable for testing against the SSS spectra in a quantitative way. Calculations with much finer energy binning, extended to other combinations of age, preshock density, and ejecta configurations are needed. Since the ionization timescales are independent of hydrodynamic timescales, a single model calculation cannot be scaled to produce consistent nonequilibrium SNR spectra for different parameters and conditions. In fact, as of this writing, no chi-squared test has been performed to compare nonequilibrium SNR spectra with SSS data.

Estimates of the iron abundance one would derive from consistent nonequilibrium fits of the SSS Tycho spectrum indicate that iron is roughly at its cosmic abundance in this remnant. For reasons unrelated to the X-ray spectrum, many researchers favor a model of Type I supernovae in which the ejecta are largely iron-group elements (e.g. Arnett 1979). If 0.3 M_0 of iron were well mixed with the X-ray emitting gas, the resulting spectrum would be dominated by iron emission features. In this case the Si and S features would be miniscule and the equivalent width of the 6.7 - 8.3 keV iron line complex would be incredible. Arnett (1980) has suggested that the iron X-ray emission features appear normal only because the reverse shock has not yet encompassed the huge mass of iron ejecta. The epoch at which it is encompassed depends on preshock density. Two lines of evidence argue against concealing iron in Tycho's ejecta this way. First, Kamper and van den Bergh (1978) show that the remnant is in the Sedov phase, so that the swept-up mass is several times the ejected mass. At such an epoch most ejecta have been reverse shocked. Second, Chevalier and Raymond determine the main shock velocity and the preshock density. Arnett's argument hinges on having a low preshock density (less than 0.1 cm^{-3}). If the preshock hydrogen density is greater than 1 cm^{-3} (the best value is 1.5 cm^{-3}, Chevalier and Raymond 1979) then any iron ejecta should have begun to pass through the reverse shock, even in a spherically symmetric model. In fully three-dimensional SNR models, the ejecta are mixed with post-main-shock gas at the epoch when swept-up mass equals ejecta mass, due to the Rayleigh-Taylor instability as the ejecta are decelerated (e.g. Cowie 1975). Thus although we do not expect mixing of layers of the presupernova star with differing abundances before or while the shock propagates out through a star, we do expect mixing at the epoch in which we see Tycho's SNR.

Acknowledgements

I would like to thank R. H. Becker, S. S. Holt, and P. F. Winkler for supplying the figures and for many rewarding discussions.

References

Arnett, W. D.: 1979, Astrophys. J. Letters, 230, pp. L37-L40.
Arnett, W. D.: 1980, "Nonuniform Abundances in Young Supernova Remnants" (preprint).
Becker, R. H., Holt, S. S., Smith, B. W., White, N. E., Boldt, E. A., Mushotzky, R. F., and Serlemitsos, P. J.: 1979, Astrophys. J. Letters, 234, pp. L73-L76.
Becker, R. H., Holt. S. S., Smith, B. W., White, N. E., Boldt, E. A., Mushotzky, R. F., and Serlemitsos, P. J.: 1980a, Astrophys. J. Letters, 235, pp. L5-L8.
Becker, R. H., Boldt, E. A., Holt, S. S., Serlemitso, P. J., and White, N. E.: 1980b, "X-Ray Spectrum of Kepler's SNR", Astrophys. J. Letters (in press).
Becker, R. H., Szymkowiak, A. E., Boldt, E. A., Holt, S. S., and Serlemitsos, P. J.: 1980c, "Is the Remnant of SN1006 Crab-like?", Astrophys. J. Letters (in press).
Chevalier, R. A. and Kirshner, R. P.: 1979, Astrophys. J., 233, pp. 154-162.
Chevalier, R. A. and Raymond, J. C.: 1978, Astrophys. J. Letters, 225, pp. L27-L30.
Chevalier, R. A. and Raymond, J. C.: 1979, private communication.
Clark, D. H., Tuohy, I. R., and Becker, R. H.: 1980, "X-Ray Line Emission from the SNR G292.0+1.8",Monthly Notices Roy. Astron. Soc. (in press).
Cowie, L. L.: 1975, Monthly Notices Roy. Astron. Soc., 173, pp. 429-436.
Holt, S. S.: 1980, "X-Ray Spectra of Supernova Remnants", in "Proceedings of the NATO Advanced Study Institute in X-Ray Astronomy", Reidel, Dordrecht, p. 35.
Itoh, H.: 1977, Publ. Astron. Soc. Japan, 29, pp. 813-830.
Itoh, H.: 1979, Publ. Astron. Soc. Japan, 31, pp. 541-562.

Kamper, K. W. and van den Bergh, S.: 1978, Astrophys. J., 224, pp. 851-856.
Mason, K. O., Pravdo, S. H., Charles, P. A., Smith, B. W., and Raymond, J. C.: 1980, "The 0.5 to 20 keV Spectrum of Cas A", Astrophys. J. (in press).
McKee, C. F.: 1974, Astrophys. J., 188, pp. 335-339.
Mewe, R. and Schrijver, J.: 1975, Astrophys. Space Sci., 38, pp. 345-352.
Pravdo, S. H. and Smith, B. W.: 1979, Astrophys. J. Letters, 234, pp. L195-L198.
Pravdo, S. H., Smith, B. W., Charles, P. A., and Tuohy, I. R.: 1980, Astrophys. J. Letters, 235, pp. L9-L12.
Raymond, J. C. and Smith, B. W.: 1977, Astrophys. J. Suppl., 35, pp. 419-439.
Winkler, P. F., Canizares, C. R., Clark, G. W., Markert, T. H., Berg, C., Jernigan, J. G., and Schattenburg, M. L.: 1980a, "High Resolution X-Ray Spectroscopy from the Einstein Observatory", in "Proceedings of the NATO Advanced Study Institute in X-Ray Astronomy", Reidel, Dordrecht, p. 61.
Winkler, P. F., Canizares, C. R., Clark, G. W., Markert, T. H., and Petre, R.: 1980b, "Observations of Puppis A with a Crystal Spectrometer", Astrophys. J. (in press).

OBSERVATIONS OF X-RAY BURSTS BY THE "HAKUCHO" SATELLITE

Minoru Oda
Institute of Space and Aeronautical Science
University of Tokyo, Meguro, Tokyo, Japan

ABSTRACT

Observational results on X-ray bursts obtained by the Japanese X-ray satellite are reviewed in conjunction with the general context of the neutron star model of the burst. Evidence indicates the existence of physically different kinds among type-I bursts which are widely believed to be from a thermonuclear flash occurring on the surface of a neutron star. There are cases which can not be accounted for by a simple nuclear flash model: The flash in a confined region of the surface and a change of the emitting area during the burst are suggested. Two type-I bursts were observed in succession within a short interval which was not of sufficient duration to accumulate the nuclear fuel for a flash. Coordinated optical/X-ray observations of bursts were performed. Optical bursts are tentatively interpreted in terms of the reprocessing of X-rays are at the accretion disk. The results will provide the basis for a discussion of the physical structure of the system. The evolution of the activity of type-II bursts from the rapid burster was observed from its onset for about two weeks and a novel mode of burst activity was discovered.

INTRODUCTION

The following talk will be confined to review what we have learned about X-ray bursts with observations performed by the Japanese X-ray satellite "Hakucho". A review of the preliminary results was presented at the ISAS X-ray astronomy symposium (Tanaka, 1979). Details of X-ray bursts and investigations before the "Hakucho" observations are described in the recent review article by Lewin and Clark (1979) and the comprehensive references listed therein. We noted, the gross common features of the type-I X-ray burst may be interpreted by means of the generally postulated model of a thermonuclear flash occurring on a weakly magnetized neutron star surface in a low-mass binary system (Woosley and Taam, 1976; Maraschi and Cavaliere, 1977; Lamb and Lamb, 1978; Joss, 1978, 1979). However, several observational facts

R. Giacconi (ed.), X-Ray Astronomy with the Einstein Satellite, 61–78.

seem to provide severe constraints to the model. Also a unique mode of type-II bursts from the rapid burster was discovered. The results may throw light on the physical structure of the system.

I will not go into the technical details of the "Hakucho" satellite (Inoue et al., 1979), its main characteristics are summarized in Table I. One of the main objectives of the mission is to perform a

Table I

X-ray Astronomy Satellite "HAKUCHO" (Cygnus in Japanese)

* Orbit : 540km - 560km, inclination = 29.9°
* Telemetry Station : Kagoshima Space Center (30°N)
* Spin-stabilized : ∿ 6 rpm
* Spin-axis : controlled by magnetic torquing
* Instrumentations :

3	detectors	(1 ∿ 30keV) ∥	to the axis with RMC
1	"	(") ∥	with tublar collimator
2	"	(") ⊥	with slats collimator
1	"	(10 ∿ 100keV) ∥	"
2	"	(0.1 ∿ 2keV) ∥	"
2	"	(") ⊥	"

Time resolutions 6 ∿ 190 msec

Panoramic Sky Watch for X-ray Bursts

* monitor and watch by modulation collimator
  over FOV 17.6° x 17.6° (FWHM)
  location ∿ 0.5°

* study of selected source
  FOV 5.8° x 5.8° (FWHM)
  location ∿ 0.1°

panoramic survey or watch for X-ray bursts over wide fields of view of the sky. Steady X-ray sources and X-ray burst sources in the 17°x17° FWHM field of view are located with an accuracy of approximately 0.5° with a pair of coarse modulation collimators. Sources in the 6°x6° FWHM field of view may be located with an accuracy of approximately 0.1° by a fine modulation collimator.

The general framework of the observational program is mainly determined by the sun angle constraint due to UV hazards to the soft X-ray counter window and the condition of sufficient electric power generation. Thus, observations of the X-ray bulge sources or the burster region are limited to the period of April to September.

Results of burst observations during the period of April-September, 1979 are tabulated in Table II. The results are compared in the table with the previous observations by SAS-3 summarized in the review article by Lewin and Clark (1979). One may note the following facts from the table.

* Ten or more burst sources were observed during the period. Among these sources the following four are new or newly confirmed burst sources: 1608-522, 1702-429, 1715-322, Cen X-4.
* 1636-536 had been very active; 1608-522 presumably associated with a recurrent nova was much more active than when the Los Alamos group watched it in 1976 (Belian et al., 1976a,b).

* 1608-522, Cen X-4 and 1730-333 are presumably associated with recurrent novae.
* Cen X-4 generated an unusually large burst (Matsuoka et al., 1980).
* 1715-322 may be the same source as the fast transient source observed by SAS-3 (Makishima et al., 1980; Hoffman et al., 1978a).
* 1730-333 (rapid burster) exhibited a novel mode of type-II burst.
* Several previously known bursters were inactive: galactic center sources, 1659-298, 1820-303 (NGC 6624), Ser X-1.
* It is clear that most of the bursters exhibit long term variation.
* No burst occurred in the Cygnus or Vela regions during approximately 30 day observations respectively: The distribution of bursters seems to show a clear boundary at a certain distance from the galactic center except 0512-401 (NGC 1851).

Table II

HAKUCHO observations of bursters (April - September, 1979)

Position & Name		Category	Before March 1979 (Lewin & Clark Review)	HAKUCHO April - September 1979
0512 - 401	4U, MXB	NGC 1851*	Out of "bulge", ~9.5 kpc	no observation
0900 - 403	4U, Vela X-1	pulsar		one very long burst (~10^3 sec)
1455 - 31	XB, Cen X-4	10yr R.T.	Vela satellite : burst?	a very large burst(Type I)~1.5kpc?
1608 - 522	4U, XB	~600d R.T.	UHURU, Vela : burst SAS-3 : obscure?	active : 22/45d FAST/SLOW
1636 - 536	4U, MXB	B.B.S.	"reliable"	active : 41/45d opt/X, FAST/SLOW
1659 - 298	MXB	transient (?)	2 ~ 2.6^h stable, slow type burst no burst for high state, ~ ~25	inactive : -/17d
1702 - 429	4U, XB		MXB 1706 - 43?	active:19/40d structures in profile
1724 - 30	4U, XB (?)	Terzan 2*	OSO-7, OSO-8 : fast transient?	inactive : -/24d
1715 - 322	4U, XB		SAS-3 : fast transient?	slow type bursts : 3/24d
1728 - 337	4U, MXB slow burster	B.B.S.	"reliable" 4 ~ 8^h, some times double peak	17/37d
1730 - 333	MXB, H rapid burster	Liller I* ~180d R.T.	type I, type II bursts	active Aug. 8 ~ 22, trapezoidal, no type I burst
1735 - 444	4U, MXB	B.B.S.	"reliable",interval irregular, opt/X	only 3/33d
1742 - 29	MXB, A?, 2S?	in GCX	ID with persistent source uncertain	
1743 - 28	MXB	in GCX	3 bursts in succession	inactive : at most a few bursts in 25 days
1743 - 29	MXB, A	in GCX nova?	ID with nova uncertain	
1820 - 303	MXB, 4U	NGC 6624*	2.2 ~ 4^h in low state no burst in high state	inactive : -/18d, high state
1837 + 049	MXB 4U Ser X-1	B.B.S.	"reliable", opt/X, intervals irregular	inactive : -/10d
1850 - 087	MXB, A, 2S	NGC 6712*	almost certain	almost no observation
1905 + 000	MXB, 4U			3/13d, oscillatory structure
1916 - 053	MXB, 4U			only 1/8d

* : in Globular cluster
B.B.S. = bright bulge source
R.T. = recurrent transient

I will describe some details of the nature of bursts, reviewing briefly works of the Hakucho team, in conjunction with the general context of the neutron star model of the burst. I am a little disturbed by the fact that the results appear to complicate rather than elaborate the tentatively postulated models of a nuclear flash for type-I bursts and the intermittent release of gravitational energy by accreted matter at the neutron star surface for type-II bursts.

(Hoffman et al., 1978b). I hope that via this process we will reach a further depth of the understanding of the burst phenomena.

FAST-TYPE AND SLOW-TYPE BURSTS

It has been noted that often a burst exhibits a rise time several seconds or longer instead of one second or less typically recognized as the rise time of a type-I burst. We found that the bursters 1608-522 and 1636-536 generate both types of bursts. In particular for 1608-522 the distinction between the two types is very clear (Murakami et al., 1980a).

Fig. 1 shows examples of the profiles of both types. Fig. 2

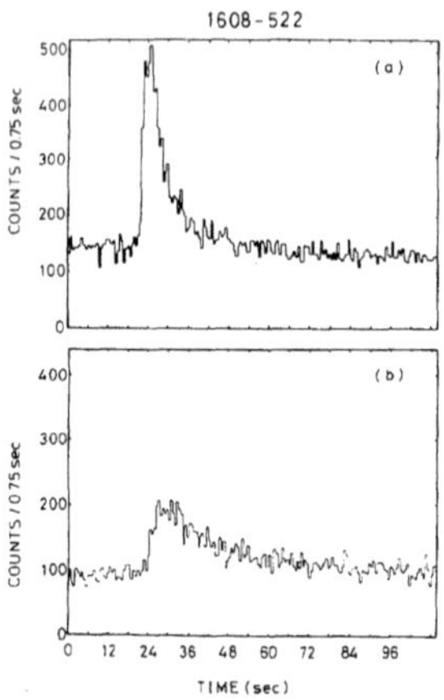

Fig. 1. Examples of the fast-type a) and slow-type b) bursts from 1608-522.

exhibits the behavior of 1608-522 for April-July, 1979. A clear correlation between the peak flux and the persistent flux is seen. Typical X-ray bursts which we now call the fast-type were observed while the persistent X-ray flux was high; when the persistent flux became low the burst profile changed to the slow-type. In Fig. 3 the relation between the peak flux and the integrated (total) flux of the fast and slow bursts are shown. It appears that for the fast bursts the peak flux varies while the duration or shape of the profile stays constant. For the slow bursts the peak flux is constant but its duration varies. There is a distinct difference between the two types. A transition in the types observed occurred between May 20 and June 20, 1979, suggesting that the source is bimodal. The typical T^4 cooling with a constant emitting area during the decay of the burst was noted for both types of bursts. The ratio α(persistent flux/ time-averaged burst flux) for the fast type burst is greater than 100 which is consistent with the He flash model (Joss, 1979). On the

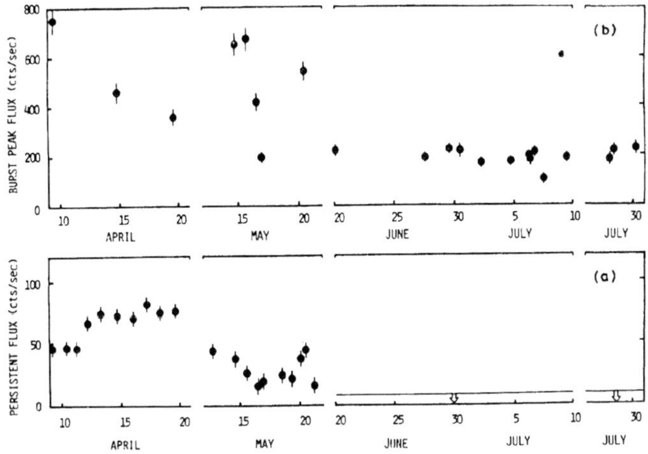

Fig. 2. a) The time history of the persistent flux of 1608-522. After June 20 the flux was low and only the upper limit was determined. b) Bursts from 1608-522. In April-May they were of fast-type and after June 20 they were of slow-type.

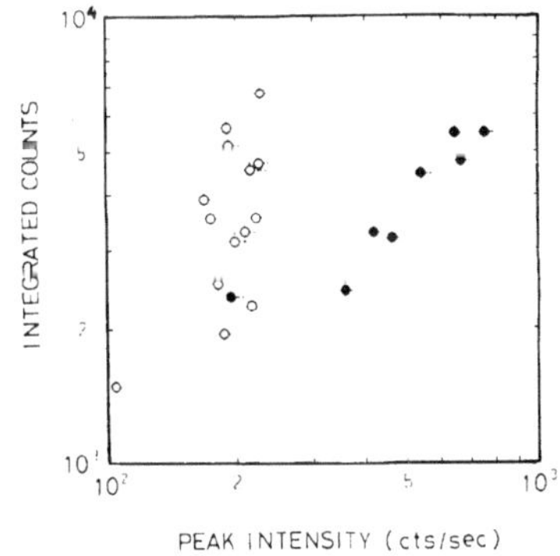

Fig. 3. Peak intensity vs. integrated counts for each individual burst in the April-May period (•) and in the June-July period (o).

other hand, for the slow type burst, α is much smaller than 100. The features of the two types of bursts are summarized in Fig. 4. We believe that the slow burst is a general phenomenon and not the exceptional case. MXB 1659-298 was noted to indicate a small α-value (Lewin and Joss, 1977): From the published profile we recognize this as being also of the slow-type. There must be a difference in the physical processes involved in the two types.

Fujimoto et al. (1980) studied semi-analytically thermal stabilities of H and He shell-burnings on an accreting neutron star. The hydrogen shell-burning leads to the ignition of the He shell by three different routes on the ΔM vs T diagram as shown in Fig. 5 depending on the accretion rate. Case 2) on the diagram corresponds to the He shell-flash as argued by Joss (1978) and Hoshi (1980). Case 1) and 3) represent combined He and H shell flashes considered by Taam and Picklum (1979) and further discussed by Fujimoto et al. It is argued

Fast Bursts and Slow Bursts from 1608-522

	Fast Bursts	Slow Bursts
Rise Time Decay Time	$\lesssim$ 2 sec $\sim$ 5 sec	$\gtrsim$ 5 sec $\sim$20 sec
Peak flux	$L \propto T^4$ cooling high : variable	$L \propto T^4$ cooling low : const.
Duration (t 1/2)	$\sim$ 6 sec : const.	8$\sim$20 sec : variable
Presistent X-ray flux	High	Low
α	$\gtrsim$ 100	< 100
Process	Pure He Flash? Joss, Hoshi	Combined He and H Flash? Taam & Picklum, Fujimoto et al.

Fig. 4.

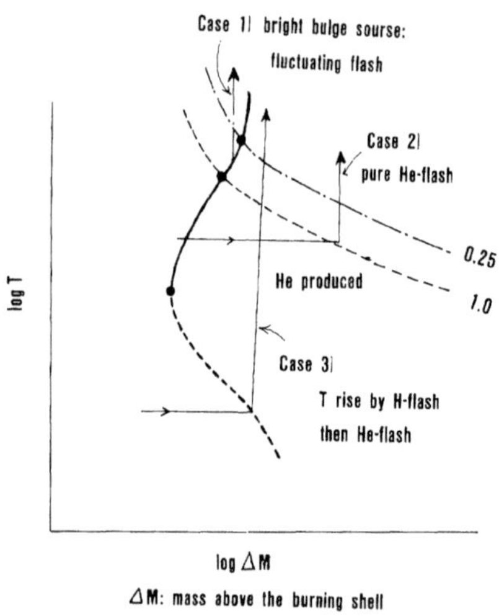

Fig. 5. Processes of the shell flash on the neutron star surface are illustrated on the ΔM-T diagram.

that the fast-type and the slow-type correspond to case 2) and case 3) respectively.

X-RAY EMISSION FROM THE NEUTRON STAR SURFACE

Next is the question whether bursts are generated from the total surface of a neutron star isotropically or from a local emitting region. Gross spectral behavior of the type-I burst is known to be consistent with the picture of a black body cooling with a constant area, but there are many cases which violate this simple view (see also Lewin and Clark, 1979).

1) Table III indicates an example for a burst from 1608-522 with the

Table III

DATE--TIME (U.T.)	PEAK FLUX CTS/S	INTEGRATED FLUX CTS	HARDNESS RATIO $\frac{F(12-30)}{F(1-12)}$	TEMPERATURE KEV	TEMPERATURE DEGREE
409-0829	750 ± 55	5470 ± 210	0.15±0.02	2.23±0.1	2.59×10^7
414-1809	461 ± 47	3200 ± 240	0.16±0.02	2.28±0.1	2.64×10^7
419-1423	358 ± 32	2440 ± 180	0.13±0.02	2.12±0.1	2.46×10^7

assumption of black body radiation. The temperature is constant while the peak flux shows considerable scatter. This indicates that the area of the emitting region changes significantly.

2) Fig. 6 shows examples of bursts from 1636-536. For case A), which

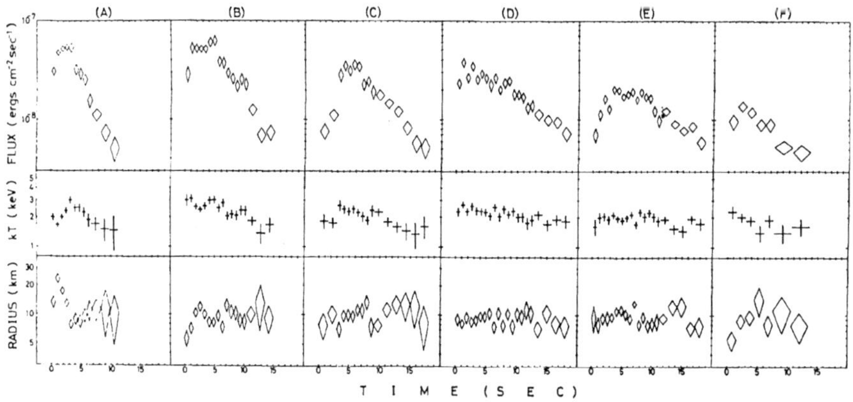

Fig. 6. Evolution of the burst from 1636-536.

is a fast type burst, the 'temperature', assigned on the assumption of black body radiation, rises and then decreases; accordingly the emitting 'area' increases and decreases to a value which is equal to that for the slow-type burst (Inoeu et al., 1979b).

3) For some bursts, ringing or oscillation in the flux and 'temperature' are noted. Among the bursters 1702-429 seems to indicate especially violent fluctuations as shown in Fig. 7. Several bursters, observed by SAS-3, have been known to exhibit double peaked structure and an irregular burst behavior (Lewin and Clark, 1979).

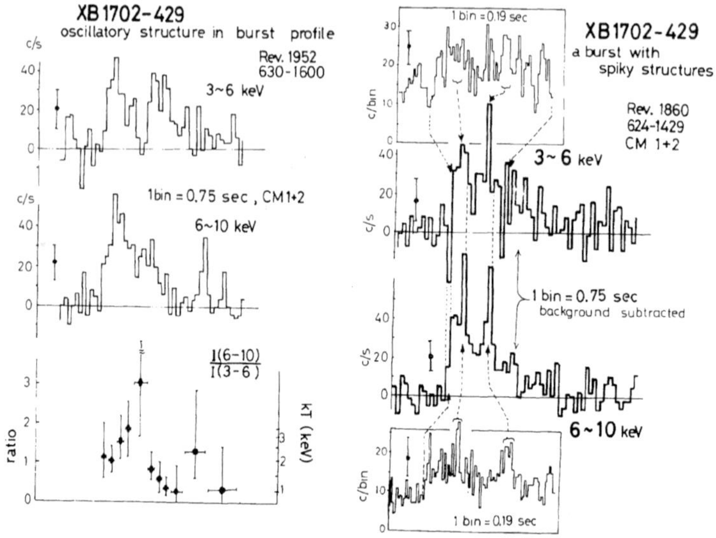

Fig. 7. Irregular burst behavior of 1702-429.

Simple-minded interpretation of the above mentioned phenomena may be a local flash on the neutron star surface and/or change of the emitting area during the flash. Simple theoretical arguments, however, appear to forbid a local nuclear flash on the surface of a weakly magnetized neutron star. On the other hand, if a strong magnetic field confines the flashing area, the question arises why we see no pulsations from bursters. Of course, the assumption of black body radiation may be an oversimplification particularly in the early phase of a burst. In any case these events are intriguing in relation to the physics of a nuclear flash on the surface of a neutron star.

INTERVALS (FREQUENCIES) OF BURSTS

No clear correlation between the burst interval and the persistent flux has been noted for 1608-522 and 1636-536. SAS-3 observed the correlation for MXB 1659-298 and MXB 1820-30, but not for MXB 1735-444, MXB 1837+05 and MXB 1636-536 (Lewin and Clark, 1979).

I stress one event of short burst interval observed for 1608-522. We detected two succeeding bursts from 1608-522 within an interval of about ten minutes (Murakami et al., 1980b). The events occurred on July 6 at 0954 UT and 1004 UT; both were slow type bursts exhibiting clear spectral softening; integrated burst sizes (counts in the 3-10 keV range) for the two bursts are 3160 ± 260 and 1690 ± 180 respectively while the size of a slow type burst from 1608-522 is generally observed to be in the range $1500 \sim 7000$. Identification of the two bursts with 1608-522 is quite certain: They were located to an accuracy of $\sim 0.1°$. Ten minutes does not appear to be long enough to accumulate enough nuclear fuel: α is smaller than 2.5 instead of $\alpha \gtrsim 100$ for the He flash and $\gtrsim 15$ for the H flash. Thus, though this is only one event, it needs to be explained.

We recall a similar case reported by the SAS-3 group for MXB 1743-28 (Lewin and Clark, 1979), which is one of the galactic center sources, where three bursts appeared in quick succession (time scale of $\sim$ 10 min): It was reported that the bursts may not have been type-I.

OPTICAL/X-RAY SIMULTANEOUS OBSERVATION

As a part of the world-wide burst watch conducted by the MIT group since 1977 the optical observations by Pedersen at ESO and the X-ray observations by "Hakucho" were coordinated by L.Cominsky, W.Lewin, G.Jernigan, J.Doty, A.Beardsly and J.van Paradijs. The observations were performed for 1636-536 during June-July, 1979 (Pedersen et al., 1979). Among 15 observed optical bursts we had 5 associated X-ray bursts. Fig. 8 shows an event with good statistics. Tentative conclusions from the data analysis are summarized as follows:

1) The parameter β relates the X-ray to the optical emission by the

$$(X_B/X_S)^{\beta} \approx Opt_B/Opt_S$$

formula, where B and S represent the burst flux and steady flux respectively. β was found to be 1/4 indicating that most (if not all) of the X-ray energy of the burst and persistent component is absorbed at the optical source and reprocessed to generate the Rayleigh-Jeans optical emission.
2) The optical burst is delayed by $\sim$ 2.5 sec: Hence, the distance of the optical source from the X-ray source is a few lightsec.
3) There is no indication of the optical burst being smeared by more than 1 sec: Therefore, the reprocessing time is shorter than 1 sec and the reprocessing region is smaller than 1 lightsec in size.

There is evidence which indicates that the steady optical emission is generated at the neutron star's accretion disk (Canizares et al., 1979). The size of the disk is typically of the order of a few light-seconds, which is consistent with conclusion 2). Thus, the evidence appears to point to a model in which the X-ray energy is absorbed at the disk, rather than the companion star, and reemitted at optical wavelengths: The question of why no smearing of the optical burst is observed remains unsolved.

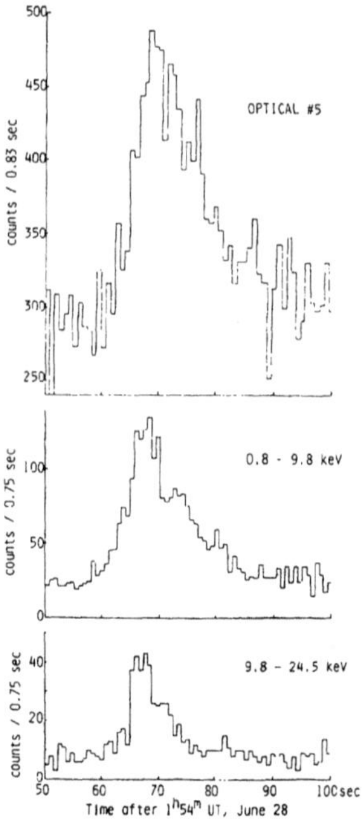

Fig. 8. Optical/X-ray observation of a burst on June 28, 1979 from 1636-536.

A NOVEL MODE OF BURSTS FROM 1730-333 (RAPID BURSTER)

We observed the activity of the rapid burster (Lewin et al., 1976; see recent review articles by Lewin, 1978 and references therein) from the beginning of its onset (Oda, 1979). The type-II bursts from the rapid burster are suggested to be generated by the release of gravitational energy of material accreted onto the neutron star surface being choked by some type of instability. The rapid burster appears to be a reccurent nova with an approximate period of 6 months. Approximately five months after SAS-3 observed the previous activity (Marshall, 1979), it started to be active at $\sim$0 hour UT of August 8, 1979.

The evolution of the burst activity is summarized in Fig. 9. This time, the rapid burster exhibited a rather unique burst (Inoue et al., 1980). For the first several hours a normal type-II burst train appeared mixed with a few flat-topped bursts. These flat-topped bursts were similar to those observed by SAS-3 (Marshall, 1979; Basinska et al., 1980) during the previous period of the rapid burster activity in March, 1979. The total energy per burst E, peak flux L_p, and average energy flux of the bursts $\langle L_B \rangle$, which may correspond to the

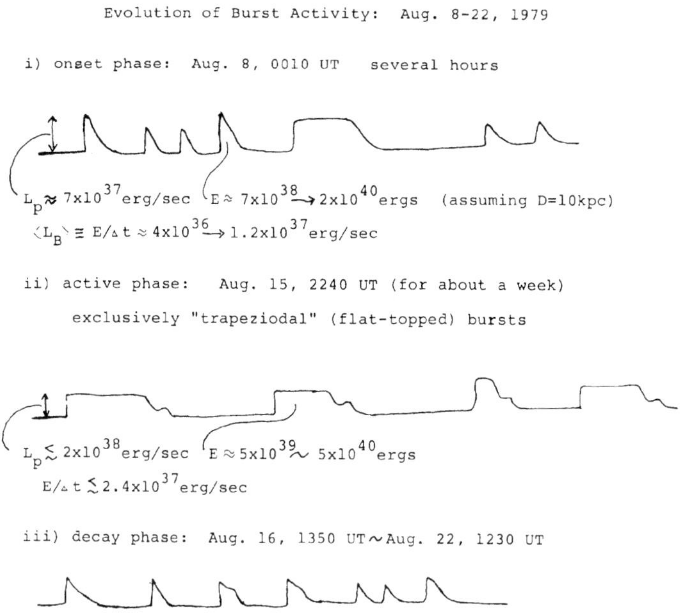

Fig. 9. Evolution of the rapid burster activity in August, 1979.

mass accretion rate $\dot{M}$, are indicated in the figure. Several hours after the onset suddenly E increased and then for about a week bursts of a flat-topped or trapezoidal profile appeared exclusively. After August 16, E and the duration of the bursts decreased and presumably they disappeared a few days after August 23, though we could not follow the source after this date due to our sun angle constraint.

Fig. 10 shows the relation between the total energy of the burst E and Δt (waiting time to the next bursts). We see an approximately linear relation and it is roughly on an extension of the line which the SAS-3 group had previously drawn on the diagram. It is noted, however, that the distribution of points is steeper than the mere proportionality: This indicates that the time-averaged burst energy $\langle L_B \rangle = E/\Delta t$ increases with E. Fig. 11 shows the evolution of these parameters with a sudden increase and gradual decay of $\langle L_B \rangle$.

Typical examples of the trapezoidal profile of the burst are shown in Fig. 12. Features of the trapezoidal bursts may be summarized as follows:

* It appeared for high $E/\Delta t$ ($\gtrsim 1.5 \cdot 10^{37}$ $(D/10\,\mathrm{kpc})^2$ erg/sec), where D is the distance to the source.
* The rise time is shorter than 1 sec, the decay time is 30-60 sec with a shoulder: It lasts for several minutes (the longest observed burst was for $\sim$700 sec).
* The level of the plateau varies by a factor of 4; the maximum peak

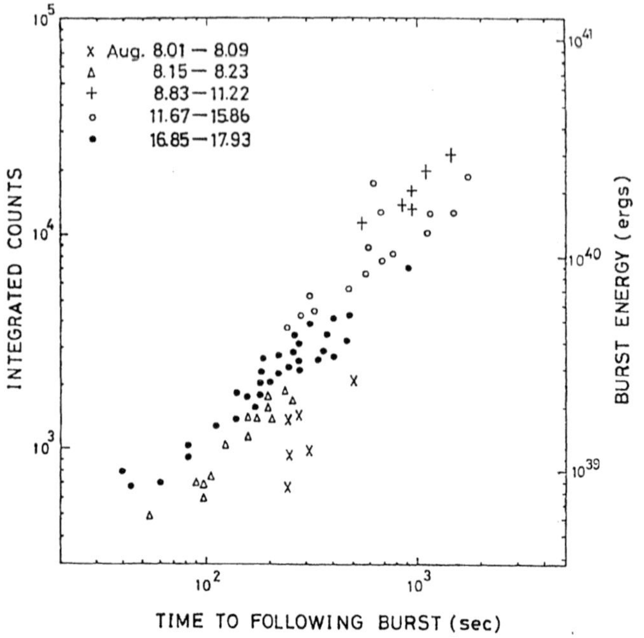

Fig. 10. Burst energy vs. time to the following burst.

flux was $2 \cdot 10^{38}$ erg/sec (hereafter D is assumed to be 10 kpc).

* No significant erratic or oscillatory structure of the plateau was observed except a few cases of interest which might be interpreted as a superposition of two structures.
* No clear spectral softening was observed during the duration of the burst.
* Dips with a duration of approximately one minute are noted before the rise and after the decay for most of the trapezoidal bursts.
* The duty cycle of the trapezoidal bursts was approximately 1/10.

Persistent emission was observed between bursts: Its flux was $L_x \approx 2 \cdot 10^{37}$ erg/sec, approximately 1/10 of the maximum peak flux of the bursts. These facts suggest that the train of trapezoidal bursts may be interpreted in terms of the transition between the high-top state and the low-top state with a one minute delay switch as illustrated in Fig. 13.

During the onset phase the peak flux L_p was 1/3 of that for the later phase. During the active phase L_p was constant $\approx 2 \cdot 10^{38}$ erg/sec for $E \lesssim 2 \cdot 10^{40}$ ergs and L_p decreases for increasing E beyond $2 \cdot 10^{40}$ ergs. The data are summarized in Fig. 14. The fact that L_p changes suggests that the simple model of the Eddington limit can not explain the flat-top burst.

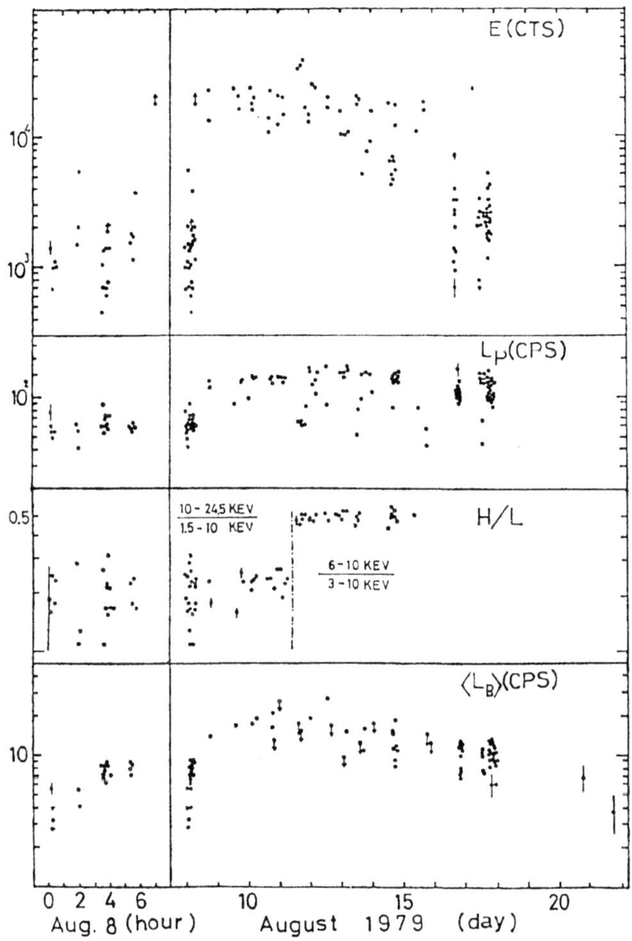

Fig. 11. Time history of the parameters of the bursts from the rapid burster throughout the period of Aug. 8-23, 1979. H/L is the hardness ratio.

The fact that L_p decreases with increasing E is an intriguing yet puzzling fact. First, this means that the amount of remaining fuel does not control the amount of energy generation, excluding the possibility that the flat-top is regulated at the emitting region (e.g. by the Eddington limit). Secondly, when the burst starts, the total energy that will be released in the burst is already determined: We do not yet know what kind of physical structure of the neutron star system may explain these features.

If there is a reservoir of material whose gravitational energy accounts for the trapezoidal burst, then the reservoir would have to

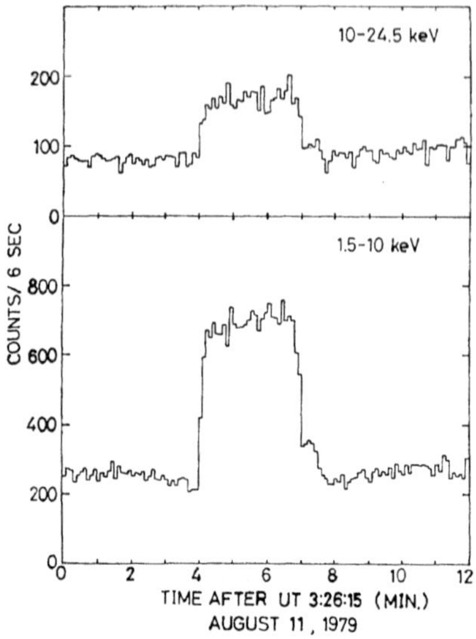

Fig. 12. An example of the trapezoidal burst. Dips in the flux or pauses of the persistent flux prior and after the burst is seen.

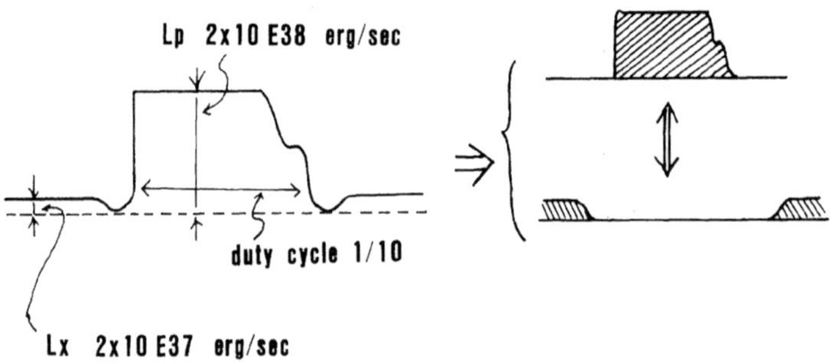

Fig. 13.

have a capacity of 10^{21} g to produce the largest observed burst. The following questions arise: What and where is the reservoir?: What sustains the flat top?: Is the peak regulated at the reservoir (e.g. is the drain rate regulated at the reservoir)?

The complex features of the trapezoidal burst may be summarized in an artificial model or toy, shown in Fig. 15. The model shows that there are two states of the reservoir: For the state a) the drain rate is smaller than (approximately 1/2 of) the in-flow rate, $\dot{M}$: Hence, the amount of material in the reservoir increases. In the state b) the drain rate is larger than (approximately 5 times) $\dot{M}$. The two drain rates for a) and b) correspond to the flux of the persistent component and the plateau of the burst respectively. The two states flip-flop with a one minute delay switch. We may play with this toy by adjusting its several parameters. But, of course, this may not necessarily represent the physical structure of the source. We hope that further analysis of data with the aid of hypothetical models and theoretical considerations combined with observations of the next recurrence of source activity will yield a better understanding of the mechanisms involved.

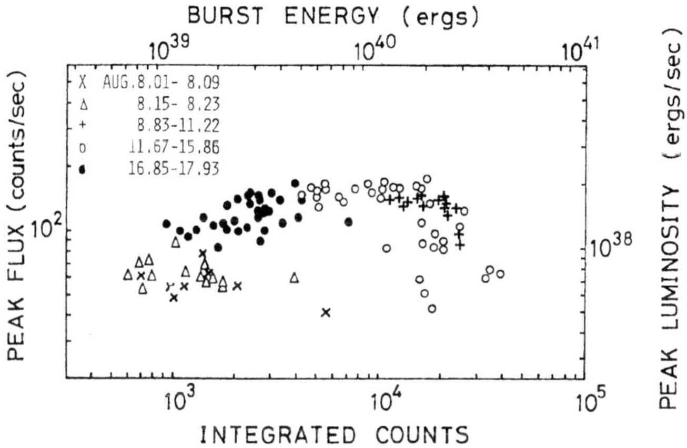

Fig. 14. The burst energy E vs. the burst peak flux L_p.

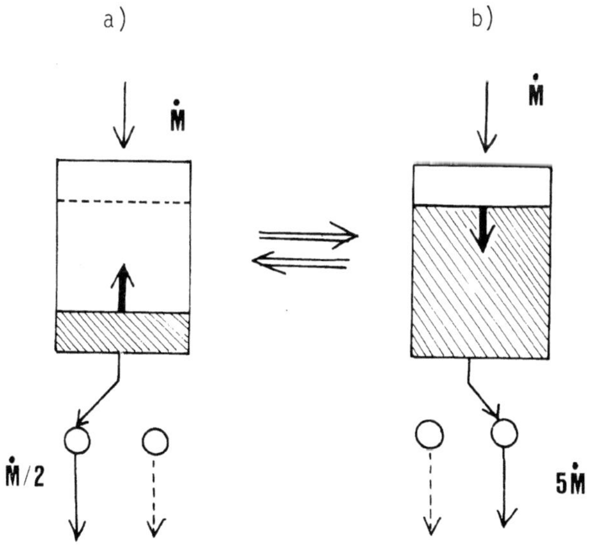

Fig. 15. Phenomenological reservoir model of the trapezoidal burst activity.

IR OBSERVATION OF THE RAPID BURSTER

Generation of remarkably bright IR bursts from the rapid burster has been reported by Apparao et al. (1979), Kulkarni et al. (1979) and Jones et al. (1979). On the other hand IR observations simultaneous with observed X-ray burst activity were performed by Okuda et al. (1979) and Glass (1979) with no success in finding IR bursts. The results are summarized in Table IV. For the observation 1), the source was known to be X-ray active on March 3-5 and not active on April 10. For the observation 2), as mentioned before, the source faded towards August 23. Thus, for both 1) and 2) IR bursts were observed when the source was presumably not active. On the other hand no simultaneous IR bursts were observed during (3) and 4)) when the source was X-ray active. In particular the IR observation at August 12, 0638 UT, when a trapezoidal burst was observed, showed no indication of unusual IR behavior of the source.

Thus, we may tentatively conclude that each of the type-II bursts, at least those observed in the X-ray are not associated with the IR burst, and the IR burst activity may appear days after the turn-off of type-II X-ray activity (though we can not be completely sure that it was off). The IR-study of the rapid burster will be the urgent subject of observations to be performed during the forthcoming active period of the source.

Table IV

A) IR Burst from the Rapid Burster, 1979

1) April 4 - 5: $\sim 2 \times 10^{-16}$ W/cm^2 μm (H-band) (Apparao, Kulkarni et al.)

*the source known to be X-ray active on March 3 - 5 but not active on April 10.

2) September 5: $\sim 10^{-16}$ W/cm^2 μm two bursts (K-band) (Jones et al.)

*the source known to be X-ray active on August 8 - 22 presumably then turned off.

B) No simultaneous IR/X-ray Bursts observed when the source was active

3) August 9, 11, 12 : total $6^h\ 12^m$ no IR burst
at Aug. 12, 0638 UT H-band (Okuda et al.)
$< 1.3 \times 10^{-17}$ W/cm^2 μm

4) August 16, 1721 - 1830 UT
August 21, 1835 - 2005 UT (Glass)

ACKNOWLEDGEMENT

I thank members of the Hakucho Team, whose works are reviewed in this talk, for their contributions to this article. I also acknowledge helps of Ms. Yoko Akiyama and Fred Marshall from MIT for the preparation of the manuscript.

REFERENCES

Apparao, K.M.V., Chitre, S.M., Ashok, N.M., Kulkarni, P.V.: 1979, IAU Circular 3344.
Basinska, E., Lewin, W.H.G., Cominsky, L., Marshall, F., and van Paradijs, J.: 1980, Ap.J. in press.
Belian, R.D., Conner, J.P., and Evans, W.D.: 1976a, Ap.J. (Letters) 206, pp.135-138.
Belian, R.D., Conner, J.P., and Evans, W.D.: 1976b, Ap.J. (Letters) 207, pp.33-36.
Canizares, C., McClintock, J.E., and Grindlay, J.: 1979, Ap.J. (Letters) 236, pp.55-59.
Fujimoto, M.Y., Hanawa, T., and Miyaji, S.: 1980, submitted to Ap.J.
Glass, I.: 1979, Private communication.
Hoffman, J.A., Lewin, W.H.G., Doty, J., Jernigan, J.G., Haney, M., and Richardson, J.A.: 1978a, Ap.J. (Letters) 221, pp.57-62.
Hoffman, J.A., Marshall, H.L., and Lewin, W.H.G.: 1978b, Nature 271, pp.630-633.
Hoshi, R.: 1980, Prog. Theor. Phys., in press.
Inoue, H., Koyama, K., Makishima, K., Matsuoka, M., Murakami, T., Oda, M., Ogawara, Y., Ohashi, T., Shibazaki, N., Tanaka, Y., Tawara, Y., Kondo, I., Hayakawa, S., Kunieda, H., Makino, F., Masai, K., Nagase, F., Yamashita, K., Miyamoto, S., Tsunemi, H., and Yoshimori, M.: 1979, Proc. 16th IUPAP Cosmic Ray Conference Vol.1, pp.5-10.
Inoue, H., Koyama, K., Makishima, K., Matsuoka, M., Murakami, T., Oda, M., Ogawara, Y., Ohashi, T., Shibazaki, N., Tanaka, Y., Tawara, Y., Kondo, I., Hayakawa, S., Kunieda, H., Makino, F., Masai, K., Nagase, F., Miyamoto, S., Tsunemi, H., and Yamashita, K.: 1979b, preprint ISAS.
Inoue, H., Koyama, K., Makishima, K., Matsuoka, M., Murakami, T., Oda, M., Ogawara, Y., Ohashi, T., Shibazaki, N., Tanaka, Y., Tawara, Y., Kondo, I., Hayakawa, S., Kunieda, H., Makino, F., Masai, K., Nagase, F., Miyamoto, S., Tsunemi, H., and Yamashita, K.: 1980, Nature 283, pp.358-360.
Jones, A.W., Selby, M.J., Mountain, C.M., Wada, R., Magro, C.S., and Munoz, M.P.: 1980, Nature 283, pp.550-551.
Joss, P.C.: 1978, Ap.J. (Letters) 225, pp.123-127.
Joss, P.C.: 1979, Ann. N.Y. Academy of Sciences. Proceedings of Ninth Texas Symposium, Munich, Germany.
Kulkarni, P.V., Ashok, N.M., Apparao, K.M.V., and Chitre, S.M.: 1979, Nature 280, pp.819-820.
Lamb, D.Q., and Lamb, F.K.: 1978, Ap.J. 220, pp.291-302.

Lewin, W.H.G., Doty, J., Clark, G.W., Rappaport, S.A., Bradt, H., Doxsey, R., Hearn, D.R., Hoffman, J., Jernigan, J.G., Li, F.K., Mayer, W., McClintock, J., Primini, F., and Richardson, J.: 1976, Ap.J. (Letters) 207, pp.95-99.
Lewin, W.H.G., and Joss, P.C.: 1977, Nature 270, PP.211-216.
Lewin, W.H.G.: 1979, Advances in Space Exploration, 3, IAU/COSPAR Symposium on X-Ray Astronomy in Innsbruck, Austria, May 1978, eds. W.A. Baity and L.E. Peterson, Pergammon Press, Oxford and New York.
Lewin, W.H.G., and Clark, G.W.:1979, ISAS Symposium on X-ray Astronomy, Tokyo, PP.3-60.
Makishima, K., Inoue, H., Koyama, K., Matsuoka, M., Murakami, T., Oda, M., Ogawara, Y., Ohashi, T., Shibazaki, N., Tanaka, Y., Hayakawa, S., Kunieda, H., Makino, F., Masai, K., Nagase, F., Tawara, Y., Miyamoto, S., Tsunemi, H., Yamashita, K., and Kondo, I.: 1980 preprint.
Maraschi, L., and Cavaliere, A.: 1977, Highlights in Astronomy 4 partI.
Marshall, F.: 1979, IAU Circular 3336.
Matsuoka, M., Inoue, H., Koyama, K., Makishima, K., Murakami, T., Oda, M., Ogawara, Y., Ohashi, T., Shibazaki, N., Tanaka, Y., Tawara, Y., Hayakawa, S., Kunieda, H., Makino, F., Masai, K., Nagase, F., Miyamoto, S., Tsunemi, H., Yamashita, K., and Kondo, I.: 1980, Ap.J. in press.
Murakami, T., Inoue, H., Koyama, K., Makishima, K., Matsuoka, M., Oda, M., Ogawara, Y., Ohashi, T., Shibazaki, N., Tanaka, Y., Tawara, Y., Hayakawa, S., Kunieda, H., Makino, F., Masai, K., Nagase, F., Miyamoto, S., Tsunemi, H., Yamashita, K., and Kondo, I.: 1980a, Ap.J. in press. 1980b, preprint ISAS RN105.
Oda, M. : 1979, IAU Circular 3392.
Pedersen, H., Oda, M., Hakucho Team, Cominsky, L., Doty, J., Jernigan, J.G., van Paradijs, J., and Lewin, W.H.G.: 1979, IAU Circular 3399.
Sato, S., Kawara, K., Kobayashi, Y., Maihara, T., Okuda, H., and Jugaku, J.: 1980, preprint KUNS535.
Taam, R.E., and Picklum, R.E.: 1979, Ap.J. 233, pp.327-333.
Tanaka, Y.: 1978, ISAS Symposium on X-ray Astronomy, Tokyo, pp.61-84.
Woosley, S.E., and Taam, R.E.: 1976, Nature 263, pp.101-103.

X-RAY SOURCES IN GLOBULAR CLUSTERS

Jonathan E. Grindlay
Harvard-Smithsonian Center for Astrophysics
Cambridge, Massachusetts

ABSTRACT

The program of x-ray observations of globular clusters being conducted with the Einstein X-ray Observatory is described. A survey of some 30 clusters has been completed using the HRI detector and yielded one new source down to luminosities $L_X > 10^{34}\ d_{10}^2$ erg s^{-1} (d_{10} is distance in 10 kpc units). Preliminary results are reported for a more sensitive survey ($L_X \gtrsim 10^{33}\ d_{10}^2$ erg s^{-1}) now being conducted with the IPC; four possible new sources have been found in the first 32 clusters observed. A related search for diffuse emission from stars in globular clusters is also in progress using much longer IPC exposures. Detailed studies of several luminous, globular cluster sources, including the x-ray burst from Terzan 2, are described. Finally, the major effort to locate precisely ($\lesssim$2 arcsec) the x-ray sources with respect to the cluster centers is discussed. Sources of systematic error and the required optical studies of globular cluster structure are reviewed. Preliminary results appear to rule out the possibility that the globular cluster sources are massive black holes, although the statistical tests are somewhat model dependent. Implications of the plausible binary systems for cluster core evolution are mentioned.

1. INTRODUCTION

X-ray observations of globular clusters have continued to be of key interest in the era of imaging x-ray astronomy introduced with the Einstein Observatory. These x-ray sources have long been recognized as especially interesting because of their x-ray similarities with the bright galactic bulge sources (Canizares 1975, Grindlay 1978), which are largely still unidentified, and because they were the first found to also emit x-ray bursts (Grindlay et al. 1976). Because globular clusters can be well-studied optically, they afford a unique opportunity in x-ray

R. Giacconi (ed.), X-Ray Astronomy with the Einstein Satellite, 79–109.

astronomy to be able (potentially) to completely specify the parameters of a compact accretion-fed source; e.g., its distance and, therefore, luminosity, mass (or at least most probable value), and possible evolutionary history. While the massive Pop I x-ray binaries may afford similar opportunities, there are usually significant complicating effects (e.g., x-ray heating of the stellar companion) which make fundamental quantities like distance (and luminosity) relatively more uncertain. Specification of source parameters is, of course, the necessary first step before developing detailed models for either persistent or burst x-ray emission. Einstein observations, with >100 times the sensitivity and $\gtrsim$10 times the angular resolution of earlier x-ray observations, are now making possible a survey of many globular clusters for the determination of "global properties" (e.g., luminosity function) as well as detailed studies of individual sources.

We shall review the progress made in Einstein observations of globular clusters, as well as related studies of bursters. Our review will incorporate some results obtained through September 1980 but will report primarily analysis completed as of the HEAD meeting in January 1980. The first preliminary results of Einstein observations of globular clusters were reported by Grindlay (1979). The review by Lewin and Clark (1979) gives a comprehensive summary of earlier observations of bursters (and previously known x-ray globular clusters). Earlier reviews by Lewin and Joss (1977), and Grindlay (1977) are also relevant, and recent burster results are given by Oda (1980a).

The plan of this review is as follows. In Section 2, we present the results of the completed high resolution imaging (HRI) detector survey of globular clusters. Eight relatively luminous sources were detected (including one new source, 47 Tuc) and upper limits were derived for 27 other clusters. We then describe preliminary results of a more sensitive survey being conducted with the imaging proportional counter (IPC). Four new low luminosity ($\sim10^{33}$ erg s^{-1}) sources are probable detections in the 32 clusters observed to date. In Section 3 we present results from some of the analysis of the x-ray properties of individual (bright) globular cluster sources. Finally, in Section 4 we describe the (still preliminary) results for the primary effort in Einstein observations of globulars--the determination of the most probable x-ray source masses from the precise x-ray position relative to the (optical) cluster center. Section 5 summarizes the current status of observations and theory for the globular cluster x-ray sources.

TABLE 1

PARAMETERS OF COMPACT X-RAY SOURCES IN GLOBULAR CLUSTERS DETECTED BY EINSTEIN*

		Structural Params.			Optical Params.			X-ray Params.		
Cluster	Dist. (kpc)	Central Density ($10^4 M_{\odot} pc^{-3}$)	Conc. ($\log r_t/r_c$)	Central Relax Time ($10^7 y$)	Spectral Type	Metals [Fe/H]	Extinct NH (10^{21} cm^{-2})	Absorp. Col. NH (10^{21} cm^{-2})	Expon. Temp. kT (KeV)	L_x (0.5-4.5 KeV) (10^{36} erg s^{-1})
NGC 104	4.6	4.8	2.0	8	G3	-0.44	0.2	-	-	0.03
NGC 1851	10.8	8.5	1.8	3.7	F7	-1.29	0.4	3.5^{+5}_{-2}	2.2	1.0
NGC 6441	10.0	13	1.7	8	G2	-0.24	3.0	≤ 0.1	6.0	3.7
NGC 6624	8.5	11	2.2	~5	G5	-0.34	1.6	0.1 ± 0.4	8.0	52
NGC 6712	7.4	0.15	1.2	35	G5	-0.43	2.3	≤ 0.1	3.0	1.2
NGC 7078	9.3	4	2.0	10	F2	-2.01	0.8	0.1 ± 1.6	5.0	3.0
Terzan 2	~9						~8	8.2^{+2}_{-5}	5.5	3.1
Liller 1†	~10						~20	32 ±5	7.9	8.8
Grindlay 1††	~9						~27	36	12.7	12

* All clusters were detected with the HRI. Cluster data are from Arp (1965), Peterson and King (1975), Peterson (1976), Grindlay (1979), and Harris and Racine (1979). Two other compact x-ray sources (both bursters) discovered by HAKUCHO (Oda 1980b) are also very likely assocated with the globular clusters Terzan 1 and Terzan 5 (Grindlay 1980). Neither has yet been observed with EINSTEIN. Thus, nine of the eleven globular cluster sources known to be compact are bursters.

† The spectrum and luminosity values given for Liller 1 (=MXB1730-335, the rapid burster) represent the integrated emission in bursts (Type II); no truly persistent emission was detected.

††This cluster was only recently discovered (Grindlay and Hertz 1980) as the counterpart of the burster MXB 1728-34. A precise cluster center or core radius will require further IR observations so that it was not included in the analysis for x-ray source mass.

2. SURVEYS FOR (PERSISTENT) X-RAY EMISSION FROM GLOBULAR CLUSTERS

A. HRI Survey

A survey of some 35 globular clusters in the Galaxy has been completed by the CFA and MIT groups using the HRI detector on the EINSTEIN Observatory. A preliminary account of the CFA portion (~20 clusters) of the observing program was given by Grindlay (1979); the complete (joint) survey will be described in detail elsewhere. In general, the clusters were selected from published lists (Arp 1965, Peterson and King 1975) to be either very centrally condensed or relatively nearby. Each cluster was observed for, nominally, 2×10^3 sec. In actual fact, exposures varied from ~$1-3 \times 10^3$ sec. Eight clusters were detected; of these, only one (47 Tuc = NGC 104) was previously unknown to contain an x-ray source. One cluster (NGC 6440) previously suspected to contain a transient x-ray source was not detected by the HRI (Grindlay 1979) although, as described below, it was subsequently detected at a much lower level by the IPC.

In Table I we list the parameters for the eight globular cluster sources detected by the HRI and used for the statistical analysis for source mass (Section 4). Spectral parameters derived from the monitor proportional counter (MPC) for seven of these eight clusters, as well as a newly discovered burster, MXB 1728-34 (Grindlay and Hertz 1980), are also given. The MPC provides spectral data and best-fit model spectra using predicted background count rates (Grindlay et al. 1980b). Spectral parameters derived from single observations on each source are given. To facilitate comparison, parameters for an exponential spectral fit (without Gaunt factor) are given even though these were not always the best fit models (e.g., NGC 6441, Terzan 2, and Grindlay 1 were best-fit by bremsstrahlung spectra with temperatures ~1.5 times those given here, and NGC 6712 was best-fit with a power law spectrum). Multiple observations are available for these sources since they were observed at several offset positions to minimize systematic errors in determining their absolute positions (see Section 4); the spectral variability derived from these observations will be reported elsewhere. We also use the optical extinction A_V for each cluster derived from Harris and Racine (1979) or our own optical (-IR) studies to derive the corresponding (Gorenstein 1975) column density N_H. Comparison with the N_H values derived from the low energy cutoffs in the x-ray spectra indicates that the latter values are not always consistent with only interstellar absorption. The spectrum of NGC 1851 appears

TABLE 2

GLOBULAR CLUSTERS SURVEYED WITH THE HRI DETECTOR: UPPER LIMITS

		Structural Params.			Optical Params.			X-ray Params.
Cluster	Dist. (kpc)	Central Density ($10^4 M_\odot pc^{-3}$)	Conc. ($\log r_t/r_c$)	Central Relax Time (10^7y)	Spectral Type	Metals [Fe/H]	Extinct NH (10^{21} cm^{-2})	L_x (0.5-4.5 KeV) (10^{34} erg s^{-1})
Kron 3	~55							<26
NGC 362	12.8	4.2	1.7	7	F8	-1.20	0.3	< 1.1
NGC 1904	13.3	0.4	1.6	14	F6	-1.58	0.06	< 2.7
NGC 2298	12.3	0.06	1.2	23	F7	-1.41	0.7	< 1.9
NGC 2419	93.3	0.002	1.4	1000	F6	-2.00	0.2	<87
NGC 2808	9.2	4.2	1.7	15	F8	-1.09	1.4	< 0.9
NGC 5634	21.6		1.6		F5	-1.50	0.4	< 4.1
NGC 5824	23.7	4	2.5	3	F5	-1.67	0.9	<13.9
NGC 6093	8.3	8	1.9	2.7	F9	-1.54	1.4	< 0.9
NGC 6218	5.3	0.13	1.3	30	F8	-1.64	1.8	< 0.7
NGC 6229	30.9	0.15	1.4	21	F8	-1.44	0.06	<10.8
NGC 6254	4.3	0.47	1.5	15	G1	-1.43	1.8	< 0.9
NGC 6266	5.9	3.6	1.6	8	G2	-1.14	3.0	< 1.3
NGC 6273	10.0	2	1.5	3.5	F3	-1.61	2.5	< 3.5
NGC 6293	7.2	5.8	1.8	1.5	F0	-1.86	2.2	< 1.7
NGC 6342	15.1		1.6		G5	-0.41	3.2	<10
NGC 6356	16.7	0.7	1.3	35	G4	-0.37	1.8	<11
NGC 6388	14.1	12	1.8	12	G3	-0.48	2.2	< 5.7
NGC 6440	3.6	46	>1.6	3.5	G5	-0.28	7	< 0.3
NGC 6517	7.8	18	2.0	1.7	G0	-1.33	7.2	< 1.9
NGC 6681	10.0	5	1.1	2.3	G3	-1.17	0.4	< 1.1
NGC 6715	21.5	5	1.9	13	F8	-1.55	1.0	<15
NGC 6760	3.8	1.4	1.5	2.1	G0	-1.06	6	< 1.1
PAL 10	8.5						8	< 2.6
NGC 6809	5.1	0.04	1.0	100	F5	-1.78	0.4	< 0.3
NGC 7006	34.7	0.01	1.4	110	F2	-1.66	0.8	<29
NGC 7089	11.3	0.8	1.6	39	F4	-1.53	0.4	< 1.3

NOTE: Cluster data are from Arp (1967), Peterson and King (1975), Peterson (1976), and Harris and Racine (1979).

cutoff at low energies when fit to any simple models. Increased spectral resolution is required to determine if this is due to local absorption or perhaps an excess at ~2 KeV from strong Si line emission. Several other clusters, on the other hand, have anomolously low NH values when their spectra are fit to simple models. It is possible that these have (variable) soft spectral components for a two-component spectrum.

Thus, for clusters too faint for MPC detection (but detectable by the IPC; see below), the optical values of N_H should provide a rough estimate of the absorption. Finally, it is noteworthy that most of the clusters are best described by either simple exponential or thermal bremsstrahlung spectra with temperatures in the range kT $\simeq$ 4-10 KeV. Thus, fainter clusters may be assumed to have exponential spectra with kT $\simeq$ 5 KeV.

The spectral parameters derived for each cluster detected by the MPC allow the luminosity to be estimated, given the distances listed by Harris and Racine (1979). The clusters span nearly three decades in luminosity ($10^{34.5}$ - $10^{37.7}$ erg s^{-1}) and are seen to have a mean luminosity of ~10^{36} erg s^{-1}. Note that these luminosities are representative values only since (as mentioned above) repeated observations of all clusters (except Liller 1) have been made and all show variability. Typical variability is a factor of ~2 about the values given. As suggested previously (Grindlay 1977), the detected cluster luminosities are significantly BELOW that obtained for the galactic bulge sources ("GX sources") if the latter are assumed to be typicaly at ~10 kpc distances. Conversely, if the globular cluster and GX sources really are similar, then most of the GX sources must be at relatively local distances of ~3 kpc--perhaps in the Sagittarius arm. A "local" distribution might partially explain the relatively broad longitude distribution of GX sources vs. the true galactic bulge sources surrounding the galactic nucleus as seen in M31 (van Speybroeck et al. 1979). The counterparts of these true galactic bulge sources in our own galaxy are then much fewer in number (perhaps because of fewer globular clusters and their subsequent decay by dynamical friction into the nucleus) and limited to only a few sources (e.g., GX 3+1, GX 5-1) truly in the galactic bulge.

Some 27 additional globular clusters were observed in either CFA or MIT programs to survey galactic globular clusters for x-ray emission using the HRI. None of these was detected and the upper limits to their ~0.5-4.5 KeV luminosities are given in Table 2. In deriving these limits, we have used the cluster distances and optical extinctions A given by Harris and Racine (1979) or Arp (1965) and have also assumed an exponential spectrum with kT ≃ 5 KeV (as justified above). Comparison of the upper limits and detected L_X values (Tables 2 and 1, respectively) shows that the HRI survey was sufficiently sensitive to detect clusters with luminosities typically ~10 times lower than the weakest cluster actually detected. This suggests (Grindlay 1979) there might, indeed, be a luminosity function cutoff below $\sim 10^{35}$ erg s^{-1} for compact x-ray sources in globular clusters, and a maximum likelihood test analysis will be used to test this hypothesis when all the data are available. However, this interpretation is complicated by the preliminary results of the IPC survey, which has greater sensitivity.

B. IPC Survey

In an effort to search for still fainter sources in globular clusters and to extend the luminosity function limits to lower L_X values, a survey of most globular clusters in the galaxy is being conducted using the Imaging Proportional Counter (IPC) on the EINSTEIN Observatory. Some ninty clusters (of the 132 clusters known) will be surveyed; each with a ~2000 sec exposure. This survey is typically a factor of $\gtrsim 10$ more sensitive than the HRI survey (this sensitivity ratio increases for increasing values of the absorption column density N_H--e.g., for clusters near the galactic plane).

Preliminary analysis of the IPC data is available for 32 clusters surveyed thus far. Of these, at least one x-ray source has been detected apparently within each of five clusters (including 47 Tuc, first detected with the HRI). The four new IPC detections are each only ~3σ detections so follow-up exposures are needed to confirm their existence. These detections and preliminary values for the upper limits for the 27 clusters not detected are summarized in Table 3. Since it is not yet possible to fit spectra to IPC data, the source spectral parameters kT and N_H are highly uncertain. Luminosities for the five detected clusters were derived from the observed counts in the 0.5-4.5 KeV band and an assumed exponential spectrum with kT = 5 KeV and low energy absorption from the "optical" values of NH. In deriving the luminosity upper limits for Table 3, we have simply assumed a scaling of 1 IPC count $s^{-1} \cong 2.5 \times 10^{-11}$ erg cm^{-2} s^{-1} (0.5-4.5 KeV) which is approximately correct for an exponential spectrum with kT - 5 KeV (as assumed above) and no low energy cutoff. An

TABLE 3

GLOBULAR CLUSTERS SURVEYED WITH THE IPC: DETECTIONS AND UPPER LIMITS

		Structural Params.						Optical Params.			X-ray Params.	
Cluster	Dist. (kpc)	Central Density[1]	Concen- tration[2]	Central Relax Time[3]	Core (arcsec)	Radius (pc)	Core Mass M_c[4]	Spectral Type	Metals [Fe/H]	Extinct NH[5]	L_x[6]	L_x/M_c[7]
NGC 104	4.6	4	2.0	8	24	3.4	3.4	G3	-0.44	0.3	6.7	2.0
NGC 1904	13.3	0.4	1.6	14	16	1.0	1.6	F6	-1.58	0.06	7	4.4
NGC 6440	3.7	46	>1.6	3.5	17	0.3	6	G5	-0.28	7	0.4	0.1
NGC 6517	7.8	18	2.0	1.7	9.5	0.4	3.4	G0	-1.33	7.2	2.6	0.8
NGC 6541	6.9	2.5	2.0	4	19	0.6	2.6	F6	-1.59	0.8	1.7	0.7
NGC 1261	13.3	0.09	1.3	40	29	1.9	2.4	F8	-1.24	0.14	< 6.6	< 2.7
NGC 2298	12.3	0.06	1.2	23	27	1.6	1.0	F7	-1.41	0.7	< 4.1	< 4.1
NGC 2808	9.2	4.2	1.8	15	15	0.7	5.1	F8	-1.09	1.4	< 3.2	< 0.6
NGC 4590	9.6	0.06	1.6	40	40	1.9	1.7	A7	-2.04	0.2	< 2.4	< 1.4
NGC 5024	17.2	0.5	1.7	100	32	2.7	3.9	F4	-1.85	0.3	< 9.7	< 2.5
NGC 5634	21.6	-	1.6	-	12	1.3	-	-	-	0.4	<13	-
NGC 5694	32.2	0.4	1.8	17	15	2.4	20	F0	-1.91	0.5	<32	< 1.6
NGC 5824	23.7	4	2.5	3	4	0.5	1.6	F5	-1.67	0.9	<17	<10.7
NGC 5904	7.6	0.5	1.8	20	30	1.1	2.7	F6	-1.25	0.2	< 1.9	< 0.7
NGC 6093	8.3	8	1.9	2.7	7	0.3	0.8	F9	-1.54	1.4	< 4.0	< 4.8
NGC 6402	10.2	0.08	1.1	170	48	2.4	3.4	G1	-1.28	3.8	< 2.1	< 0.5
NGC 6528	7.3	3.3	1.5	3.5	12	0.4	1.0	G5	-0.43	4	< 1.9	< 1.9
NGC 6715	21.5	5	1.9	13	6	0.7	5.6	F8	-1.55	0.9	<17	< 3.0
PAL 10	8.5	-	-	-	-	-	-	-	-	8	< 2.9	-
NGC 6809	5.2	0.04	1.0	100	104	2.6	0.7	F5	-1.78	0.4	< 8.6	<11.8
NGC 6838	4.0	0.11	1.2	8.2	48	0.9	0.4	G5	-0.28	1.8	< 0.5	< 1.4
NGC 6934	14.7	0.16	1.4	30	24	1.7	3.2	G0	-1.38	0.8	< 6.6	< 2.1
NGC 6981	17.3	0.02	1.2	70	34	2.8	1.6	G3	-1.27	0.2	< 9.1	< 5.7
NGC 7006	34.7	0.01	1.4	110	16	2.7	1.0	F2	-1.66	0.8	<39	<39
NGC 7089	11.3	0.8	1.6	39	24	1.3	7.2	F4	-1.53	0.4	< 3.6	< 0.5
UKS1751-241	~10	-	-	-	-	-	-	-	-	-	< 4.4	-
Terzan 3	~10	-	-	-	-	-	-	-	-	-	< 5.9	-
Terzan 7	~10	-	-	-	-	-	-	-	-	0.8	< 5.6	-
Terzan 10	~10	-	-	-	-	-	-	-	-	-	< 5.0	-
Terzan 12	~10	-	-	-	-	-	-	-	-	-	<12	-

[1] (10^4 $M_\odot pc^{-3}$)
[2] (log r_t/r_c)
[3] (10^7 y)

[4] (10^4 $M_\odot$)
[5] (10^{21} cm^{-2})
[6] (0.5-4.5 KeV) (10^{33} erg s^{-1})
[7] (10^{33} erg $s^{-1}/10^4$ $M_\odot$)

FIGURE 1

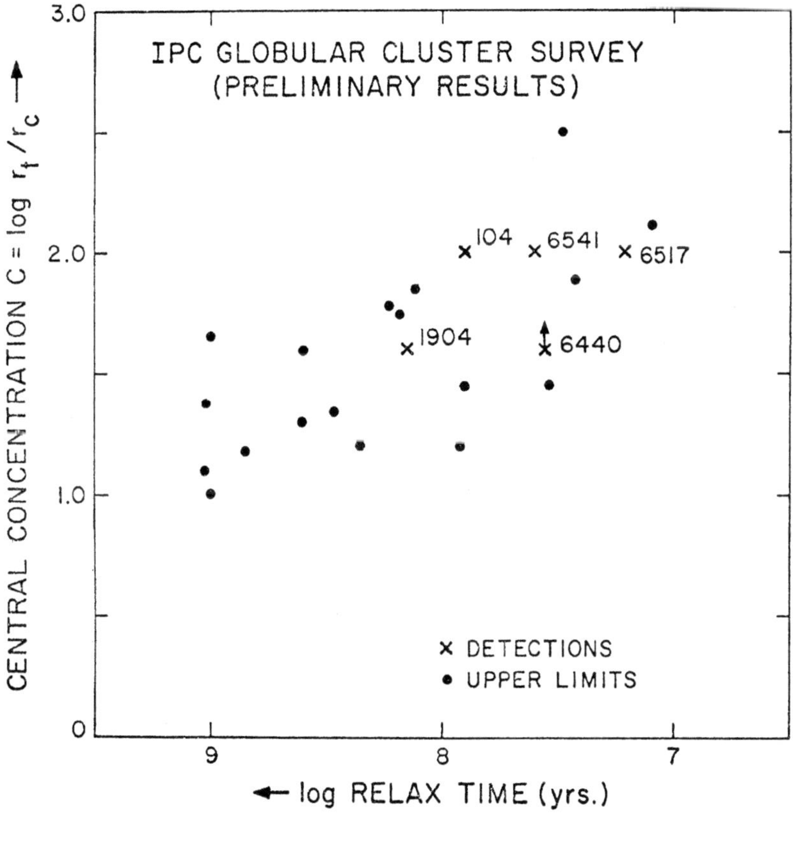

The probable new globular cluster x-ray sources are located in clusters of relatively short central relaxation time and high central concentration.

approximate correction for the cutoff due to the optical extinction NH values given in Table 3 has also been applied.

Several interesting results are apparent from these preliminary results. First, the detected x-ray sources are all of similar luminosity: $L_x \sim 1 - 10 \times 10^{33}$ erg s^{-1}. This is just below the typical upper limits derived in the HRI survey (Table 2). Two clusters (NGC 1904 and NGC 6440) were detected in the IPC survey but not by the HRI; the detected L_x values are well below (factors of ~3-10) the corresponding upper limits. Note that the apparent luminosity of NGC 104 (47 Tuc) as detected by the IPC is a factor of ~4 lower than that detected a year before by the HRI (cf. Table 2). Some, but not all, of this difference may be due to the uncertainty in the spectral shape assumed. Second, the newly detected clusters all have relatively large central concentration parameter values $c = \log(r_t/r_c)$, where r_t and r_c are the tidal and core radii of the cluster (Peterson and King 1975) and Peterson (1976). They also have relatively short central relaxation times T_R, given their moderately high central densities. This is apparent in Figure 1 where we plot the IPC survey clusters for which c and T_R are available (Peterson 1976). The detected clusters all are at the upper right of the diagram, just as is the case for the high luminosity clusters in the similar diagram of escape velocity (proportional to c) vs. T_R plotted by Bahcall and Ostriker (1975). Finally, the detected clusters have similar values of their luminosity per unit mass: $L_{33}/M_4 \simeq 0.8 - 4.4$ for all detected clusters except NGC 6440 (with $L_{33}/M_4 \simeq 0.1$, though this is more uncertain given the large extinction).

From the IPC survey results alone we are unable to determine the nature of the newly detected sources. That is, they may be either the lowest luminosity end of a continuous luminosity function of compact sources in globulars or they may be diffuse sources due to the combined stellar emission from the cluster core. The ~1 arcmin IPC resolution is not sufficient to resolve these sources if they are diffuse given the ~15 arcsec (typical) cluster core radii (cf. Table 3). Obviously HRI observations, with appropriately long integration times, are needed.

The fact that the luminosities of the new cluster sources are comparable with that of NGC 104 (47 Tuc), known (from the HRI) to be a point source, suggests they, too, are point like. In addition, the apparent preference (Figure 1) for centrally condensed, evolved cluster suggests these sources have a similar origin (see

Section 5). However, their similar L/M values (Table 3) are expected if they are diffuse, and the fact that the detected clusters are all relatively close and have relatively massive cores also supports this interpretation. The diffuse emission origin would require a typical luminosity of $\gtrsim 10^{28.5}$ erg s^{-1} for a $\sim 0.5\ M_\odot$ main sequence dwarf in the cluster core. Present measurements of K and M dwarf stellar x-ray luminosities (Vaiana et al. 1980) are based on detections of only two Pop II stars with luminosities only $\sim 10^{27.5}$ erg s^{-1}. Although diffuse emission may therefore be unlikely, a larger stellar (Pop II) sample is needed.

A program of deep IPC observations of several globulars with large, well-resolved (in the IPC) core radii is also being conducted with EINSTEIN. The motivation is to look for diffuse emission from either stars (CFA observations) or stellar winds (guest observation). Analysis of these data is incomplete but will, of course, be crucial to understanding the new low luminosity sources.

3. STUDIES OF INDIVIDUAL GLOBULAR CLUSTER SOURCES

Detailed study of the spectral and temporal properties of the individual globular clusters detected by the MPC (Table 1) is in progress. At present, the analysis of the Terzan 2 data is most complete and has been reported by Grindlay et al. 1980. This analysis concentrated on the single x-ray burst observed and lead to the important conclusion that the burst luminosity must greatly exceed (by a factor of $\gtrsim 10$) the Eddington limit if the emission is pure blackbody with emissivity = 1 from the surface of a neutron star with mass and radius consistent with neutron star models. It was shown that in general the factor k between the true luminosity and Eddington value is given by

$$k = \text{const}\ (R^2 T_a^4)/(Mg^3)$$

where T_a is the apparent blackbody temperature, M and R are the true (intrinsic) source mass and radius, and g is gravitational redshift factor $g = (1 - 2GM/c^2r)^{\frac{1}{2}}$.

Marshall (1980) has generalized this relation for k to consider also the source emissivity, ε, and the composition (i.e., helium fraction) of the gas. For the simplifying assumption that $\varepsilon = k = 1$, then the maximum apparent temperature $kT_a \simeq 1.65$ KeV cannot be exceeded for <u>any</u> neutron star models (M and R) if the luminosity does <u>not</u> exceed the Eddington value. If the distance to the source is known, then Marshall (1980) also shows that the fraction f of the available surface area for emission may also be specified. This quantity is also an "apparent" quantity

FIGURE 2

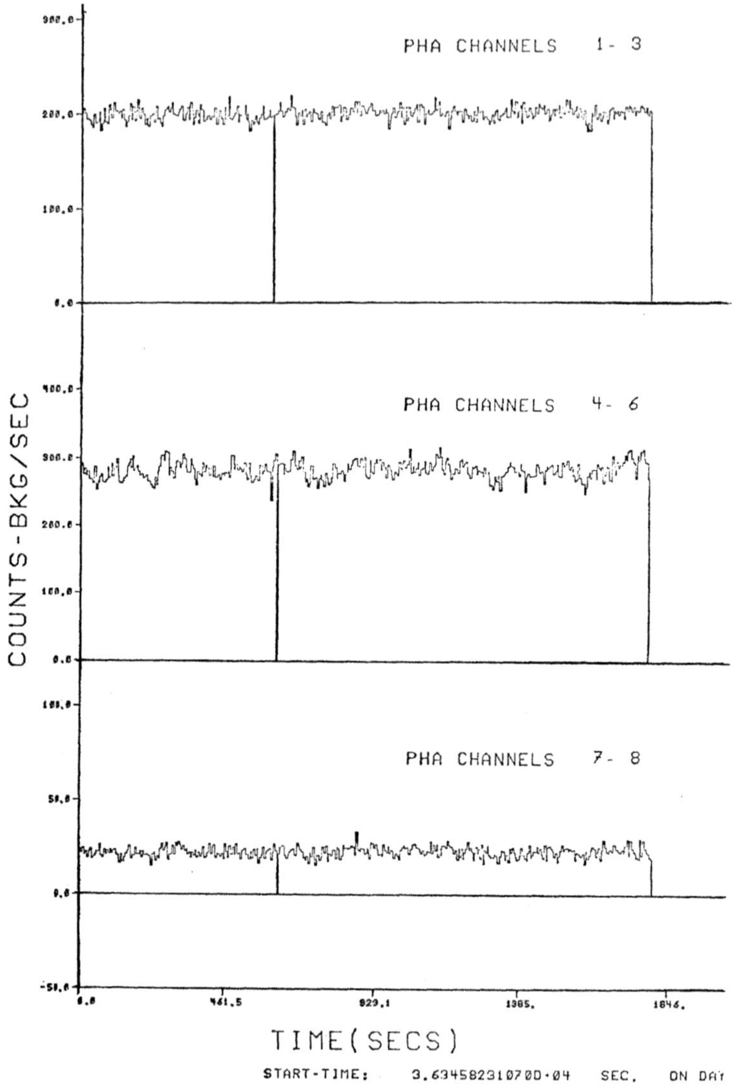

Typical energy-dependent variability observed from the bright x-ray source in the globular cluster NGC 6624. These are net counts (background subtracted) from the source in the energy ranges 1.1-4.5, 4.5-9.7, and 9.7-21 KeV. The momentary decrease to zero counts (between times 461 and 923 seconds) is a telemetry drop-out.

(f_a) which must be corrected for gravitational effects for the true surface area fraction f; the correction is complicated and depends on the angle between the emitting fractional surface area and the line of sight. Nevertheless, using the Terzan 2 data as well as published burst data for NGC 6624 (Clark et al. 1976), NGC 6712 (Hoffman, Cominsky and Lewin 1980) and MXB 1728-23 (Hoffman, Lewin and Doty 1977) it appears likely that k > 1 and f < 1 and that $\varepsilon \gtrsim 0.5$. It is also clear that the quantity εf_a as derived from spectra of individual bursts is not constant and that the emission region changes in either height or area or emissivity from burst to burst. Hoshi (1980) also shows that bursts from several sources exceed the Eddington limit by factors of $\gtrsim$5 if the sources are ~1 $M_\odot$ neutron stars.

Analysis of the persistent emission from the bright globular cluster source is also being conducted. Several clusters have had long exposures with EINSTEIN (e.g., ~6 x 10^4 sec total on NGC 6624) and are thus well suited for variability analysis and detailed spectral analysis studies. The temporal analysis will search for periodicities with both epoch folding and power spectral analysis. The results should be considerably more sensitive (given the larger detector area and long pointings) than achieved in earlier studies (e.g., Parsignault and Grindlay 1978, Cominsky et al. 1980). Preliminary results are now available for several clusters for which only about a third of the data have been analyzed. These results are from analysis of the TIP data (photon arrival times) from the MPC on EINSTEIN (see Grindlay et al. 1980b). Pulsation limits of $\lesssim$3 percent for periods ~0.5 - 100 sec should be possible; analysis will be conducted for both shorter and longer periods. For periods longer than ~5-10 sec, an additional sensitivity increase can be achieved by using the binned (in 2.56 sec integrations) spectral data (PHA data; 8 channels from 1.1-21 KeV) instead of the TIP data since, unlike the TIP (with memory overflow limits), all counts are recorded. This will also allow analysis for energy-dependent variability. This has not yet been conducted for the globular clusters and bursters as a class.

An example of the temporal data still to be analyzed is shown in Figure 2 where we show a segment of data from NGC 6624. These are subtracted for energy-dependent background (predicted from a correlation technique--see Grindlay et al. 1980b) and are plotted for three energy ranges: 1.1-4.5 KeV, 4.5-9.7 KeV, and 9.7-21.0 KeV. Two types of temporal behavior can be seen: relatively short duration (~30 sec) flares, with amplitudes of ~10 percent, and smoother longer term variations with time scales of ~300 sec. The fast variations may be related to the burst phenomenon (no bursts are observed, however, as NGC 6624 was in a "high state" in all data examined so far). The longer term variations may be quasi-periodic--detailed analysis (using autocorrelation techniques and phase-shifted folding) is in progress to test this hypothesis. If indeed the ~200-300 sec variations in NGC 6624 are quasi-periodic, this would be an important link with 4U1626-67 which shows quasi-periodic (~10^3 sec) variations (Rappaport and Joss 1980). Since these variations, in turn, are very likely related to the recently discovered (Middleditch et al. 1980) 2400 sec binary period, quasi-periodic variability in NGC 6624 might be the first direct x-ray evidence that the globular cluster sources are also, indeed, compact binaries with comparably short periods.

4. GLOBULAR CLUSTER X-RAY SOURCE MASSES

The principal scientific objective in the x-ray imaging observation of globular clusters conducted with the EINSTEIN Observatory is to estimate the x-ray source mass from the positional offsets between the x-ray source and the cluster center. The bright x-ray sources in globulars (most of the sources in Table 1) were located by SAS-3 (Jernigan and Clark 1979) to be well within the cluster tidal radii (typically several arcmin) but not necessarily within the much smaller core radii (typically ~10 arcsec). We shall show that the (preliminary) EINSTEIN x-ray positions together with new measurements of the cluster centers and core radii indicate that all of these sources (with one possible exception NGC 1851) are well within one core radius of the cluster centers.

An x-ray source in a globular cluster may be considered to be a single and identifiable test particle in a self-gravitating swarm of test particles. As such a precise x-ray position allows the test particle to be compared with that expected for an object in "thermal" equilibrium within the cluster core. If the cluster core is indeed isothermal with constant total energy "kT" per particle (as the surface brightness distribution usually suggests), then the more massive the test particle the lower its mean velocity (mV^2 = const) and the closer it will be to the cluster center. In fact, as Bahcall and Wolf (1976) originally showed (and see Grindlay 1979 for a simplified derivation due to Tremaine), the expectation value for the mass ratio of the test particle (x-ray source) is inversely proportional to the positional offset squared, or

$$q = M_x/\langle m\rangle = 0.9\ (r_c/R_x)^2$$

where M_x is the x-ray source mass, <m> is the mean mass of an ambient cluster (core) star, R_x is the x-ray positional offset and r_c is again the core radius. Thus two quantities must be measured to determine q: R_x and r_c. The mean stellar mass may be taken as ~0.5 $M_\odot$ from studies of cluster stellar evolution though this assumption (as well as that of a uniform isothermal spherical gravitational potential well) will be examined in Section D below.

A. Precise X-ray Position Measurements

Since the core radii r_c of most of the x-ray globular clusters are as small as ~6 arcsec (cf. Table 4 below), the positional accuracy of the x-ray position measurements should be a fraction of this inherent angular scale of the problem. Thus, x-ray positions must (ideally) be measured to ~1 arcsec for a minimal contribution to the "smearing" (from measurement errors) of the probability distribution for the mass ratio q (see Section D below). The design of the EINSTEIN Observatory (Giacconi et al., 1979) permits ~1 arcsec positions to be obtained. Realization of this capability has been extremely difficult, although ultimately possible.

In the course of normal analysis of HRI observations of bright (galactic) x-ray sources with known positions (i.e., optical counterparts), several major difficulties became apparent. The 1σ position error was found to be 4.5 arcsec, much larger than expected, and images were (in some cases) smeared by as much as 10 arcsec. In addition, the angular separation between guide stars in the two star trackers used to determine the aspect was found to vary smoothly (again only for certain well-defined cases) around the satellite orbit. After extensive analysis, these difficulties were found to be due to two major problems with the star trackers: first, from anomolous star separations within the individual trackers themselves, one of the two functioning trackers was found to have a distortion calibration error; second, the variable star separation between trackers was found to correlate well with the predicted (ephemeris) value of the imposed geomagnetic field along the tracker axis. Since the star trackers are magnetically focussed image-dissector tubes, this magnetic distortion could be understood as a consequence of local induced fields (possibly from magnetic material in the support structure of the x-ray telescope mirrors) with larger than expected values.

These problems have now been solved with extensive analysis of the magnetic distortions and derivation of the functional form and numerical values of correction terms which, when applied a posteriori to the data, remove the magnetic field effects to within $\lesssim 0.5$ arcsec. The absolute distortion of the one tracker was remeasured in a special sequence (~2 days) of observations of bright stars in the tracker. All of these corrections have been included in their preliminary form in deriving the globular cluster positions reported here. Final correction terms have now been derived so that the final 1σ positional error can be derived when the final aspect system is implemented and many sources with precise optical positions have been processed. These numbers will be included in our final analysis (Grindlay et al., 1980a). The sigma values reported here are only approximate--they are the r m s scatter in repeated (~4 or 5) measurements of the source position where an attempt was made to randomize remaining systematic errors by offsetting the pointing directions on a 8 x 8 arcmin square centered on the cluster position.

The preliminary x-ray positions for the eight globular clusters detected with the HRI (see Table 1) are listed in Table 4 below, together with the cluster centers and core radii.

B. Optical Studies: Precise Cluster Centers and Core Radii

Precise positions for the cluster centers and core radii are just as important as precise x-ray positions in the determination of source masses. Approximate core radii (with accuracy of typically 30 percent) have been determined photometrically (and through star counts) for all of these clusters (e.g., Peterson 1976, Canizares et al., 1978, and references in Grindlay 1979). Clearly more accurate values are needed. There also have been only approximate positions for the cluster centers. Jernigan and Clark (1979) derived approximate (± 3 arcsec) cluster centers for comparison with the SAS-3 data and Shawl and White (1980) have derived more accurate centers (± 1 arcsec claimed) for six of our eight clusters. The latter study, while nearly the required accuracy, may have been biased by bright giants in the cluster core (the study used Palomar and ESO B images) or, conceivably, nonconcentricity between the cluster halo (actually measured) and core. Instead, the optical center (and cluster core radius) is desired for the underlying diffuse emission from main sequence stars in the core since they contribute most of the mass and thereby determine the gravitational potential felt by the x-ray source. Note also that the method of Shawl and White

(1980) is not able to derive the precise core radius value for each cluster which is just as important as a precise cluster center.

Accordingly, we have undertaken a program to measure the cluster centers and core radii to the required accuracy and with a minimum bias. A preliminary account of this work is given by Grindlay (1979) and final results will be reported by Hertz and Grindlay (1980). Our analysis is based on PDS photometry of a series (short and long exposure) of high quality 4m plates (obtained by JEG at CTIO) of the clusters in the U and B bands. Final results are based on the U plates but the (identical) analysis of the B plates provides an estimate of the systematic difference due the presence of individual cluster stars (giants), which are much more prominent on the B plates.

The plates are scanned with a PDS in 1 arcsec bins (0.5 arcsec steps) out to about 30-50 core radii (using published values) from an approximate visual center of the cluster. A grid of small raster scans is performed over the entire plate to measure background variations, and scans are done on each of the 16 CTIO spot sensitometer images which were simultaneously recorded on the end of each plate. The sensitometer spots provide the density/intensity calibration curve for each plate. A 2-dimensional surface fit to the background rasters enables the total background (density variations, fog, and field stars) to be predicted accurately within the raster of the cluster itself and the density-intensity calibration curve to be applied.

For some clusters, even the U plates contain images of very bright individual cluster stars. These are "removed" by two methods which give virtually identical results for the final cluster centers: (a) the stars are truncated to have the mean intensity of surrounding pixels or (b) a two-dimensional Fourier transform (FFT) of the cluster image is performed and the high frequency quadrant of the transformed image is smoothly truncated. Since the detectable individual stars contribute the high spatial frequencies, only the smooth image of the underlying cluster remains when the inverse transform is then made back to a real image. The center (of symmetry) is not altered by this FFT procedure but the cluster shape (and core radius) is affected. Thus, core radii are determined (see below) using method (a) although fits to equivalently FFT-smoothed King (1966) curves would yield equivalent results.

An example of a digitized U band image of the cluster

FIGURE 3

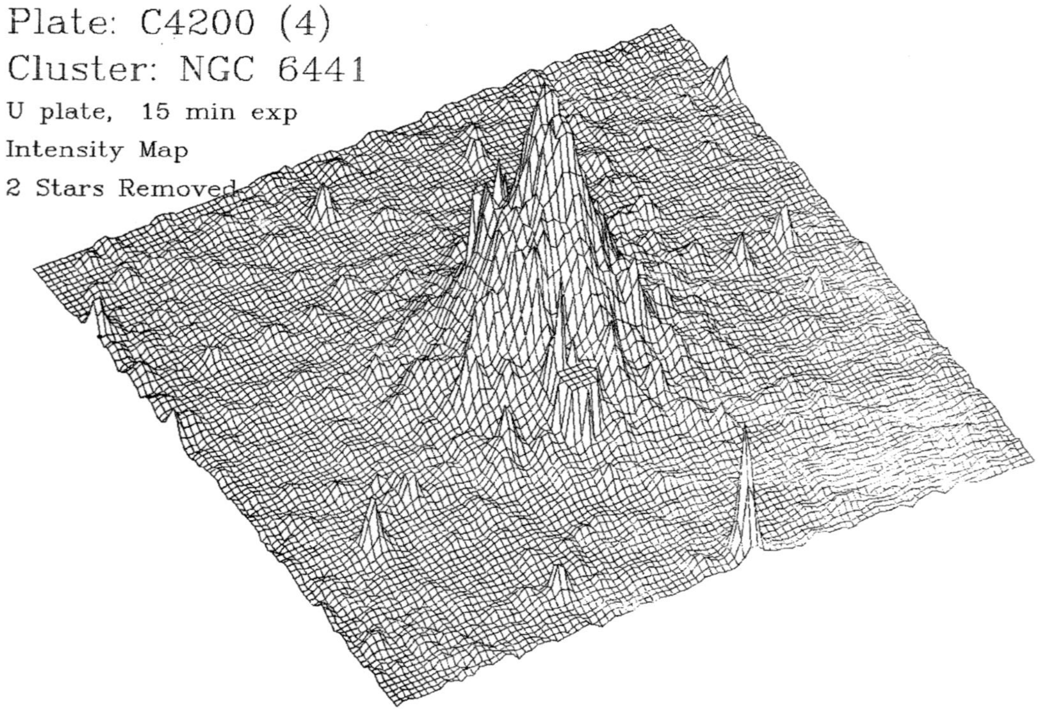

Digitized CTIO 4m plate (U-band) image of NGC 6441.

FIGURE 4

Cluster: NGC 6441
Plate: C4200 (5)
U plate, 15 min exp
Measure of Symmetry

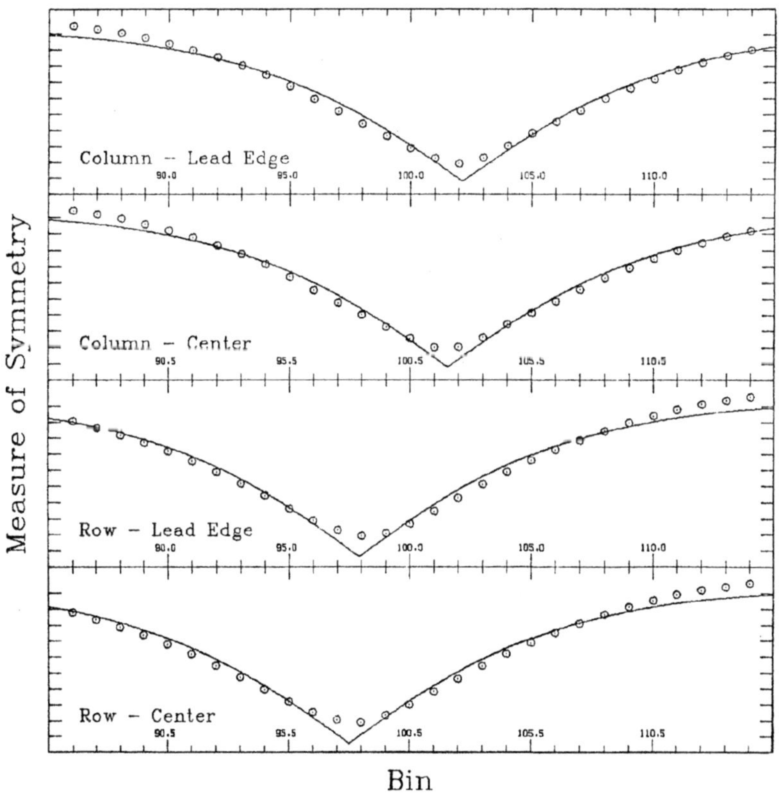

Results of folding analysis (in "rows" and "columns") for the center of symmetry of the digitized image of NGC 6441. Each bin is 0.9 arcsec.

NGC 6441 is shown in Figure 3. This image was smoothed by method (a) above. The cluster center is measured from this image by defining the center of symmetry in each of the two orthogonal axes as the minimum of the quantity.

$$s_i = \sum_{j=1}^{N/2} f_{i+j} - f_{i-j}$$

where f is the intensity measured in all the jth bins (i.e., entire row or column) on either side of the "test bin" s and where i ranges over the central N/2 x N/2 bins of the N x N pixels in the digitized image. The results of this image-"folding" procedure for the NGC 6441 image of Figure 3 are shown in Figure 4. The minimum is seen to be sharply defined in a cusp, which can be accurately determined by either fitting a parabola to the data or, more accurately (as done here), by fitting a folded King curve with the published cluster core radius. The center of symmetry and its formal uncertainty is then derived from "sliding" the model curve along the data until a minimum χ^2 fit is obtained.

The core radius r_c of each cluster is then derived by forming concentric annuli around the derived cluster center and summing the intensities within. The fitted background is subtracted from each annulus and the resulting surface brightness distribution is then fitted to a series of King models generated in steps of 0.25 of the concentration parameter c [= log (r_t/r_c), where r_t is the cluster tidal radius]. For each trial value of c, the best fit (χ^2 minimization) core radius r_c is derived; the overall minimum χ^2 model then defines r_c and c.

Preliminary results for the cluster centers are given in Table 4 along with the preliminary x-ray positions. Core radius values are also given but are taken from Grindlay (1979) rather than the analysis just described. This analysis is nearing completion after delays in digitizing the plates with sufficient dynamic range to determine r_c and c. Final results of this optical analysis will be described by Hertz and Grindlay (1980) and used by Grindlay et al. (1980a).

C. X-ray vs. Optical Positions

The preliminary results for the x-ray positional offsets from the cluster centers are plotted in Figure 5. Both the position offsets and sigmas are scaled to be in units of their respective core radius values. The error bars are the errors on the mean x-ray position (see discussion above) combined with the (smaller) error of the optical center determination as well as the astrometry.

TABLE 4

PRELIMINARY X-RAY POSITIONS vs. OPTICAL CLUSTER CENTERS

Cluster	No. Obs.	X-ray Position (ra, dec)	Approx. Sigma (arcsec)	Cluster Center (ra, dec)	Core Radius (arcsec)	$\frac{R_X}{r_C}$
NGC 104	4	00 21 51.44 -72 21 33.5	3.5	00 21 52.27 -72 21 32.1	24	0.28±.15
NGC 1851	5	05 12 27.94 -40 06 01.8	2.1	05 12 28.06 -40 06 11.1	7	1.34±.33
NGC 6441	4	17 46 48.36 -37 02 15.6	0.9	17 46 48.56 -37 02 14.7	9	0.29±.15
NGC 6624	3	18 20 27.74 -30 23 15.1	1.8	18 20 27.44 -30 23 14.9	6	0.65±.34
NGC 6712	5	18 50 21.12 -08 46 04.2	1.5	18 5.0 20.47 - 8 46 7.7	49	0.21±.04
NGC 7078	3	21 27 33.03 +11 56 49.6	1.1	21 27 33.40 +11 56 48.7	10	0.60±.15
Liller 1	1	17 30 06.70 -33 21 15.0	1.9	17 30 6.95 -33 21 20.6	7	0.93±.30
Terzan 2	4	17 24 19.81 -30 45 40.3	2.1	17 24 19.89 -30 45 38.8	7	0.26±.33

NOTES: All positions are epoch 1950.0. The sigma values are the combined sigma in the x-ray vs. optical center positions, where the x-ray position uncertainty is approximated as just the r m s scatter in the multiple observations and the optical center uncertainty is derived from the χ^2 fitting procedure described in the text. An additional ~1 arcsec uncertainty in the core radii is also included. The final column of R_X/r_C values are the offsets of the x-ray sources from the cluster centers expressed in units of the core radius.

FIGURE 5

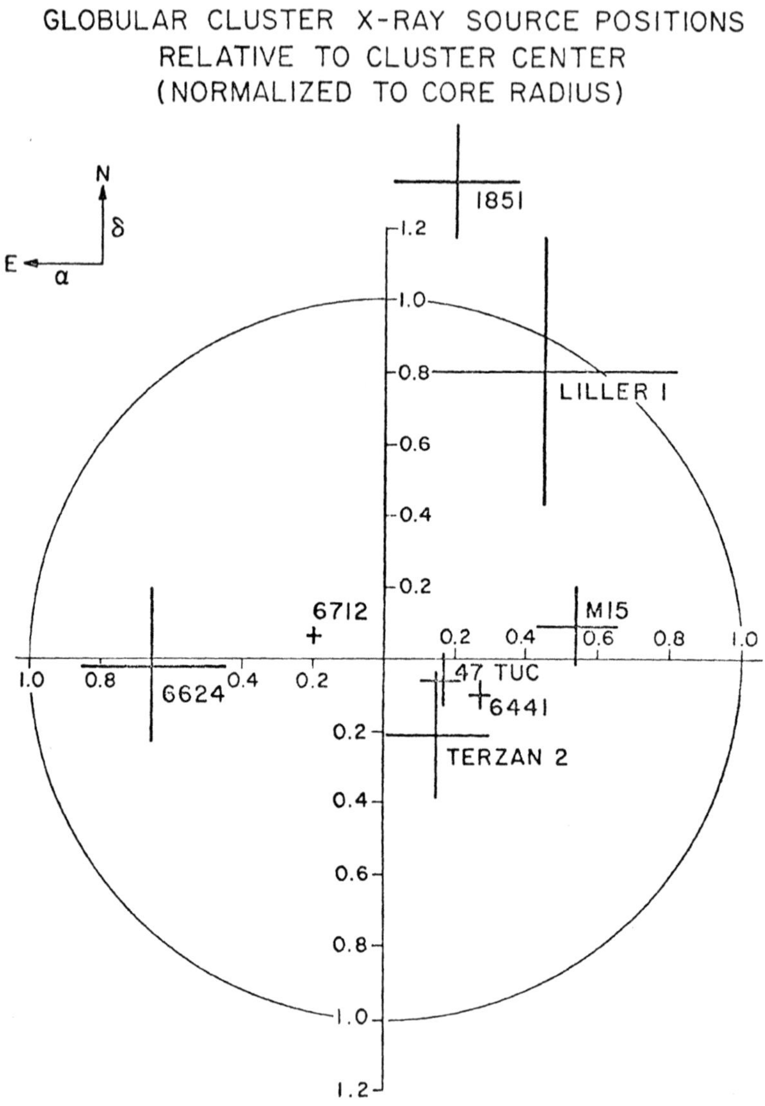

Preliminary EINSTEIN results for location of x-ray sources in globular clusters. The circle defines one core radius; the error bars are only approximate 1σ values (see text).

These error bars are thus only approximate and will be refined in the final analysis (Grindlay et al., 1980a).

It is clear that most of the sources are probably located within ~0.5 core radius of the cluster centers. This is as expected for sources with mass ratio $q \simeq 4$, or an x-ray source mass of ~2 $M_\odot$ given the mean cluster star mass of 0.5 $M_\odot$ and the expression for q given above. For only one cluster (NGC 1851) is the x-ray source quite possibly outside the cluster core radius. This apparently larger positional offset is consistent with the distribution in offsets that would be expected for $q \simeq 4$ as we shall now show.

D. Absolute Likelihood Analysis for Mass Ratio q

The observed distribution of positional offsets (Figure 5) must be compared with that expected for trial values of q as well as the cluster potential. This could be done by conducting extensive Monte Carlo simulations and counting the number of occurences in which the statistical realization of an assumed q, potential (or cluster core "shape"), and measurement errors gave rise to the observed distribution. A much less tedious, but essentially equivalent, analysis method has been developed by Lightman, Hertz and Grindlay (1980). We have called this an "absolute" likelihood method, as opposed to a relative likelihood analysis (Jernigan and Clark 1979), in that it yields an absolute probability that the observed offsets are a particular realization of a given q and cluster model (i.e., the functional form of the potential can be tested). The most significant feature of our analysis is that by using a generalized Kolmogorov-Smirnov test it includes the measurement errors explicitly. The relationships between the errors, size of the data set (number of clusters) and width of the derived probability distribution for q are also derived.

In Figure 6, we show the absolute likelihood distribution of q, L(q), derived from the eight preliminary values for the positional offsets and errors plotted in Figure 5 and listed in Table 4. The peak value of L(q) is greater than 0.5 indicating a good "fit" of the data to the hypothesis that a single value of q and a simple cluster potential (King model) describes all the cluster sources. The value of $q \simeq 4.0$ at the peak is the most probable mean ratio for the globular cluster sources; since the mean stellar mass $\langle m \rangle$ in the core may be taken as ~0.5 $M_\odot$, the x-ray source mass is the $M_X = q \cdot \langle m \rangle \simeq 2.0\ M_\odot$. Thus, the most probable mass is, in fact, the value expected if the globular cluster sources are compact binary systems

FIGURE 6

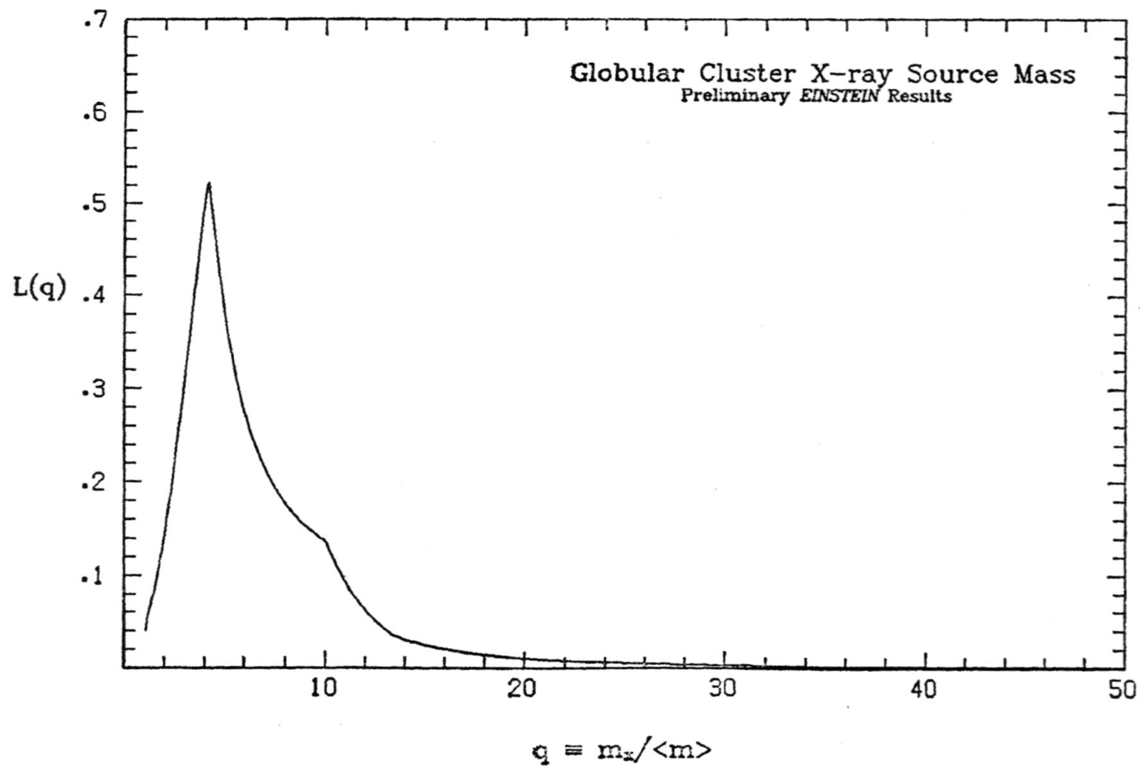

Absolute likelihood distribution L(q) derived for the preliminary results of the x-ray positions in eight globular clusters and assuming a single King potential and mean stellar mass <m>. The peak at $q \simeq 4$ indicates the data are consistent with a common value of $q \simeq 4$ (or $M_x \simeq 2M_\odot$) for the source masses. NOTE THAT THESE RESULTS ARE PRELIMINARY AND THAT BOTH THE POSITION OF THE PEAK L(q) AND THE DISTRIBUTION WIDTH MAY CHANGE SOMEWHAT IN THE FINAL ANALYSIS.

consisting of a ~1.4 $M_\odot$ neutron star and a ~0.5-0.7 $M_\odot$ companion for the total mass of ~2 $M_\odot$. Note, however, that the data are consistent with q in the range 1.8-11.0 (or ~0.9-5.5 $M_\odot$) with a 10 percent absolute likelihood, and that at the ~1 percent level (~3σ) essentially any (large) value of q can be accommodated. The long tail towards large q is primarily due to the still significant measurement errors. As Lightman, Hertz, and Grindlay (1980) show, a value of σ (normalized to the core radius) for the (mean) combined measurement error in the data set restricts the analysis to be sensitive to q only if $\sigma^2 << 1/(3q-1)$. Thus, since the data indicate $q \simeq 4$, the mean σ must be less than 0.30 (as is, indeed, the case in Table 4) to have allowed the measurement of the peak. The L(q) distribution of Figure 6 becomes flat for $q \gtrsim 12$ simply because σ is too large (i.e., σ^2 >> .03) to allow the sources to be measured within the small range of radii expected about the cluster centers.

Further insight into the interpretation of Figure 6 for L(q) can be had by comparison with the simulations of Lightman, Hertz, and Grindlay (1980). These simulations were carried out with eight clusters (in ancitipation of our actual data set), with several values of σ as well as several models for the cluster potential. In Figure 7 we reproduce (a) the L(q) curves for several σ values, including the critical value described above, for data simulated with King potentials and a fized q = 5; (b) the L(q) curves for data generated with fixed σ but two different assumed cluster potentials approximately corresponding to a sub-core within the cluster core; and (c) the L(q) curves for data generated with fixed σ and a single King potential but a mixture of two actual q values (see Figure caption). Comparison of the L(q) curves in Figures 7a and 7b indicates that the amplitude of the L(q) curve is substantially reduced at small q for the "core-within-a-core" model. Comparison with the actual data of Figure 6, in turn, indicates that the clusters are probably well described by a simple single core King potential. Further discussion of this question will be given by Hertz and Grindlay (1980) since the detailed cluster measurements allow this question to be explored directly. (In fact, two of the eight clusters-- NGC 6624 and M15--do appear to have significant central cusps or "sub-cores." However, two clusters are apparently not enough to bias the sample.) Similarly, Figure 7c indicates that with only eight clusters mixed q models are difficult to distinguish from a single q model except at small values of q.

FIGURE 7

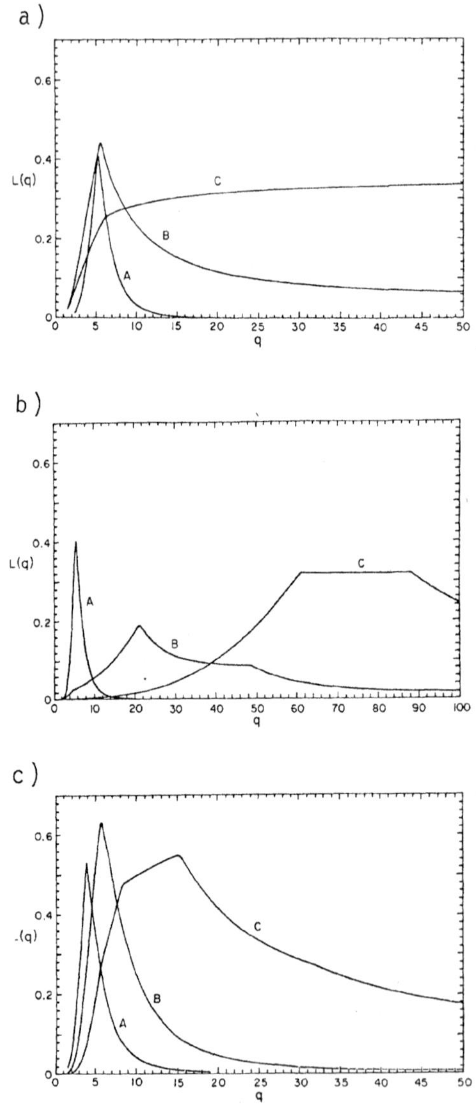

Absolute likelihood distribution from Lightman, Hertz, and Grindlay (1980) for simulated data on eight clusters showing

a) the sensitivity to σ for a fixed q (= 5) with σ^2 = 0, 0.07 (the critical value for q = 5) and 0.14 for Curves A, B, and C, respectively;

b) the sensitivity to the cluster potential model when the data are tested against the hypothesis of a simple King potential but the data are actually the stochastic realization of a cluster with a subcore within the core with relative core radius 0.0 (Curve A), 0.1 (Curve B), or 0.2 (Curve C) and fixed q = 5, σ = 0; and

c) the sensitivity to mixed values of q (but σ = 0) where the three curves show all sources with q = 5 (Curve A), half (= 4) sources with q = 5 and half with q = 10 (Curve B), and half with q = 5 and half with q = 40 (Curve C).

5. CONCLUSIONS

Although the work described here is still undergoing final analysis, several interesting new results are already apparent. Foremost among these, of course, is the analysis for the source masses which provide strong support for the hypothesis that the globular cluster X-ray sources are binaries (e.g. Clark 1975, Lewin 1980) and not massive ~1000 $M_{\odot}$ black holes (e.g. Bahcall and Ostriker 1975) or intermediate mass (~10-100 $M_{\odot}$) black holes (Grindlay 1978). While supermassive (isolated) black holes can almost certainly be ruled out, stellar mass (e.g. < 5 $M_{\odot}$) black holes cannot (yet) be disregarded. However, the restriction on mass and the current limits on free gas in cluster cores (see Durison and Burns 1980 for a recent discussion) would almost certainly require the X-ray sources to be members of binary systems. The generally successful predictions of the nuclear flash model for X-ray bursts (Joss 1978) provide strong evidence that these binary systems contain neutron stars. Although our preliminary results provide the first direct determination of the mass of globular cluster X-ray sources (and bursters), similar conclusions had previously been reached from a variety of arguments (see Lewin and Clark 1980 for a comprehensive summary).

The second key result of this study, also still only preliminary, is that a number of new low luminosity globular cluster X-ray sources have been found. It is likely that four new cluster sources are detected with fluxes near the IPC threshold and luminosities $L_x \simeq 10^{33}$ erg s^{-1}. It is especially interesting that the globular clusters containing these sources are, on the whole, centrally condensed and evolved (cf. Figure 2) as is the case with most of the clusters containing high luminosity sources. It thus seems likely that these sources are the lowest luminosity end of a luminosity distribution of compact sources, presumably binaries, with $10^{32} < L_X < 10^{37}$ (cf. Tables 1 and 3). It is also possible that the ~10^{33} erg s^{-1} sources are diffuse and due to summed emission from the stars in the high stellar density cores detected. We regard this as a less likely interpretation for two reasons: 1) two of the four low luminosity sources are known to be point sources (NGC 104, from the HRI - Table 1 - and NGC 6440, detected as a bright X-ray transient source[1] by UHURU [Forman et al. 1976] and

[1]Note that the detection of NGC 6440, if indeed the quiescent emission from the bright transient, is the first such detection and suggests a luminosity increase in the transient outburst by a factor of $\gtrsim 10^5$.

OSO-7 [Markert et al. 1975]) and 2) the mean X-ray luminosity per unit mass in the core (L_x/M_c in Table 3) of the three possibly diffuse cluster sources (NGC 1904, NGC 6517, and NGC 6541) are not significantly below many of the upper limits in Table 3. The diffuse vs. point source nature of these sources might be established in analysis of much deeper exposures of several other globulars, which should also show diffuse emission (i.e. they are of similar concentration, metallicity; spectral type, etc.) if the three clusters in question here are also diffuse sources. Follow up HRI observations will also be carried out if Observatory operations permit.

What conclusions can be drawn regarding the origin of the globular cluster sources? First, we note from Tables 1 and 3 that there are now at least fifteen globular clusters in the Galaxy with detectable x-ray sources (we include Terzan 1 and Terzan 5--see footnote in Table 1. Most have very high central densities and concentration and all but four of these are within the galactic bulge. Most of these are moderately metal rich; the five clusters in the bulge for which metallicity is not yet known (Liller 1, Terzan 1, 2, 5, and Grindlay 1) are probably also metal rich, given their locations near the galactic center (Harris and Racine 1979). It also appears significant that of the eleven globular cluster sources known to be compact (Table 1), all but two are known to burst. Neither of these two clusters (NGC 104 and NGC 7078) is in the galactic bulge.

It is thus clear that the formation of x-ray sources in globular clusters is favored if the cluster is in the galactic bulge and that most, if not all, of these sources are x-ray bursters. Of course, a number of x-ray bursters have been optically identified with stellar-like objects not in globular clusters (of the ~30 bursters presently known [Lewin and clark 1980, Oda 1980a], nine are in globulars, six have stellar-like optical counterparts, and the rest are unidentified). However, again, most of these "field bursters" are also in the galactic bulge. Since we have concluded above that the globular cluster sources are most probably compact binaries as had been concluded previously by Lewin and Clark (1980) for globular clusters, bursters, and galactic bulge sources in general, a common origin might be possible.

The preference for compact binary x-ray sources to be in the galactic bulge and metal rich globular clusters might possibly be understood as follows: a metal rich cluster has possibly undergone more stellar evolution and nuclear processing than a metal poor cluster--perhaps because of a larger initial fraction of massive stars. [Alternatively, the initial composition of the gas from which the cluster formed was more metal rich.] Thus, a metal rich cluster might be expected to contain a higher density and total number of collapsed stellar remnants (white dwarfs, neutron stars, or black holes). It would then be more likely that tidal capture of these remnants by stars in the high density cluster core and the consequent formation of compact binary systems would occur.

An additional effect favoring the galactic bulge for the production of the cluster sources could be involved. That is, since the density of field stars and field binaries is higher in the galactic bulge, there will be more interactions between these stars and the compact stellar remnants in the cluster cores. Compact binary systems might be formed in the cluster from exchange reactions (Spitzer and Mathieu 1980) between field binaries and compact remnants in the cluster core as the cluster passes through the galactic plane. Binary-binary interactions (either between cluster binaries or a cluster binary and a field binary) could also be important and might lead to the ejection of one of the binaries from the cluster. This might be the origin of the x-ray bursters not in globular clusters and might also explain why only one compact x-ray source has been detected in each cluster (see Grindlay 1979). Alternatively, these bursters in the field may have evolved from cataclysmic varible type binary systems where accretion on the white dwarf causes collapse to a neutron star. However, this does not explain the strong preference for the occurrence of bursters in globular clusters unless primarily white dwarfs are produced from stellar evolution in globulars and they, in turn, form capture binaries in which the white dwarfs evolved to neutron stars.

ACKNOWLEDGMENTS

I thank my collaborators in these studies, particularly Paul Hertz, for their help and our MIT collaborators for permission to include the MIT HRI survey results. This work was partially supported by NASA Contract NAS8-30751 as well as the Alfred P. Sloan Foundation.

REFERENCES

Arp, H.C. 1965, in Galactic Structure (Vol. 5, Stars and Stellar Systems), A. Blaauw and M. Schmidt, ed., Univ. Chicago Press, p. 401.

Bahcall, J. and Ostriker, J. 1975, Nature, 256, 23.

Bahcall, J.N. and Wolf, R.A. 1976, Ap. J., 209, 214.

Canizares, C.R. 1975, Ap. J., 201, 589.

Canizares, C., Grindlay, J., Hiltner, W., Liller, W. and McClintock, J. 1978, Ap.J., 224, 39.

Clark, G.W. 1975, Ap.J.(Letters), 199, L-143.

Clark, G.W. et al. 1976, Ap.J.(Letters), 207, L-105.

Cominsky, L. et al. 1980, Ap.J., in press.

Durisen and Bursts (1980), M.N.R.A.S., in press.

Forman, W., Jones, C., and Tananbaum, H. 1976, Ap.J.(Letters), 207, L-25.

Giacconi, R. et al. 1979, Ap.J., 230, 540.

Gorenstein, P. 1975, Ap.J., 198, 95.

Grindlay, J., Gursky, H., Schnopper, H., Parsignault, D., Heise, J., Brinkman, A., and Schrijver, J. 1976, Ap.J.(Letters), 201, L-127.

Grindlay, J.E. 1977, in Highlights of Astronomy, Vol. 4, 111.

Grindlay, J.E. 1978, Ap.J., 221, 234.

Grindlay, J.E. 1979, in Galactic X-ray Sources, P.W. Sanford, ed., Cambridge Univ. Press (Proc. NATO A.S.I. on X-ray Astron., Cape Sounion, Greece), in press.

Grindlay, J.E. 1980, IAU Circular No. 3506.

Grindlay, J., Marshall, H., Hertz, P., Soltan, A., Weisskopf, M., Elsner, R., Ghosh, P., Darbro, W., and Sutherland, P. 1980b, Ap.J.(Letters), 240, L-121.

Grindlay, J.E. and Hertz, P. 1980, Ap.J.(Letters), submitted.

Grindlay, J., Hertz, P., Steiner, J., Murray, S., and Lightman, A. 1980a, in preparation.

Harris, W. and Racine, R. 1979, Ann. Rev. Astron. Astrophys., 17, 241.

Hertz, P. and Grindlay, J. 1980, in preparation.

Hoffman, J., Lewin, W., and Doty, J. 1977, M.N.R.A.S., 179, 57P.

Hoffman, J., Cominsky, L., and Lewin, W. 1980, Ap.J.(Letters), 240, L-27.

Hoshi, R. 1980, preprint.

Jernigan, G. and Clark, G. 1979, Ap.J.(Letters), 231, L-125.

Joss, P.C. 1978, Ap.J.(Letters), 225, L-123.

King, I.R. 1966, A.J., 71, 64.

Lewin, W. 1980, in Globular Clusters (D. Hanes and B. Madore, eds.), Cambridge Univ. Press, Chapt. 18.
Lewin, W. and Joss, P. 1977, Nature, 270, 211.
Lewin, W. and Clark, G. 1980, Annals N.Y. Acad. Sci., 336, 451. (Proc. Ninth Texas Symposium on Relativistic Astrophysics, Munich, West Germany).
Lightman, A., Hertz, P., and Grindlay, J. 1980, Ap.J., 241, 367.
Lin, D.N. and Tremaine, S. 1980, preprint.
Markert, T., Backman, D., Canizares, C., Clark, G., and Levine, A. 1975, Nature, 257, 32.
Middleditch, J., Mason, K., Nelson, J., and White, N. 1980, Ap.J., in press.
Oda, M. 1980a, invited talk at H.E.A.D. Meeting, A.A.S., Cambridge, Massachusetts (Jan., 1980); see contribution in this volume.
Oda, M. 1980b, IAU Circular No. 3506.
Parsignault, D. and Grindlay, J.E. 1978, Ap.J., 225, 970.
Peterson, C.J. and King, I.R. 1975, A.J., 80, 427.
Peterson, C.J. 1976, A.J., 81, 617.
Rappaport, S. and Joss, P. 1980, invited talk at H.E.A.D. Meeting, A.A.S., Cambridge, Massachusetts (Jan., 1980); see contribution in this volume.
Shawl, S.J. and White, R.E. 1980, Ap.J.(Letters), 239, L-61.
Spitzer, L. and Mathieu, R. 1980, preprint.
van Speybroeck, L., Epstein, A., Forman, W., Giacconi, R., Jones, C., Liller, W., and Smarr, L. 1979, Ap.J.(Letters), 234, L-45.

MODELS FOR GALACTIC X-RAY SOURCES

Paul C. Joss
Department of Physics
and
Center for Theoretical Physics
Massachusetts Institute of Technology

This discussion will be highly circumscribed in a few respects. First, I shall limit my attention to those compact galactic X-ray sources whose X-ray luminosities are considerably in excess of the solar luminosity, and in particular I shall exclude from my discussion the weak X-ray emission that has been detected from many apparently isolated stars (Linsky 1980; Vaiana 1980). I shall also exclude any detailed discussion of individual peculiar objects and shall instead concentrate on the general trends among the majority of the bright galactic sources.

The key breakthrough in the development of our understanding of compact galactic X-ray sources was the discovery of X-ray pulsars with the *Uhuru* satellite (Schreier *et al.* 1972; Tananbaum *et al.* 1972). There is by now overwhelming evidence that these objects are neutron stars in close binary stellar systems (see Rappaport and Joss 1980 for a recent review). The X-ray pulsations are thought to be thermal emission from the magnetic polar caps of a neutron star that is accreting matter from a companion star and whose magnetic field is misaligned with its rotation axis. This interpretation receives strong support from (1) the very high stability of the pulse periods, (2) the secular rates of change of the pulse periods which, according to the theory of accretion torques on accreting collapsed stars (Pringle and Rees 1972; Lamb, Pethick, and Pines 1973), are consistent with the underlying objects being neutron stars but not larger objects, such as degenerate dwarfs (Rappaport and Joss 1977), (3) the periodic eclipses of the X-ray flux and periodic variations in pulse period, which can be quantitatively understood as due to orbital motion of the X-ray star about its companion (see Rappaport and Joss 1980 for references), and (4) the photometric and spectroscopic variability of the optical companion stars which, in some cases, confirm the orbital period and phase as deduced from the X-ray observations (Bahcall 1978a and references therein).

Among the compact galactic X-ray sources that are not X-ray pulsars, some still show direct evidence of binary membership, such as

R. Giacconi (ed.), X-Ray Astronomy with the Einstein Satellite, 111–122.

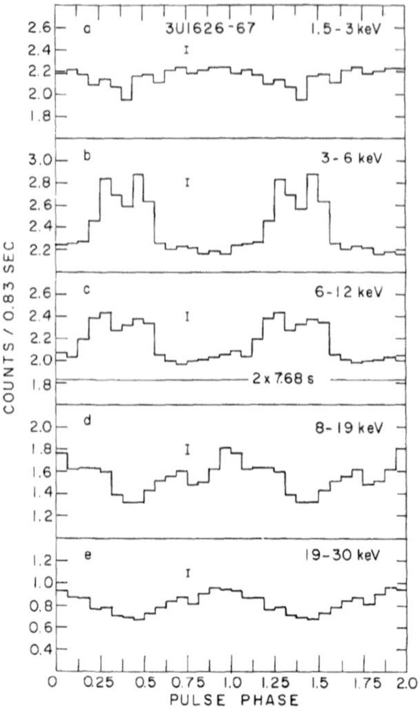

Figure 1. Pulse profiles of 4U1626-67 (=3U1626-67) for several ranges of X-ray energies (from Rappaport et al. (1977). The characteristics of these profiles are well within the range observed in other X-ray pulsars (see Fig. 1 of Rappaport and Joss 1980).

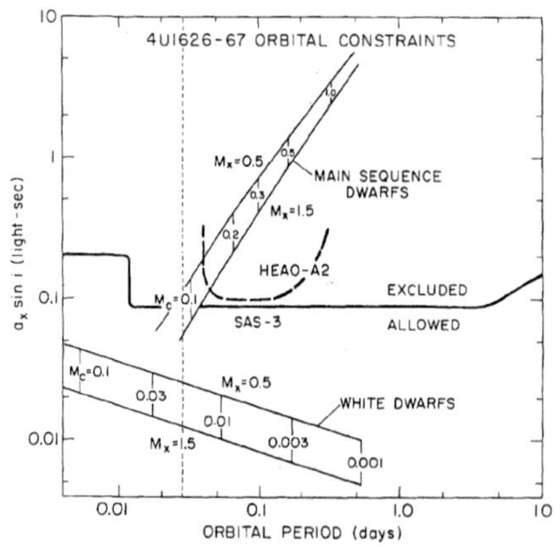

Figure 2. Orbital constraints for the 4U1626-67 system (adapted from Li et al. 1980). The 95% confidence upper limits for $a_x \sin i$ are derived from SAS-3 data (heavy curve) and HEAO-1 data (dashed curve; Pravdo et al. 1979). Under the assumptions that 4U1626-67 is a compact star of mass M_x, in a circular orbit (with sin i=1) about a companion star that rotates synchronously with the orbit and fills its Roche lobe, each point in the $a_x \sin i$ - P_{orb} plane uniquely corresponds to a mass and radius for the companion (Joss, Avni, and Rappaport 1978). With these assumptions, the loci of parameters for main-sequence dwarf and degenerate dwarf companions are shown for a range of possible masses for the X-ray star. All indicated masses are in units of solar masses. Very low mass main-sequence dwarf companions, as well as degenerate dwarf companions, are consistent with the limits on $a_x \sin i$. For values of sin i less than unity, the loci of companion-star parameters are shifted vertically downward by a factor of sin i. The dotted vertical line represents the orbital period of 2500 s, as recently discovered by Middleditch et al. (1980) (see text).

X-ray eclipses (as in 4U1700-37; Jones *et al.* 1973) or periodic variability of the optical counterpart (as in Cygnus X-1; see Bahcall 1978a for references). In total, more than a dozen bright galactic sources have now been found to display some direct evidence of binary character (see Bradt, Doxsey, and Jernigan 1979). However, many other sources show little or no evidence of binary membership. In some cases, this lack of evidence may simply reflect the low apparent brightnesses and/or the relative lack of observations of the source, but such explanations fail for some brighter sources that have been thoroughly scrutinized for evidence of binary membership without success.

The best example of a class of sources that fail to show any binary characteristics is the galactic-bulge sources (Salpeter 1973; Canizares 1975; and references therein). These sources characteristically have relatively soft X-ray spectra, inconspicuous optical counterparts, and little X-ray variability (with some notable exceptions that will be discussed below). In particular, there is no evidence of X-ray eclipses, periodic optical photometric variability, or any other direct indication of binary membership.

In the following paragraphs, I shall argue that there are now a few indirect but collectively compelling lines of evidence that the galactic-bulge sources are, in fact, close binary stellar systems. I shall thereby conclude that the great majority of bright galactic X-ray sources, with only a tiny handful of exceptions (such as the Crab and Vela pulsars), are likely to be binaries.

The first clue concerning the underlying nature of the galactic-bulge sources is supplied by the X-ray pulsar 4U1626-67 (Rappaport *et al.* 1977). The intrinsic properties of this pulsar (pulse period of $7\overset{s}{.}7$, X-ray spectrum, and so forth) are well within the range of properties found in other X-ray pulsars (see Fig. 1). However, this source shows no evidence of X-ray eclipses or regular variations in the pulse arrival times that would be expected from orbital motion (Rappaport *et al.* 1977). By now, very severe constraints on any orbital motion have been obtained (Joss, Avni, and Rappaport 1978; Pravdo *et al.* 1979; Li *et al.* 1980; see Fig. 2). These X-ray data strongly suggest that the source is in a highly compact binary (orbital period $\lesssim 0.3$ days) with a low-mass ($\lesssim 0.1\ M_{\odot}$) main-sequence dwarf or degenerate dwarf companion (Fig. 3). The apparent absence of orbital motion by the neutron star presumably results from a combination of the small dimensions of the binary system and the low mass of the companion, which permits the neutron star to remain almost at rest near the center of mass of the system.

The optical counterpart of 4U1626-67 is a faint, extremely blue object (McClintock *et al.* 1977), which displays $7\overset{s}{.}7$ pulsations (Ilovaisky, Motch, and Chevalier 1978) that evidently result from the reprocessing of some of the X-radiation into optical light. The very faintness of the optical counterpart rules out giant, supergiant, or

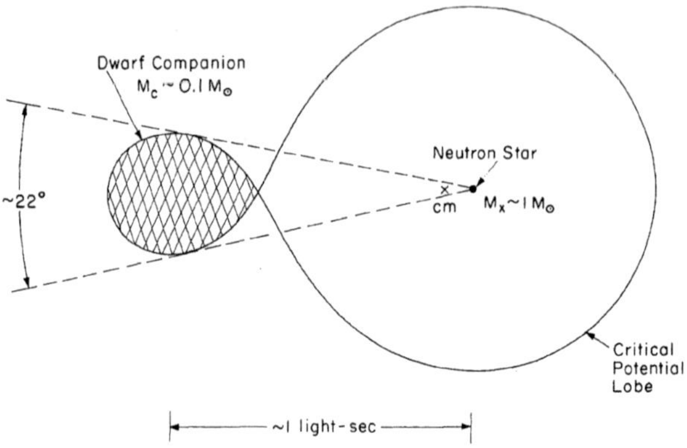

Figure 3. Highly-compact-binary model for 4U1626-67 (adapted from Joss and Rappaport 1979). The figure is drawn to scale for the indicated illustrative values of M_c and M_x. The indicated orbital separation of ~1 light-second corresponds to a main-sequence dwarf companion; for a degenerate dwarf companion of the same mass, the orbital separation would be reduced to ~0.3 light-seconds.

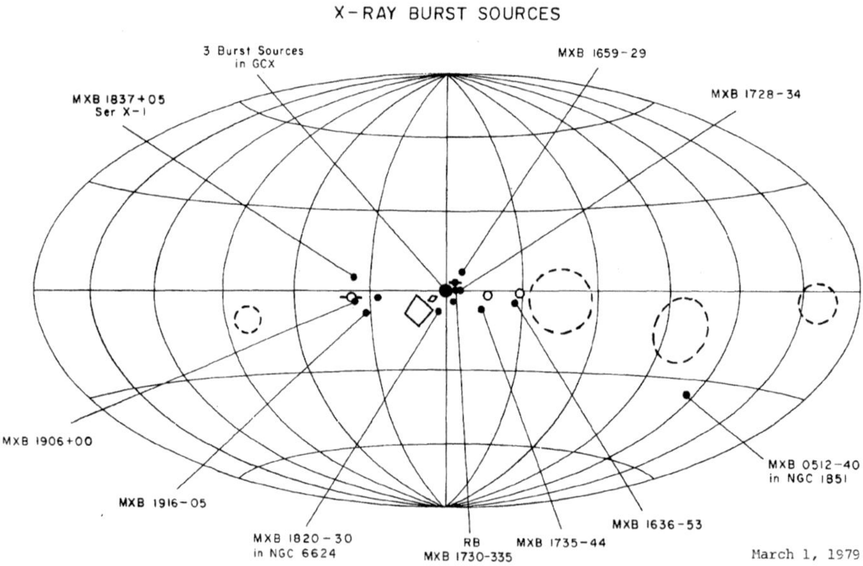

Figure 4. Map of 25 X-ray burst sources with accurately known positions as of March 1, 1979, in galactic coordinates (from Lewin and Clark 1980). The concentration of these sources in the direction of the galactic center, together with the association of several of them with globular clusters, strongly suggests an identification with an old stellar population and is one of several lines of evidence linking the burst sources to the broader class of galactic-bulge sources (see text and Lewin and Clark 1980).

early-type main-sequence stellar companions and thus supports the picture of a highly compact binary system with a low-mass or degenerate companion (see Rappaport and Joss 1980 for a further discussion). However, the crucial evidence in favor of this picture was the recent discovery (Middleditch *et al.* 1980) of a sideband in the power density spectrum of the optical pulsations (see Fig. 13 of Rappaport and Joss 1980). The sideband can be understood as resulting from the difference between the synodic and sidereal rotation periods of the neutron star and its X-ray emission pattern as seen from the companion star, on whose surface some of the reprocessing into optical pulsations takes place. The fact that the sideband is at a lower frequency than the X-ray pulsations indicates that the orbital motion is in the same sense as the rotation of the neutron star, and the magnitude of the frequency shift implies an orbital period of $\sim$2500 s. This period is entirely consistent with the highly-compact-binary model (see Fig. 2); the principal and down-shifted optical pulsations are shifted in phase by an average of $\sim$1 s, which implies an orbital separation of $\sim$1 light-second, also supporting this model.

In the context of the highly-compact binary model, several other aspects of the 4U1626-67 system can be readily understood. The *a priori* probability for detecting eclipses is small because the size of the companion star is small compared to the overall dimensions of the system (Joss and Rappaport 1979; see Fig. 3); thus, the apparent absence of eclipsing behavior is not surprising. The relatively small size of the companion may also allow it to be largely obscured from the X-ray emission by an accretion disk surrounding the neutron star (see Milgrom 1978), so that much of the optical flux results from reprocessing of X-rays in the disk rather than on the star itself; this effect can account for the low level of optical photometric variability at the orbital period (McClintock *et al.* 1977; Middleditch *et al.* 1980). A $\sim 10^3$ s quasi-periodic oscillation in the observed X-ray intensity (Joss, Avni, and Rappaport 1978; Li *et al.* 1980; see Fig. 12 of Rappaport and Joss 1980) may be understood as variable obscuration by material at the outer edge of the accretion disk; although this explanation is far from secure, the timescale of the oscillations is remarkably consistent with that expected from the orbital period and the system geometry (Li *et al.* 1980).

An X-ray luminosity of $\gtrsim 10^{36}$ ergs s^{-1} can be driven by mass transfer resulting from the decay of the orbit due to gravitational radiation, possibly augmented by a self-excited stellar wind and/or the evolution of the companion star. This estimate is consistent with the luminosity of $\sim 7 \times 10^{36}$ ergs s^{-1} deduced from the observed spin-up rate of the neutron star (Joss, Avni, and Rappaport 1978; Li *et al.* 1980; see also Rappaport and Joss 1980 for a review).

In view of the remarkable similarity of the optical counterpart of 4U1626-67 to the counterparts of several of the galactic-bulge sources, it seems reasonable to regard the emerging model for the former source as a prototype for the galactic-bulge sources as well.

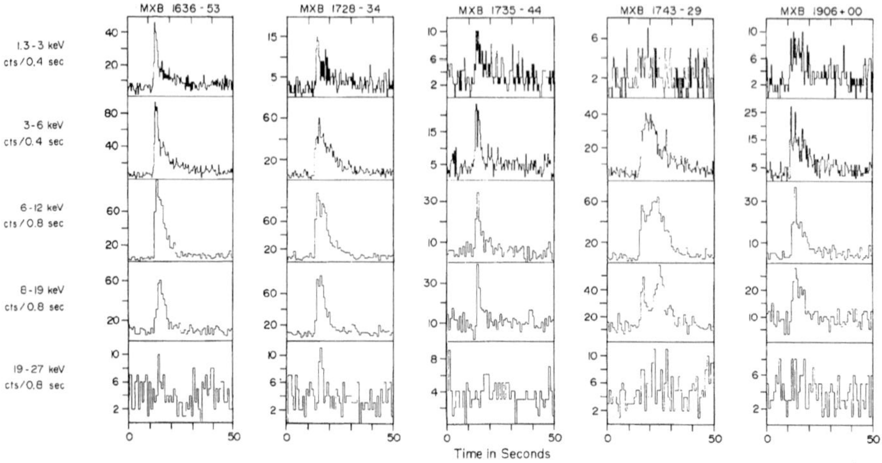

Figure 5. Profiles, in five energy channels, of X-ray bursts from five different burst sources (from Lewin and Joss 1977).

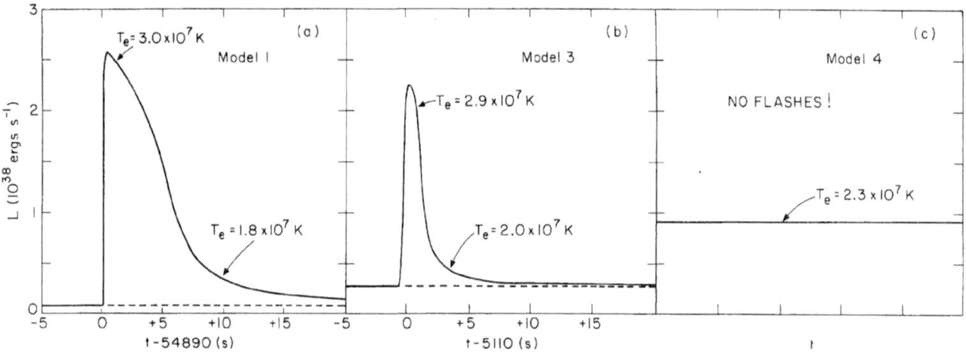

Figure 6. X-ray bursts resulting from helium-burning flashes in the surface layers of an accreting neutron star, as computed by Joss (1978). (a) Burst resulting from a low accretion rate and core temperature for the neutron star ($\dot{m}=3\times10^{16}$ g s^{-1} and $T_c=2.5\times10^8$ K, respectively). L is the bolometric luminosity from the neutron-star surface, and time t=0 is at the start of accretion onto the surface. The dashed line denotes the level of persistent accretion-driven luminosity, and the effective blackbody temperature (T_e) is indicated at a few points. (b) Same as (a), but for a higher accretion rate and core temperature ($\dot{m}=1\times10^{17}$ g s^{-1} and $T_c=3.9\times10^8$ K). The behavior of both of these models is in remarkably good agreement with the typical properties of observed X-ray bursts (see text). (c) Results for a model with still higher accretion rate and core temperature ($\dot{m}=3\times10^{17}$ g s^{-1} and $T_c=5.7\times10^8$ K). In this case, no flashes occur, so that the nuclear energy generation rate is roughly constant and never produces more than a small perturbation on the accretion-driven luminosity. This behavior may correspond to the properties of the bright galactic-bulge sources, which do not display bursts (see text). (From Joss 1979.)

In the case of 4U1626-67, we are confronted with an accreting neutron star that either is relatively young or has, in any event, retained a strong surface magnetic field. The galactic-bulge sources, on the other hand, are probably very old systems (ages $\gtrsim 10^{10}$ yr), as evidenced by their spatial distribution and the association of some of them with globular clusters (Salpeter 1973; Canizares 1975). If they contain accreting neutron stars, their surface magnetic fields may have had time to decay away (see Ruderman 1972 and references therein; Flowers and Ruderman 1977), or perhaps some difference in the process of formation of neutron stars in earlier stellar generations led to the production of relatively weak surface fields. It is also conceivable that the accreting objects are black holes of roughly solar mass, formed when neutron stars accreted sufficient matter from their binary companions to exceed the maximum stable mass for a neutron star (which most current theoretical estimates place at $\sim 2\ M_\odot$).

The second line of evidence concerning the physical character of the galactic-bulge sources comes from the X-ray burst sources (Figs. 4 and 5). These sources form a subset of the galactic-bulge sources, but they have a distinctive property in that they display occasional intense bursts of X-ray emission (see Lewin and Clark 1980 for a thorough review). Calculations of the properties of thermonuclear flashes in the surface layers of an accreting neutron star (see, e.g., Joss 1978; Taam and Picklum 1979; Fujimoto, Hanawa, and Miyaji 1980; Joss and Li 1980; Taam 1980) show remarkably good agreement between the expected properties of such flashes and the observed features of X-ray bursts from most burst sources, including their rise times ($\lesssim 1$ s), peak X-ray luminosities ($\sim 10^{38}$ ergs s^{-1}), decay time scales (~ 10 s), "tails" of softer X-ray emission lasting $\sim 10^2$ s, and typical recurrence intervals between bursts ($\sim 10^4 - 10^5$ s) (see Fig. 6). Moreover, the spectra of the X-ray bursts can generally be well fitted by blackbody emission from a region of ~ 10 km scale size, as expected for thermal emission from a neutron star following a thermonuclear flash beneath its surface; and the ratio of the average persistent X-ray luminosity to time-averaged burst luminosity is usually $\sim 10^2$ (see van Paradijs *et al.* 1979), which is consistent with the idea that the persistent emission results from accretion onto a neutron star (liberating gravitational potential energy of ~ 100 MeV per nucleon) and that the burst emission results from thermonuclear flashes (releasing ~ 1 MeV per nucleon).

Aside from the existence of bursting behavior, there seems to be little difference between the burst sources and the remainder of the galactic-bulge sources. That the brightest of the galactic-bulge sources do not display bursts can be understood as the consequence of a decrease in the instability of the nuclear burning shells of the neutron star at higher accretion rates (Joss 1978; see Fig. 6). Moreover, the lack of bursting behavior in the X-ray pulsars can be understood as resulting from a decrease in the instability of the burning shells due to the intense magnetic field and the resultant funneling of the accretion flow onto the magnetic polar caps of the neutron star (Joss and Li 1980).

If we accept the ideas that most X-ray burst sources result from thermonuclear flashes on accreting neutron stars and that these sources form a physical subset of the galactic-bulge sources, then the conclusion seems inescapable that the galactic-bulge sources are contained in highly compact binary stellar systems. The persistent X-ray luminosities of these systems are as high as $\sim 10^{38}$ ergs s^{-1}, implying accretion rates ($\sim 10^{18}$ g s^{-1}) that almost certainly require the presence of a mass-losing companion star to supply the accreting matter. On the other hand, the low luminosities of the optical counterparts of these sources exclude all but intrinsically faint optical companion stars, which would presumably be of low mass (Joss and Rappaport 1979). The absence of X-ray eclipses among the galactic-bulge sources and the lack of large-scale periodic photometric variability among their optical counterparts also seems to require highly compact binary systems with low-mass companion stars, wherein the size of the companion is sufficiently small compared to the overall dimensions of the system that it can be largely or completely shielded from the X-ray emission by the accretion disk (Milgrom 1978). In this picture, the X-ray source is visible only if we are sufficiently far from the orbital plane that eclipses are not observed, and much or all of the optical light emanates from the disk (as appears to be the case in 4U1626-67; see McClintock *et al.* 1977 and the above discussion).

I shall conclude with some brief remarks concerning the advances in our overall understanding of compact galactic X-ray sources during the past several years.

A few years ago, it seemed that binary X-ray sources could be divided into two distinct categories: those with massive ($\gtrsim 20$ $M_\odot$) optical companion stars, and those whose companions had masses of the order of a few solar masses (i.e., Hercules X-1). It seemed plausible that these two categories represented two distinct modes of mass loss from the companion star and that binary systems with intermediate-mass companion stars might be excluded from being luminous X-ray sources. However, it has now become apparent that nature is more clever than astrophysicists in finding ways to effect mass transfer in close binary stellar systems. The gap between roughly solar-mass companions and very massive companions has been filled by systems wherein the companion is an emission-line B star, of mass $\sim$5-10 $M_\odot$, which is evidently undergoing spontaneous mass loss; the best documented case of this type of behavior is that of 4U0115+63 (Rappaport *et al.* 1978). At the other extreme, 4U1626-67 evidently has a very low-mass ($\lesssim 0.1$ $M_\odot$) companion star; the mass transfer in such a system is most likely effected by gravitational radiation resulting from the orbital motion (see above). Cygnus X-3 (Parsignault *et al.* 1972) and 4U2129+47 (Thorstensen *et al.* 1979) have also been demonstrated to have short orbital periods of $4^h\!.8$ and $5^h\!.2$, respectively; if we make the simple (and speculative) assumptions that the companion stars in these systems are main-sequence stars that fill their Roche lobes, we find that their masses are again likely to be well under a solar mass. A summary of the estimated masses of the companion stars in several binary X-ray sources is given in Figure 7; we find that there now appears to be a fairly continuous distribution of companion masses from $\sim 0.1\ M_\odot$ to $\sim 30\ M_\odot$, and we expect the remaining gaps in this distribution to eventually be filled in as additional orbital determinations of X-ray binaries are obtained.

The binary X-ray sources can be broken into broad categories, according to (1) whether the total system mass M_{tot} is dominated by a massive companion star (i.e., $M_{tot} \gtrsim 10\ M_\odot$) or is not very much greater than the mass of the collapsed object (in which case usually $M_{tot} \lesssim 1\ M_\odot$), and (2) whether the accreting object is a degenerate dwarf, neutron star, or black hole. An attempt to make such a breakdown is shown in Figure 8. There are no good candidates for X-ray binaries containing degenerate dwarfs and massive companion stars; there are, however, several classes of low-mass X-ray binaries containing degenerate dwarfs, including AM Herculis and a few similar systems (wherein the degenerate dwarf is evidently highly magnetized) (see, e.g., Hearn and Richardson 1977), Sco X-1 and a few similar systems (though this interpretation is still uncertain; see Kylafis and Lamb 1979), and some of the cataclysmic variables, such as SS Cygni (see Cordova *et al.* 1980 for references) (though the X-ray luminosities of these systems are marginal in the context of this discussion). The best candidate for a system containing a black hole is still Cygnus X-1 (see Bahcall 1978b), but perhaps this interpretation should still be regarded as tentative, in view of the enormous theoretical significance of the existence or non-existence of black holes. It is conceivable that a fraction of the galactic-bulge sources are low-mass binaries containing black holes, but there is no real evidence to support this conjecture.

In contrast, there are many X-ray binary systems for which the accreting object has been well-identified as a neutron star; these include the X-ray pulsars, most of the burst sources and, according to the above arguments, many or all of the remaining galactic-bulge sources. Other systems, such as Cygnus X-3 (Lamb *et al.* 1977) and 4U1700-37 (see Bahcall 1978a for references), may also contain neutron stars, though the evidence is less compelling. At present, then, it seems probable that the great majority of the high-luminosity compact X-ray sources in our Galaxy contain neutron stars, with perhaps only a handful of exceptions.

I am grateful to S. Rappaport for helpful comments on the manuscript. This work was supported in part by the National Science Foundation under grant AST78-21993, by NASA under grant NSG-7643 and contract NAS5-24441, and by the United States-Israel Binational Science Foundation (BSF). The author is an Alfred P. Sloan Research Fellow.

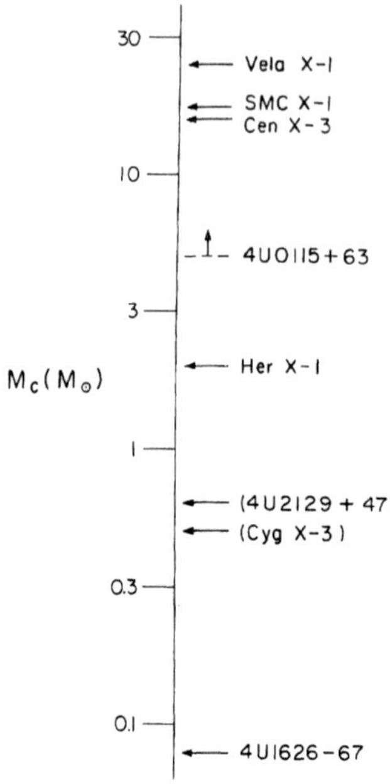

Figure 7. Estimated masses of the companion stars in some selected galactic X-ray sources. The indicated value for the companion to 4U0115+63 is a lower limit, but a mass of ~5-10 $M_\odot$ is plausible in view of its spectral type (Be). The estimates for Cyg X-3 and 4U2129+47 are especially speculative (see text).

	Low-Mass Binaries ($M_{tot} \lesssim 1\ M_\odot$)	High-Mass Binaries ($M_{tot} \gtrsim 10\ M_\odot$)
White Dwarfs	(1) Cataclysmic Variables (SS Cyg, etc.) (2) AM Herculis-type (3) Sco X-1, Cyg X-2?	?
Neutron Stars	(1) 4U1626-67 (2) Cyg X-3? (3) Galactic Bulge?	(1) X-Ray Pulsars (2) Others? (4U1700-37, etc.)
Black Holes	Galactic Bulge??	Cyg X-1?

Figure 8. Breakdown of galactic X-ray sources into six categories (see text for explanation).

REFERENCES

Bahcall, J.N. 1978a, in "Physics and Astrophysics of Neutron Stars and Black Holes," ed. R. Giacconi and R. Ruffini (Amsterdam: North Holland), p. 63.

Bahcall, J.N. 1978b, *Ann. Rev. Astron. Astrophys.*, *16*, 241.

Bradt, H.V., Doxsey, R.E., and Jernigan, J.G. 1979, in "X-Ray Astronomy," ed. W.A. Baity and L.E. Peterson (Oxford: Pergamon Press), p. 3.

Canizares, C.R. 1975, *Astrophys. J.*, *201*, 589.

Cordova, F.A., Chester, T.J., Tuohy, I.R., and Garmire, G.P. 1980, *Astrophys. J.*, *235*, 163.

Flowers, E., and Ruderman, M. 1977, *Astrophys. J.*, *215*, 302.

Fujimoto, M.Y., Hanawa, T., and Miyaji, S. 1980, preprint.

Hearn, D.R., and Richardson, J.A. 1977, *Astrophys. J. (Letters)*, *213*, L115.

Ilovaisky, S.A., Motch, Ch., and Chevalier, C. 1978, *Astron. Astrophys.*, *70*, L19.

Jones, C., Forman, W., Tananbaum, H., Schreier, E., Gursky, H., Kellogg, E., and Giacconi, R. 1973, *Astrophys. J. (Letters)*, *181*, L43.

Joss, P.C. 1978, *Astrophys. J. (Letters)*, *225*, L123.

Joss, P.C. 1979, *Comments on Astrophys.*, *8*, 109.

Joss, P.C., Avni, Y., and Rappaport, S. 1978, *Astrophys. J.*, *221*, 645.

Joss, P.C., and Li, F. 1980, *Astrophys. J.*, *238*, 287.

Joss, P.C., and Rappaport, S.A. 1979, *Astron. Astrophys.*, *71*, 217.

Kylafis, N.D., and Lamb, D.Q. 1979, *Astrophys. J. (Letters)*, *228*, L105.

Lamb, F.K., Pethick, C.J., and Pines, D. 1973, *Astrophys. J.*, *184*, 271.

Lamb, R.C., Fichtel, C.E., Hartman, R.C., Kniffen, D.A., and Thompson, P.J. 1977, *Astrophys. J. (Letters)*, *212*, L63.

Lewin, W.H.G., and Clark, G.W. 1980, *Ann. N.Y. Acad. Sci.*, *336*, 451.

Lewin, W.H.G., and Joss, P.C. 1977, *Nature*, *270*, 211.

Li, F., Joss, P.C., McClintock, J.E., Rappaport, S., and Wright, E. 1980, *Astrophys. J.* (in press).

Linsky, J. 1980, this volume.

McClintock, J.E., Canizares, C.R., Bradt, H.V., Doxsey, R.E., Jernigan, J.G., and Hiltner, W.A. 1977, *Nature*, *270*, 320.

Middleditch, J., Nelson, J., Mason, K.O., and White, N. 1980, to be submitted to *Astrophys. J.*

Milgrom, M. 1978, *Astron. Astrophys.*, *67*, L25.

Parsignault, D.R., Gursky, H., Kellogg, E.M., Matilsky, T., Murray, S., Schreier, E., Tananbaum, H., Giacconi, R., and Brinkman, A.C. 1972, *Nature Phys. Sci.*, *239*, 123.

Pringle, J.E., and Rees, M.J. 1972, *Astron. Astrophys.*, *21*, 1.

Pravdo, S.H., White, N.E., Boldt, E.A., Holt, S.S., Serlemitsos, P.J., Swank, J.H., and Szymkowiak, A.E. 1978, *Astrophys. J.*, *231*, 912.

Rappaport, S., Clark, G.W., Cominsky, L., Joss, P.C., and Li, F. 1978, *Astrophys. J. (Letters)*, *224*, L1.

Rappaport, S., and Joss, P.C. 1977, *Nature*, *266*, 683.

Rappaport, S., and Joss, P.C. 1980, this volume.

Rappaport, S., Markert, T., Li, F.K., Clark, G.W., Jernigan, J.G., and McClintock, J.E. 1977, *Astrophys. J. (Letters)*, *217*, L29.
Ruderman, M. 1972, *Ann. Rev. Astron. Astrophys.*, *10*, 427.
Salpeter, E.E. 1973, in "I.A.U. Symposium No. 55, X- and Gamma-Ray Astronomy," ed. H. Bradt and R. Giacconi (Dordrecht: Reidel), p. 135.
Schreier, E.J., Levinson, R., Gursky, H., Kellogg, E., Tananbaum, H., and Giacconi, R. 1972, *Astrophys. J. (Letters)*, *178*, L71.
Taam, R.E. 1980, preprint.
Taam, R.E., and Picklum, R.E. 1979, *Astrophys. J.*, *233*, 327.
Tananbaum, H., Gursky, H., Kellogg, E.M., Levinson, R., Schreier, E., and Giacconi, R. 1972, *Astrophys. J. (Letters)*, *174*, L143.
Thorstensen, J., Charles, P., Bowyer, S., Briel, U.G., Doxsey, R.E., Griffiths, R.E., and Schwartz, D.A. 1979, *Astrophys. J. (Letters)*, *233*, L57.
van Paradijs, J., Joss, P.C., Cominsky, L., and Lewin, W.H.G., *Nature*, *280*, 375.
Vaiana, G. 1980, this volume.

BINARY X-RAY PULSARS

Saul Rappaport and Paul C. Joss
Department of Physics and Center for Space Research
Massachusetts Institute of Technology
Cambridge, MA 02139

I. INTRODUCTION

As X-ray astronomy enters the Einstein era when faint, distant, and exotic X-ray sources are being investigated, we thought it might be useful to review some of the progress made over the past decade on a class of relatively nearby objects that have contributed richly to our understanding of astrophysics -- the binary X-ray pulsars. We shall touch briefly on three general topics related to X-ray pulsar research: (1) pulse profiles, (2) studies of pulse periods, and (3) orbital determinations and the evaluation of binary system parameters. We shall also discuss three special topics of current interest: (1) the apsidal motion test in X-ray binaries; (2) 4U1626-67 and the nature of highly compact X-ray binary systems; and (3) searches for new X-ray pulsars in the Einstein era. Most of the data that we shall invoke were obtained with the Uhuru, Copernicus, Ariel 5, SAS-3, OSO-8, and HEAO-1 satellites.

II. X-RAY PULSAR PHENOMENOLOGY

A. Pulse Profiles

The pulse periods of the known binary X-ray pulsars range over three decades in period, from 0.7 s to 835 s. The apparent "gap" in the distribution of pulsar periods between 10 and 100 s has recently been closed with the discoveries of two pulsars with periods of 17 and 38 s (Kelley et al. 1980a; White and Pravdo 1979). Sample pulse profiles for 14 of the ~18 known pulsars are presented in Figure 1. The pulse profiles are characterized by (i) large "duty cycles" of ≳50% (compared to the ~3% duty cycles typical of radio pulsars; Manchester and Taylor 1977), (ii) modulation factors that range from 25 to 90%, (iii) a range from symmetric to highly asymmetric shapes, and (iv) no obvious trend in pulse morphology as a function of pulse period or X-ray luminosity. For several of the sources (e.g., 4U1626-67, A0535+26, and 4U0900-40) the pulse profiles vary radically

R. Giacconi (ed.), X-Ray Astronomy with the Einstein Satellite, 123–152.

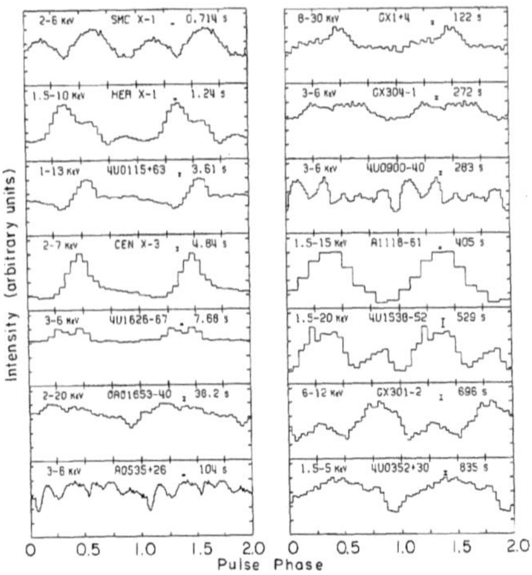

Figure 1: Sample pulse profiles for 14 binary X-ray pulsars (from Joss and Rappaport 1980). The profiles for Her X-1, 4U0115+63, Cen X-3, 0A01653-40, A1118-61, and 4U1538-52 are from Joss et al. (1978), Johnston et al. (1978), Ulmer (1976), White and Pravdo (1979), Ives et al. (1975) and Becker et al. (1977), respectively; the remaining profiles are from SAS-3 data. (For other references see Rappaport and Joss 1977a.) In each case, the data are folded modulo the pulse period and plotted against pulse phase for two complete cycles. The approximate pulse periods and energy intervals are indicated for each pulsar. Non-source background counting rates have been subtracted. A typical $\pm 1\sigma$ error bar, derived from photon counting statistics, is indicated for each pulse profile.

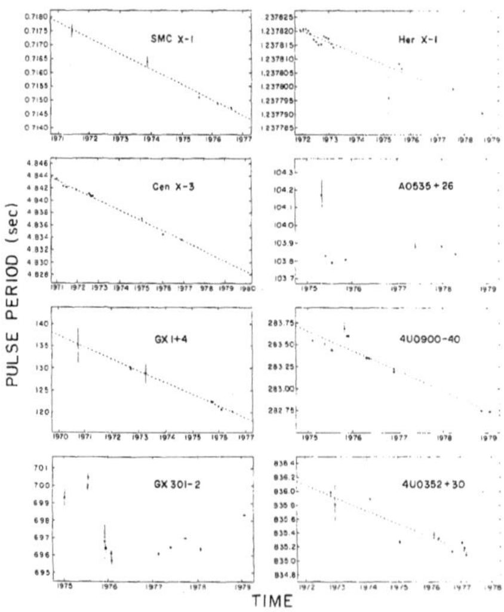

Figure 2: Pulse period histories for 8 binary X-ray pulsars. The heavy dots are individual measurements of pulse period; the vertical bars represent the 1σ uncertainties in the period determination. The data are from the Uhuru, Copernicus, Ariel 5, SAS-3, OSO-8, and HEAO-1 satellites, the Apollo-Soyuz Test Project, and a balloon and sounding rocket experiment. In general, the original references are quoted elsewhere in the text (see Rappaport and Joss 1977a,b; White, Mason and Sanford 1977; and Lamb, Pines and Shaham 1978 for additional references). The dashed lines are minimum chi-squared fits of a straight line to the data points. (In some cases, where the pulse period was measured with very great accuracy, a larger error bar, of fixed size, was assigned in the fitting procedure to take into account the known systematic deviations of the pulse period variations from a simple linear time dependence.)

with energy, while for others (e.g., Cen X-3 and 4U0352+30) the basic pulse profile is retained over a wide range of energies (see Joss and Rappaport 1980 for references).

At present, we are far short of a complete theoretical understanding of the complex pulse shapes (Fig. 1) or their dependence on energy. It seems clear, however, that the pulsations result from an X-ray beam pattern that is misaligned with the rotation axis of an accreting, magnetized neutron star and viewed from different directions as the star rotates (Pringle and Rees 1972; Davidson and Ostriker 1973; Lamb, Pethick and Pines 1973; Baan and Treves 1973) (see note 1). The X-ray luminosity is generated by the release of the gravitational potential energy of the accreting matter, which is, in turn, presumably supplied by a close binary stellar companion. The existence of the companion star is, in many cases, demonstrated by direct observational evidence. The X-ray beam pattern is determined by the external magnetic field of the neutron star, the resultant funneling of the accretion flow, and the complex radiative transfer processes for X-rays propagating from the neutron-star surface through regions of accreting, magnetized plasma.

Early attempts to derive theoretical pulse profiles (see, e.g., Lamb, Pethick and Pines 1973; Gnedin and Sunyaev 1973; Basko and Sunyaev 1975; Tsuruta 1975; Wang 1975) utilized highly simplifying approximations for the actual physical processes. These efforts suggest emission patterns ranging from pencil beams directed along the magnetic axes of the neutron star to fan beams perpendicular to the magnetic axes.

Recently, a few workers (Yahel 1979; Pravdo and Bussard 1980) have used Monte Carlo techniques to simulate the generation of the pulse profiles and X-ray spectra. These calculations have taken into account many of the important scattering processes for X-rays propagating through the accretion column. However, such calculations have not yet been carried out in a self-consistent manner (i.e., the effects of the radiation upon the plasma and the emission of radiation by the plasma are not considered). In particular, it is not even known where in the accretion column or neutron star the X-rays are generated. Moreover, any complete model of the emission mechanism must explain the gross similarity among the pulse profiles of the various X-ray pulsars, whose rotation periods and X-ray luminosities range over three and four orders of magnitude, respectively. An in-depth understanding of this problem is a goal that can be pursued during the next few years while we await a new X-ray astronomy satellite designed to carry out temporal studies.

B. Studies of Pulse Periods

The pulse periods for a number of X-ray pulsars have been sufficiently well measured over the past decade to provide important information regarding the torques exerted by the accreting material.

Nearly all of the available pulse period measurements (as of January 1980) for the eight best-measured X-ray pulsars are shown in Figure 2. When we examine the pulse period histories of these eight pulsars, it is apparent that the "spin-up" trend first noted in Her X-1 and Cen X-3 (Giacconi 1974; Gursky and Schreier 1975; Schreier and Fabbiano 1976) is very prominent in at least six of them.

The trend in most of the X-ray pulsars toward a secular decrease in pulse period can be understood in terms of torques exerted by the matter accreting onto the neutron star. Calculations of these torques (Pringle and Rees 1972; Lamb, Pethick and Pines 1973; Ghosh and Lamb 1979 and references therein) show that over a wide range of conditions, the rate of change, $\dot{P}$, of the intrinsic pulse period is related to the X-ray luminosity and the parameters of the neutron star:

$$\frac{\dot{P}}{P} \simeq -3 \times 10^{-5} \left(\frac{\xi v_r}{v_{ff}}\right)^{1/7} \left(\frac{M_x}{M_\odot}\right)^{-10/7} \left(\frac{R}{10\mathrm{km}}\right)^{6/7} \left(\frac{R_g}{10\mathrm{km}}\right)^{-2} \left(\frac{\mu}{10^{30}\mathrm{Gcm}^3}\right)^{2/7} \left(\frac{L}{10^{37}\mathrm{ergs\ s}^{-1}}\right)^{6/7} \left(\frac{P}{1\mathrm{s}}\right) \mathrm{yr}^{-1}. \quad (1)$$

Here, ξ is the fractional solid angle subtended at the neutron star by the infalling matter at the magnetopause; v_r/v_{ff} is the ratio of the average radial infall velocity of a particle to its free-fall velocity just outside the magnetopause; M_x, R, R_g, μ, and P are the mass, radius, gyroradius, magnetic dipole moment, and rotation period of the neutron star, respectively; and L is the accretion-driven luminosity. The quantity $(\xi v_r/v_{ff})^{1/7}$ is not expected to differ greatly from unity (see, e.g., Lamb, Pethick and Pines 1973).

The overall minus sign in equation (1) is explicitly for the case where the sense of the orbital angular momentum in the accreting matter is the same as that of the rotation of the neutron star. This is, in fact, to be expected if (1) the rotation of the progenitor star was in the same sense as the binary orbital motion, (2) the neutron star retained this sense of rotation during its formation, and (3) the sense of the orbital angular momentum of the matter leaving the companion star is retained as it enters the accretion disk that surrounds the neutron star.

We note that even for the six sources in Figure 2 that exhibit a "spin-up" trend, the change in pulse period is not always monotonic on short timescales (see, e.g., Giacconi 1974; Fabbiano and Schreier 1977; Ögelman _et al_. 1977). Furthermore, for two of the sources (GX301-2 and A0535+26; Kelley, Rappaport and Petre 1980; Li _et al_. 1979) there are long intervals where the pulse period is nearly constant or even increasing with time.

It is possible that A0535+26 becomes a "fast pulsar" (Elsner and Lamb 1976) when in the low part of its transient cycle and thereby undergoes periods of spin-down. The pulse period behavior of GX301-2 may possibly be explained by occasional reversals in the sense of rotation of its accretion disk (cf. Shapiro and Lightman 1976). (For further discussions of such effects, see Ghosh, Lamb and Pethick 1977; Ghosh and Lamb 1978, 1979; Davies, Fabian and Pringle 1979; and references therein.)

For simplicity, we can rewrite equation (1) as

$$\frac{\dot{P}}{P} = -3 \times 10^{-5} f \left(\frac{P}{1s}\right)\left(\frac{L}{10^{37} \text{ergs s}^{-1}}\right)^{6/7} \qquad (2)$$

(Rappaport and Joss 1977b), where the dimensionless function f is expected to be of order unity for a neutron star and contains parameters that are not yet measurable for most or all of the X-ray pulsars.

Following Rappaport and Joss (1977b), we have estimated the spin-up rate for each of ten X-ray pulsars in the following way. For the six sources in Figure 2 that show a clear secular trend toward decreasing pulse period, we have adopted the average value of $\dot{P}$ (i.e., the dashed lines in Fig. 2). In the cases of the other two sources shown in Figure 2, we have adopted the average value of $\dot{P}$ during the longest interval with generally decreasing pulse period as a representative measure of the accretion torques during times when equations (1) and (2) are most likely to be valid (see Elsner and Lamb 1976; Ghosh and Lamb 1979). For two other pulsars (4U0115+63 and 4U1626-67), not indicated in Figure 2, a reliable value of $\dot{P}$ is available (Rappaport _et al_. 1977, 1978), but the distance to the source, and hence the X-ray luminosity, is not known.

The empirical data relating the spin-up rate and the quantity $PL^{6/7}$ (see equation 2) are shown for 10 X-ray pulsars in Figure 3. The solid line is the relation expected from equation (2) if the X-ray stars are neutron stars with commonly accepted values of mass, radius, and magnetic moment as indicated in equation (1) (i.e., if f = 1). The dashed line, lying far below the data points, is the expected relation for accreting degenerate dwarfs. This result (see also Mason 1977) provides a compelling quantitative argument that the X-ray pulsars are, in fact, accreting neutron stars.

An alternative method of displaying the pulsar spin-up data is to plot $\dot{P}$ vs $PL^{3/7}$ (Ghosh and Lamb 1979). This has the advantage of sorting the pulsars according to their "fastness" (i.e., the ratio of the rotational angular velocity of the neutron star to that of the orbiting matter in an accretion disk just outside the magnetopause), and may best reveal possible systematic deviations from a simple linear relation.

Important additional information concerning the internal structure and dynamics of neutron stars may eventually be gleaned from studies of the observed smaller-scale fluctuations in pulse period

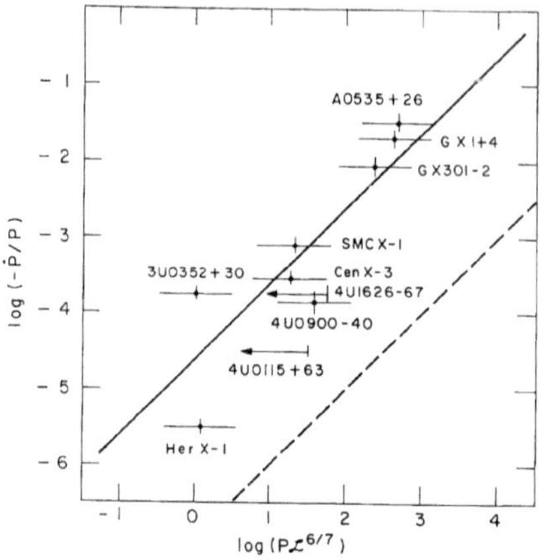

Figure 3: The empirical relation between the fractional rate of change of pulse period, $\dot{P}/P$, and the parameter (pulse period x luminosity $^{6/7}$) for 10 binary X-ray pulsars (adapted from Rappaport and Joss 1977b and Joss and Rappaport 1980). The units of $\dot{P}/P$, P, and L are yr^{-1}, s, and 10^{37} ergs s^{-1}, respectively. The solid line is the best fit to a straight line with logarithmic slope 1.0 (equation (2)); the data point for Her X-1 and the limits for 4U1626-67 and 4U0115+63 were excluded from the fit (see Elsner and Lamb 1976, Ghosh and Lamb 1979, and the text of §II.B for discussions of the rather special case of Her X-1). The intercept at log $(\dot{P}/P)$ = -4.6 is in good agreement with the expected value if the X-ray emitting objects are accreting neutron stars (see text and equation (2)). The dashed line is the expected relation for ~1 $M_\odot$ degenerate dwarfs.

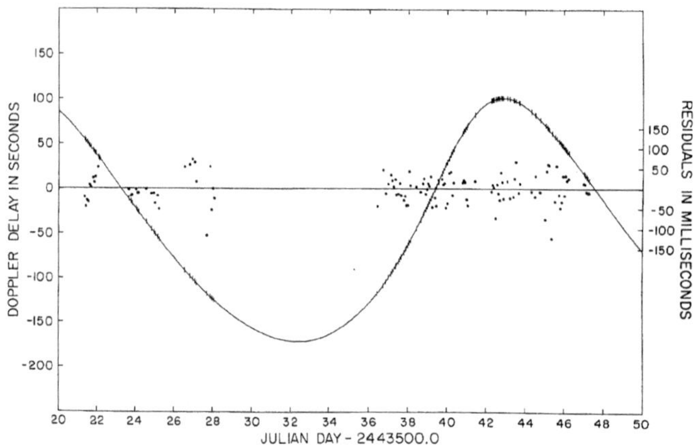

Figure 4: Doppler delay data for 4U0115+63 (from Rappaport et al. 1978). The vertical bars are the measured delays in pulse arrival time (left-hand scale); the length of each bar is considerably greater than the uncertainty in the measurement. The solid curve represents the expected delays for a Keplerian orbit with the best-fit orbital parameters (see Table 1). Small circles (right-hand scale) indicate the residual differences between the measured delays and the best-fit curve.

(see, e.g., Lamb, Pines and Shaham 1978; Boynton and Deeter 1979; Lamb and Boynton 1980).

C. Orbital Determinations and the Evaluation of Binary System Parameters

Measurements of the pulse arrival times from several binary X-ray pulsars have been used very successfully to determine the orbits of these systems. As will be discussed below, the binary orbital parameters constitute the key ingredients that allow a determination of many of the physically interesting system parameters. In the early 1970s the relatively tight orbits of the luminous X-ray pulsars Cen X-3 and Her X-1 were measured with the Uhuru satellite (Schreier et al. 1972a; Tananbaum et al. 1972); sources like 4U0115+63 and GX301-2 have lower average X-ray luminosities and relatively wide orbits, and the orbital parameters of these systems were measured only recently (Rappaport et al. 1978; White, Mason and Sanford 1978; Kelley, Rappaport and Petre 1980).

The measured Doppler delay curve for the spectrally "hard" transient source 4U0115+63 is shown in Figure 4 (Rappaport et al. 1978). These data were obtained with SAS-3 in observations that spanned an interval of 26 days. Analysis of these data revealed a short pulse period ($P \simeq 3.6$ s; Cominsky et al. 1978) a long orbital period ($P_{orb} \simeq 24$ d), a large projected semi-major axis ($a_x \sin i \simeq 140$ lt-sec), and a moderate eccentricity ($e \simeq 0.34$). The companion is thought to be a B emission-line star that does not fill its critical potential lobe; the episodes of X-ray flaring are probably attributable to periods of enhanced spontaneous mass loss from the companion (Rappaport et al. 1978). This measurement ended speculation about the basic nature of 4U0115+63 and, by inference, the other hard X-ray transients as well.

Another example of an X-ray pulsar in a wide binary orbit is provided by the GX301-2 system. Unlike 4U0115+63, this pulsar has a long pulse period (695 s; White et al. 1976) that undergoes relatively large intrinsic variations (see Fig. 2). In this case, a tentative orbital determination was made possible by combining the data from two extended Ariel V observations (White, Mason and Sanford 1978) and a 30-day SAS-3 observation. The Doppler delay curves and the best orbital fit are shown in Figure 5. The orbital parameters are $P_{orb} \simeq 35$ d, $a_x \sin i \simeq 304$ lt-sec, and $e \simeq 0.44$ (Kelley, Rappaport and Petre 1980). The companion star, WRA 977 (Vidal 1973; Dower et al. 1978), has an estimated mass and radius of 30 $M_\odot$ and 45 $R_\odot$, respectively (Hammerschlag-Hensberge et al. 1976; Parkes et al. 1980). In this system, the companion star may fill its critical potential lobe near periastron passage but not during most of the orbital cycle. With this preliminary picture, one can now begin more detailed calculations of the mass transfer process (e.g., stellar wind vs. critical-lobe overflow; see, e.g., Davidson and Ostriker 1973; Okuda and Sakashita 1977; Kelley, Rappaport and Petre 1980).

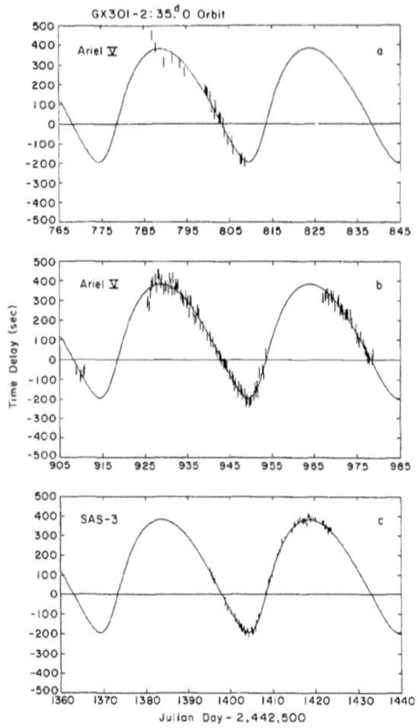

Figure 5: Doppler delay data for GX301-2 (from Kelley, Rappaport and Petre 1980). The vertical bars are the measured delays after the best-fit second-order polynomial (representing the changes in intrinsic pulse period) has been subtracted from the data (see Kelley, Rappaport and Petre 1980 for a discussion). The length of each bar represents $\pm 1\sigma$ uncertainties in the arrival times: ~10 s for the SAS-3 data and ~26 s for the Ariel V data (Kelley, Rappaport and Petre 1980; White, Mason and Sanford 1978; White 1979). The solid curves represent the delays predicted by the best-fit orbital parameters (see Table 1).

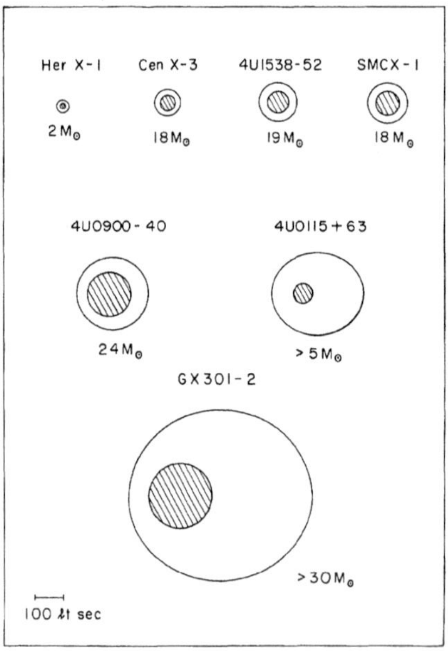

Figure 6: Schematic sketch, to scale, of the orbits and companion stars of seven binary X-ray pulsars (from Joss and Rappaport 1980). The approximate mass of the companion star is indicated below each orbit. See text (§II.C) for a discussion of the derivation of the orbital parameters, stellar masses, and stellar radii.

All the measured orbits of binary X-ray pulsars are shown to scale in Figure 6, in order of increasing size of the semi-major axis (see also Table 1). Nominal values for the masses and radii of the companion stars are indicated in the figure; the derivation of these parameters will be discussed below. A few other X-ray binaries that lack detectable X-ray pulsations (e.g., LMC X-4, 4U1700-37, and Cyg X-2) have been sufficiently well studied in optical light (Chevalier and Ilovaisky 1977; Hutchings et al. 1978; Hutchings 1976; van Paradijs et al. 1978; Cowley et al. 1979) to yield important information about the binary system parameters; however, the information is much less complete than for the systems where the orbit of the X-ray star can be measured by pulse timing observations.

There are five binary X-ray pulsars for which sufficient information is now available to allow a determination of many of the system parameters. In particular, for four of these systems, the following three key quantities (in addition to the orbital period) have been measured: (i) the projected semi-major axis of the X-ray star; (ii) the amplitude of the Doppler velocity curve for the optical companion; and (iii) the duration of the X-ray eclipse. These systems are SMC X-1, Cen X-3, 4U0900-40, and 4U1538-52 (Primini, Rappaport and Joss 1977; Hutchings et al. 1977; Fabbiano and Schreier 1977; Hutchings et al. 1979; Rappaport, Joss and Stothers 1980; van Paradijs et al. 1977; Becker et al. 1977; Davison et al. 1977; Crampton et al. 1978).

The orbital period, P_{orb}, and the projected semi-major axis of the X-ray star, $a_x \sin i$, yield the mass function:

$$f(M) = \frac{4\pi^2 (a_x \sin i)^3}{G\, P_{orb}^2} = \frac{M_c \sin^3 i}{(1+q)^2} , \qquad (3)$$

where M_c is the mass of the companion star, $q \equiv M_x/M_c$ is the mass ratio, and G is the universal gravitational constant.

The mass ratio q can be determined directly from the ratio of the velocity of the companion star to that of the X-ray star:

$$q = \frac{a_c \sin i}{a_x \sin i} = \frac{K_c\, P_{orb} \sqrt{1-e^2}}{2\pi\, a_x \sin i} , \qquad (4)$$

where $a_c \sin i$ is the projected semi-major axis of the orbit of the companion, K_c is the semi-amplitude of the optical Doppler velocity curve, and e is the orbital eccentricity.

The final ingredient for determining the system parameters is the orbital inclination angle, i. This can be computed approximately,

Table 1

Orbital Parameters for Seven Binary X-ray Pulsars

Source	$a_x \sin i$* (lt-s)	P_{orb} (d)	f(M) ($M_\odot$)	e*	References
Her X-1	13.206 ± 0.006	1.700	0.85	< 0.002	1
Cen X-3	39.792 ± 0.005	2.087	15.5	0.0008 ± 0.0001	2
4U1538-52	55.2 ± 3.7	3.730	13	. . .	3
SMC X-1	53.46 ± 0.03	3.892	10.8	< 0.0007	4
4U0900-40	113.0 ± 0.8	8.965	19.3	0.092 ± 0.005	5
4U0115+63	140.13 ± 0.1	24.31	5.00	0.3402 ± 0.0002	6
GX301-2	304 ± 3	34.99	25	0.44 ± 0.01	7

* Wherever possible, we have given 1σ confidence limits for the error bars. The quoted upper limits on e are 95% confidence limits.

1. Tananbaum *et al.* (1972); Schreier and Fabbiano (1976); Fechner and Joss (1977); Joss *et al.* (1980).
2. Schreier *et al.* (1972a); Fabbiano and Schreier (1977).
3. Becker *et al.* (1977); Davison, Watson, and Pye (1977).
4. Primini, Rappaport, and Joss (1977).
5. Rappaport, Joss and McClintock (1976); Rappaport, Joss and Stothers (1980).
6. Rappaport *et al.* (1978).
7. Kelley, Rappaport and Petre (1980); White, Mason and Sanford (1978).

Table 2

Parameters Used in the Monte Carlo Analysis

Source	$a_x \sin i$* (lt-sec)	K_c* (km s^{-1})	θ_e + (degrees)	References
SMC X-1	53.46 ± 0.03	19 ± 2	26.5-29	1
Cen X-3	39.79 ± 0.01	24 ± 6	35-40	2
4U0900-40	113.0 ± 0.8	21.8 ± 1.2	31-40	3
4U1538-52	55.2 ± 3.7	33 ± 7	25-33	4

* The quoted uncertainties are ±1σ error limits.

\+ The extreme range of eclipse half-angles allowed by the observations.

1. Schreier *et al.* (1972b); Primini *et al.* (1976); Primini, Rappaport and Joss (1977); Hutchings *et al.* (1977).
2. Pounds *et al.* (1975); Pounds (1976); Fabbiano and Schreier (1977); Hutchings *et al.* (1979).
3. Forman *et al.* (1973); van Paradijs *et al.* (1977); Watson and Griffiths (1977); Rappaport, Joss and Stothers (1980).
4. Davison, Watson and Pye (1977); Becker *et al.* (1977); Crampton, Hutchings and Cowley (1978).

with the aid of a simple model. If we replace the companion star by a sphere of radius R_c whose volume equals the actual volume of the critical potential lobe, then the relation between inclination angle and eclipse half-angle, Θ_e, is

$$R_c \simeq D\left[\cos^2 i + \sin^2 i \sin^2 \Theta_e\right]^{1/2} , \quad (5)$$

where D is the stellar separation. This simple approximation is surprisingly accurate, even for stars that suffer appreciable tidal or rotational distortion (see, e.g., Avni 1976; Rappaport 1979). Moreover, the size and shape of the critical potential lobe can be computed from the mass ratio and the rotation rate of the companion star; its effective radius, R_L, has been found (Plavec 1968) to be reasonably well fitted by the expression

$$R_L \simeq D\,(a - b \log q) \quad , \quad (6)$$

where a and b are constants that depend upon the ratio, Ω, of the rotational frequency of the companion star to the orbital frequency (Plavec 1968; Avni 1976; Bahcall 1978b; Rappaport 1979). If we define the radius of the companion as some fraction, β, of the radius of the critical potential lobe, then equations (5) and (6) may be combined to yield an expression for the inclination angle:

$$\sin i \simeq \left[1-\beta^2(a-b \log q)^2\right]^{1/2}/\cos \Theta_e \ . \quad (7)$$

From the above relations, we see that if K_c, $a_x \sin i$, and Θ_e are measured, and if we make reasonable assumptions about the rotation rate and fraction of the critical lobe occupied by the companion star, then all of the system parameters mentioned above may be determined. The amplitude of the observed ellipsoidal light variations in several of these systems indicates that the companion stars nearly fill their critical potential lobes (i.e., $\beta \gtrsim 0.9$; see, e.g., Avni and Bahcall 1975a,b). It can then be argued that tidal dissipation should force these systems into approximately synchronous rotation (i.e., $\Omega \simeq 1$; see, e.g., Lecar, Wheeler, and McKee 1976). Moreover, measurements of the rotational velocities of the companion stars in several of the X-ray binaries (see, e.g., Conti 1978; Crampton et al. 1978; Hutchings et al. 1978) are consistent with $0 \lesssim \Omega \lesssim 1.5$. (For other discussions of these issues, see van den Heuvel 1975; Avni and Bahcall 1976; Rappaport and Joss 1977c; Avni 1978; Bahcall 1978a,b; Conti 1978; Petterson 1978.)

The best-fit values for the system parameters are easily found from equations (3), (4), (5), and (7) by inserting the most probable values for K_c, $a_x \sin i$, and Θ_e and a best guess for Ω and β. The values that we used for K_c, $a_x \sin i$, and Θ_e in our analysis are given in Table 2. A difficulty arises in propagating both the experimental errors in these quantities and the theoretical and experimental uncertainties in Ω and β. We have therefore evaluated the system parameters and their uncertainties by means of a Monte Carlo propagation technique (Rappaport, Joss and Stothers 1980). In 2 x 10^4 trial evaluations, $a_x \sin i$ and K_c were chosen randomly with respect to Gaussian distributions of the appropriate widths, while the values of Θ_e were chosen randomly and uniformly in the appropriate range, to reflect the experimental uncertainties. To simulate the theoretical and observational uncertainties in the values of β and Ω, we chose values of β randomly and uniformly over the range 0.9-1.0 (see note 2) and Ω randomly and uniformly between 0.0 and 1.5.

The ranges of companion star masses and radii for SMC X-1, Cen X-3, 4U0900-40, and 4U1538-52, as determined by the Monte Carlo error propagation technique, are shown in Figure 7 as contours of constant probability. The most probable parameter values are in good agreement with previously reported values (Primini, Rappaport and Joss 1977; Hutchings et al. 1979; van Paradijs et al. 1977; Rappaport, Joss and Stothers 1980; Becker et al. 1977; Davison et al. 1977). The outer contour, which contains at least 98% of the Monte Carlo events, represents reasonably secure error limits for the masses and radii. These uncertainties are important to bear in mind when trying to fit the companion stars into an evolutionary scenario involving mass-losing supergiant companions (see, e.g., van den Heuvel 1976, 1977; Conti 1978).

In the case of the Her X-1/HZ Her system, a reliable optical Doppler velocity curve is not yet available; however, extensive studies of optical pulsations (Middleditch and Nelson 1976), which are apparently produced by the reprocessing of X-ray pulsations that strike the surface of the companion star, have provided additional information needed to estimate the system parameters (Bahcall and Chester 1977; Chester 1978). Even if the optical pulsations are excluded, there is sufficient observational information to place some constraints on the system parameters (see, e.g., Joss and Rappaport 1976; Bahcall 1978b).

In Figure 8, we show the probability distributions for additional system parameters of SMC X-1, Cen X-3, 4U0900-40, and 4U1538-52, as determined by the Monte Carlo error propagation technique.

The measured neutron-star masses for these four systems, are shown in Figure 9 (cf. Joss and Rappaport 1976; Rappaport and Joss 1977c; Avni 1977; Bahcall 1978b). The most probable masses (indicated by the filled circles) and the corresponding uncertainties are obtained directly from the probability distributions in Figure 8. The shorter error bars shown in Figure 9 represent the uncertainties in the neutron star masses for the less conservative assumptions that $\beta = \Omega = 1$. Also included in the figure are the masses of Her X-1, analyzed in a somewhat different but related manner (see above), and the binary radio pulsar PSR1913+16 (Taylor et al. 1979), added for completeness. These six cases represent the only reliable mass measurements for objects known to be neutron stars. Not included are the masses of the collapsed objects in such systems as LMC X-4 (Hutchings

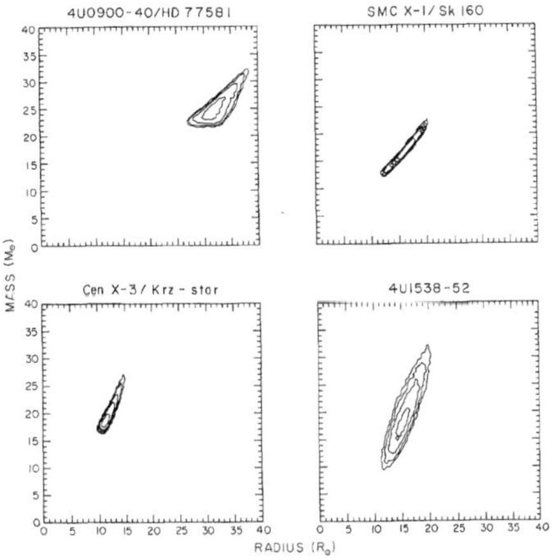

Figure 7: Computer generated error contours for the mass and radius of four companion stars in binary X-ray pulsar systems (from Joss and Rappaport 1980). The input data are the orbital period, the projected semi-major axis of the neutron-star orbit, the X-ray eclipse duration, and the amplitude of the optical Doppler velocity curve (see text). The error contours are derived from a Monte Carlo analysis (Rappaport, Joss and Stothers 1980; see text) and are generated directly from the raw output of the Monte Carlo analysis. The various contours represent arbitrary confidence levels; however, the outer contour contains ~98% of the Monte Carlo events. For simplicity, we neglected the orbital eccentricity of 4U0900-40 in computing sin i from equation (7). We estimate that this approximation results in an error of less than 10% in M_c, R_c, and M_x (see also Avni 1976).

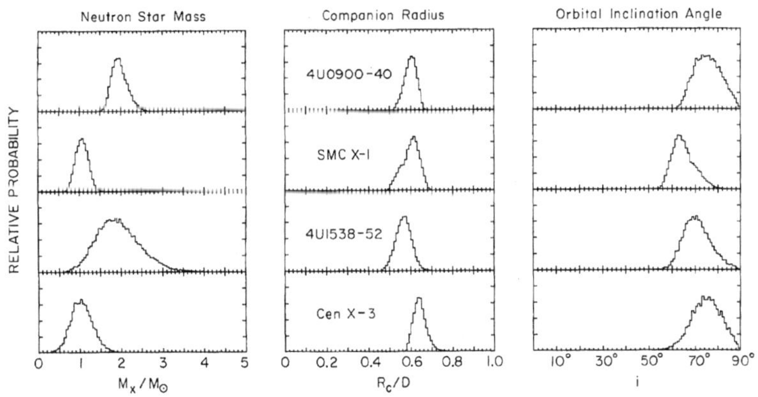

Figure 8: Several of the binary system parameters and their uncertainties, for the SMC X-1, 4U0900-40, Cen X-3, and 4U1538-52 systems, as determined by the Monte Carlo analysis technique described in the text. The quantities M_x, R_c/D, and i are the neutron star mass, the ratio of the radius of the companion to the orbital separation, and the orbital inclination angle, respectively. Note that the inclination-angle distribution plotted here will differ from the distribution of sin i by a factor of cos i (so that, for example, the latter distribution may peak at sin i = 1 even though the distribution shown here goes to zero as i approaches 90°).

et al. 1978) and Cyg X-2 (Cowley et al. 1979), where, in view of the lack of detectable pulsations, we do not know whether the objects are neutron stars.

Note that the masses given in Figure 9 are all consistent, within the uncertainties, with a mass of 1.4 ± 0.2 $M_\odot$ (shaded region on Fig. 9). This is the range of masses that might be expected for neutron stars if they are formed in dynamical events resulting from the collapse of the degenerate cores of highly evolved stars (Arnett and Schramm 1973; Iben 1974 and references therein) or from the collapse of accreting degenerate dwarfs in close binary stellar systems (see, e.g., Canal and Schatzman 1976). Furthermore, all of the masses are consistent with neutron-star and quark-star models based on conventional many-body nuclear and high-energy physics (Arnett and Bowers 1977; Fechner and Joss 1978; and references therein).

III. SOME CURRENT PROBLEMS OF SPECIAL INTEREST

A. The Apsidal Motion Test in X-Ray Binaries

Almost all of our knowledge of stellar interiors is derived from theoretical studies. At present, very few observations can be performed to test any theoretical predictions. Two direct measurements are, however, possible: the measurement of the neutrino flux from the sun (see Davis 1978 for a recent review) and the apsidal motion test for binary stellar systems (Russell 1928). In the former case, one attempts to gain information about the thermal and chemical structure of the solar interior, while in the latter, a one-parameter measure of the mass distribution within a star may be obtained. Reliable measurements of apsidal motion, which results from the tidally and rotationally induced gravitational quadrupole moment of a star, have been made for about 16 close binary systems (Kopal 1965; see Stothers 1974 for more recent references), wherein both stars are nondegenerate and are usually on the main sequence. The rate at which the longitude of periastron, ω, of an eccentric orbit advances can then be related to the "apsidal motion constant" k, which is calculated for various stellar model mass distributions under standard theoretical assumptions (see Schwarzschild 1958). In general, the measured values of k are smaller than the standard calculated values by factors of $\sim$ 2. In all cases, however, the interpretation is complicated by the fact that both stars of the binary system contribute to the apsidal motion.

Binary X-ray pulsars, on the other hand, provide a potentially less ambiguous means of carrying out these measurements. In such systems one of the stars is a neutron star, which acts essentially as a point mass and has a completely negligible apsidal motion constant compared to that of the companion star. Of the well studied X-ray binaries, Her X-1 (Fechner and Joss 1977), Cen X-3 (Fabbiano and Schreier 1977), and SMC X-1 (Primini, Rappaport and Joss 1977) have highly circular orbits, and 4U1538-52 is not sufficiently well mea-

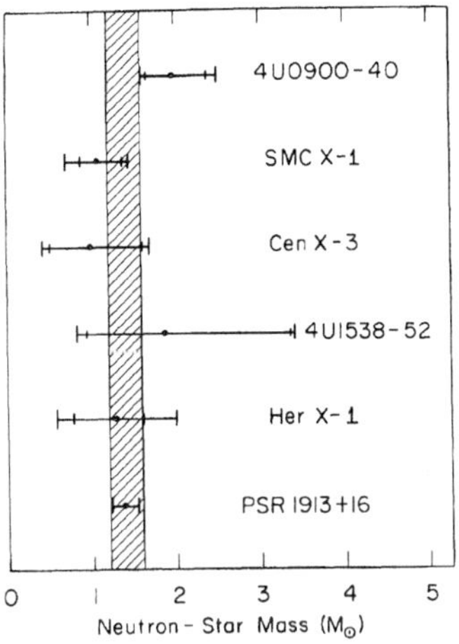

Figure 9: Empirical knowledge of neutron star masses (from Joss and Rappaport 1980). Five of the neutron-star masses are derived from observations of binary X-ray pulsars (see text). PSR1913+16 is a binary radio pulsar (Taylor et al. 1979) and is added for completeness. The most probable value for the mass of each neutron star is indicated by the filled circle. For the X-ray binary systems, an inner set of error limits is also shown, corresponding to the less conservative assumptions that $\beta = 1$ and $\Omega = 1$ (see text). The hatched region represents the range of neutron-star masses (1.2-1.6 $M_\odot$) that might be expected on the basis of current theoretical scenarios for neutron-star formation (see text).

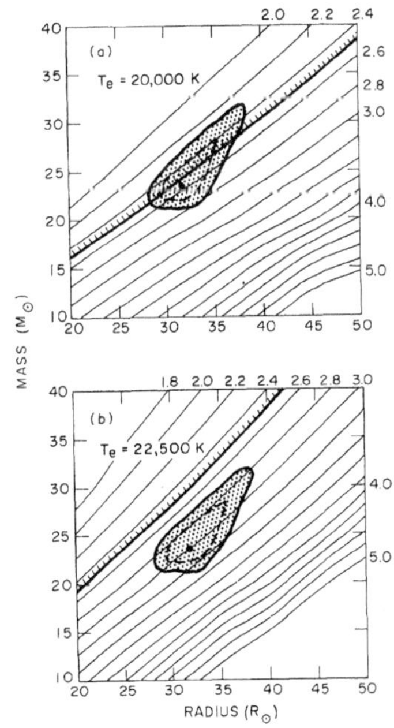

Figure 10: Contours of apsidal motion constant, k, in the mass/radius plane, calculated for two assumed values of the effective temperature, T_e. (a) T_e = 20,000 K; (b) T_e = 22,500 K (from Rappaport, Joss, and Stothers 1980). The numbers to the top and to the right of each figure denote the value of -log k for the various contours. The shaded zone represents the error region for the companion star HD 77581 in the 4U0900-40 binary system (see also Fig. 7). The heavy curve is the limit on k as measured with SAS-3 and separates the allowed (lower) from the excluded (upper) region.

sured to significantly constrain the eccentricity (Becker et al. 1977; Davison, Watson and Pye 1977). Only 4U0900-40 (Rappaport, Joss and McClintock 1976), GX301-2 (White, Mason and Sanford 1978; Kelley, Rappaport and Petre 1980), and 4U0115+63 (Rappaport et al. 1978) have measurably eccentric orbits. The orbit of GX301-2 has only recently been measured, and the observational baseline is too short to allow a determination of the rate of apsidal motion, $\dot{\omega}$. The 4U0900-40 and 4U0115+63 systems are therefore the best candidate systems in which to attempt measurements of $\dot{\omega}$.

The orbit of 4U0900-40 was first measured in July 1975 (Rappaport, Joss and McClintock 1976) and found to have a small but detectable eccentricity of $e \simeq 0.1$. Subsequent measurements of the orbit by Ögelman et al. (1977) and Becker et al. (1978) confirmed the eccentricity but had neither the precision nor the temporal baseline to attempt a measurement of apsidal motion. The pulse phase of 4U0900-40 was again tracked for one orbital cycle with SAS-3 during November 1978 (Rappaport, Joss and Stothers 1980). An analysis of these observations showed, in particular, that the apparent eccentricity has a statistical significance of $\sim 20\sigma$. In order to obtain constraints on the rate of apsidal motion in the 4U0900-40 system, we performed an orbital fit to the joint 1975/1978 data set (Rappaport, Joss and Stothers 1980). We obtained only an upper limit to the apsidal motion: $\dot{\omega} < 3\overset{\circ}{.}8\ \mathrm{yr}^{-1}$ (97% confidence).

For a binary system where the orbital and stellar parameters are known, limits on $\dot{\omega}$ can be used to constrain the allowed values of k through the relation

$$\dot{\omega} \simeq \left(\frac{2\pi k}{P_{orb}}\right)\left(\frac{R_c}{D}\right)^5 \left\{15qg(e) + \Omega^2(1+q)h(e)\right\} \tag{8}$$

(Cowling 1938; Sterne 1939). The factors g(e) and h(e) in this expression are slightly greater than unity and incorporate the contributions to $\dot{\omega}$ from the non-infinitesimal eccentricity of the orbit. (The contribution to apsidal motion from the rotational distortion of the primary was neglected in Rappaport, Joss and Stothers (1980), but in fact its effect can be significant. However, this omission did not affect any of their results or conclusions.) The contribution to $\dot{\omega}$ from general relativistic effects (see, e.g., Landau and Lifshitz 1962) is negligible in this system. We have evaluated the quantity $(R_c/D)^5(15qg + \Omega^2(1+q)h)$ and its uncertainties for the 4U0900-40 system by means of the Monte Carlo analysis described in § II.C. Our limit on apsidal motion then yields a limit on log k ($\lesssim -2.5$) that is a direct measure of the structure of the companion star HD 77581.

We have calculated values of k for a wide range of stellar models (Rappaport, Joss and Stothers 1980). Companion masses in the range 10-40 $M_\odot$, radii of 20-50 $R_\odot$, and effective temperatures, T_e, of 17,000-28,000 K were considered. (For discussions of the effective temperature of the B0.5 Ib supergiant HD 77581, see Avni and Bahcall 1975b; Rappaport, Joss and Stothers 1980; and references therein). However, only combinations leading to luminosities that are greater than or equal to the zero-age main-sequence values and less than the Eddington limit were used. The method of generation of the stellar models is described by Rappaport, Joss and Stothers (1980). The results of some of these calculations are shown in Figure 10 as contours of constant log k in the M_c - R_c plane; only the results for T_e = 20,000 K (Fig. 10a) and T_e = 22,500 K (Fig. 10b) are shown.

We have evaluated M_c, R_c, and the corresponding uncertainties for HD 77581 by the methods discussed in §II.C (Rappaport, Joss and Stothers 1980). The results are shown superposed on the contours of -log k in Figure 10. The shaded error region for M_c and R_c and the inner region enclosed by the dashed curve contain 98% and 70% of the Monte Carlo events, respectively. The filled circle denotes the most probable parameter values (i.e., M_c = 24 $M_\odot$ and R_c = 32 $R_\odot$).

The heavy curve in each of Figures 10a and b separates the allowed from the excluded regions, based on our limit for k. The experimental limit on k and the theoretical calculations of k are inconsistent for $T_e \lesssim$ 18,000 K (not shown in Fig. 10). Thus, these results in principle constrain the effective temperature of HD 77581 (cf. Nandy and Schmidt 1975, who suggest that the effective temperature of the B0 Ia star ε Ori could be as low as ~18,000 K). For T_e = 20,000 K, the lack of observed aspidal motion is just consistent with theoretical expectations. For $T_e \gtrsim$ 23,000 K, aspidal motion should not have been observed.

We anticipate that another measurement of $\dot{\omega}$, a few years in the future, should be able to detect the apsidal motion of 4U0900-40. This will provide a unique measure of the internal mass distribution of an evolved supergiant. Such a measurement will, in turn, yield a direct check on theoretical stellar models and on our understanding of the evolution of close binary stellar systems.

The transient source 4U0115+63 has been detected in X-rays on only two occasions separated by ~ 7 years (Cominsky et al. 1978). The orbit of this system was measured with high accuracy (§II.C) during the outburst in 1978 (Rappaport et al. 1978); the longitude of periastron at this time was ω = 47°.7 ± 0°.1. Significant pulse timing data are also available (Kelley et al. 1980b) from the original *Uhuru* discovery observation (Giacconi et al. 1972).

A recent analysis (Kelley et al. 1980b) of the 1971 *Uhuru* data on 4U0115+63 has yielded a value for the longitude of periastron seven years prior to the SAS-3 observation. The limited timing data, which extend over only ~7 days, were insufficient to yield a complete orbital solution. If we assume, however, that the values of P_{orb}, a_x sin i, and e obtained in 1978 were not significantly different from the corresponding 1971 values, then the *Uhuru* data can be used to determine the longitude of periastron in 1971. The *Uhuru* data were fitted with an eccentric orbit with P_{orb}, $a_x \sin i$, and e fixed at the accurately measured 1978 values, while the X-ray pulse period and phase, time of periastron passage, and ω were left as free parameters. The best-fit orbital solution yielded $\omega = 51° \pm 19°$ (95% confidence) and is shown superposed on the *Uhuru* Doppler delay data in Figure 11. This result implies an upper limit to a positive apsidal motion in the 4U0115+63 system of $\dot{\omega} \lesssim 2°$ yr^{-1} (95% confidence; Kelley *et al*. 1980b).

There have been very few optical observations of the companion star in this system (see, e.g., Johns *et al*. 1978); however, the available evidence indicates that it is probably a B emission-line star of moderate mass and radius (Hutchings 1980). A main-sequence B star is expected to have k ~ 0.01 (see, e.g., Schwarzschild 1958; Kopal 1965; Stothers 1974). In such a system, the value of $\dot{\omega}$ computed from equation (8) is ~ 1 arc min yr^{-1} for a non-rotating companion star with a nominal radius of ~8 $R_{\odot}$; for rapid rotation rates the apsidal motion increases as Ω^2 (see eq. 8). Thus, even though the limits on apsidal motion in this system are more stringent than for the 4U0900-40 system, the implied constraints on the structure of the companion star are less significant. This results mainly from the factor $(R_c/D)^5$ in equation (8), whose value for the 4U0115+63 system is small. A rate of apsidal motion as small as 1 arc min yr^{-1} is detectable, however, if 4U0115+63 can be observed for a complete orbital cycle during a future outburst. This will lead to an accurate determination of k for a seemingly unevolved companion star in an X-ray binary system.

B. 4U1626-67 and the Nature of Highly Compact X-Ray Binaries

The X-ray source 4U1626-67 is a 7.68 s pulsar with a hard energy spectrum and a pulse profile that varies dramatically with energy (Rappaport *et al*. 1977; Pravdo *et al*. 1979). The celestial position of 4U1626-67 was determined by Bradt *et al*. (1977), and the source was subsequently identified with a faint blue star (V ≃ 18.7, B-V ≃ 0.0, U-B ≃ -1.2) by McClintock *et al*. (1977). The pulse period is decreasing at a rate $\dot{P}/P \simeq -1.8 \times 10^{-4}$ yr^{-1} which, from the theory of accretion torques (see §II.B), implies an X-ray luminosity of ~7 x 10^{36} ergs s^{-1}. This luminosity, together with the measured X-ray flux of ~2 x 10^{-9} ergs cm^{-2} s^{-1}, yields a distance of ~3-12 kpc (Joss, Avni and Rappaport 1978). The absolute magnitude of McClintock's star is therefore $M_V \gtrsim 3$. This is consistent with a system containing a late-type dwarf companion with negligible intrinsic optical luminosity, wherein most of the light results from reprocessed X-radiation. Giant, supergiant, and B emission-line stellar companions are apparently ruled out.

Severe constraints have been placed on the size of the orbit of the X-ray star. The lack of Doppler shifts in the pulse arrival times has yielded limits on $a_x \sin i$ of 0.1-0.2 lt-sec for orbital periods in the range 10 m $\lesssim P_{orb} \lesssim$ 20 d (Rappaport *et al*. 1977; Joss, Avni and

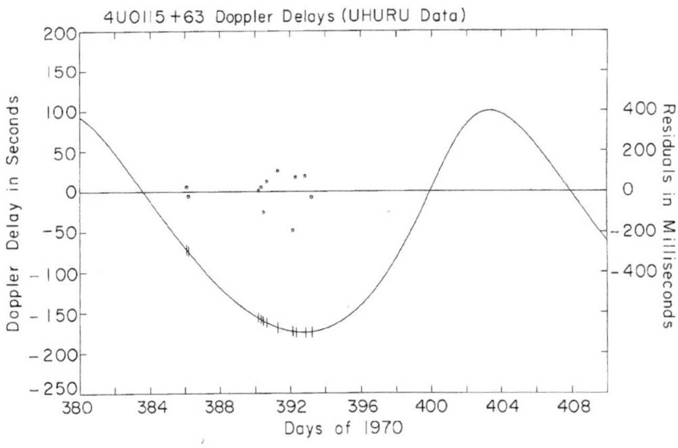

Figure 11: Uhuru Doppler delay data for 4U0115+63. The vertical bars are the measured delays in pulse arrival time; the length of each bar is considerably greater than the uncertainty in the measurement. The solid curve represents the expected delays for the best-fit orbital parameters given by Kelley et al. (1980b). In the fit to the arrival times, P_{orb}, $a_x \sin i$, and e were held fixed at the values measured in 1978 (Rappaport et al. 1978), while the pulse period, time of periastron passage, and ω were treated as free parameters (see text). The small circles indicate the residual differences between the measured delays and the best-fit curve.

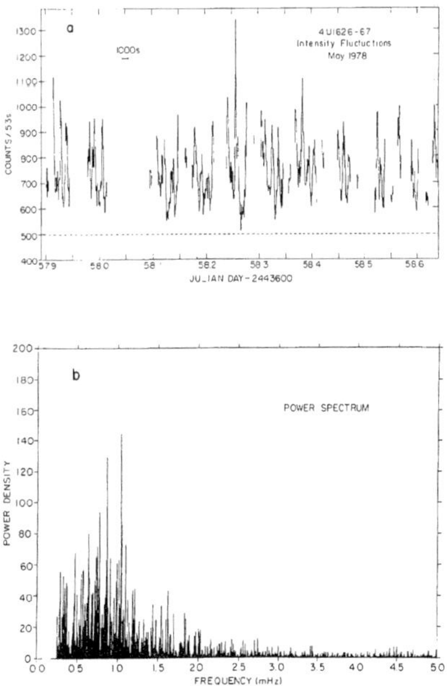

Figure 12: (a) SAS-3 counting rate data from 4U1626-67 in the 1.5-12 keV X-ray energy band (from Li et al. 1980). The dashed line indicates the non-source background level. The large flares in intensity have a characteristic recurrence time of $\sim 10^3$ s. Gaps in the data train are times when the source was occulted by the earth, the satellite was in the South Atlantic Anomaly, or data were lost in transmission. (b) Power density spectrum for 4U1626-67 computed from the X-ray counting rate data. Note the enhanced power density for frequencies near 1 mHz ($\sim 10^3$ s period). The plot is terminated for frequencies below ~0.25 mHz because of excessive noise near the ~95 min orbital period of the satellite.

Rappaport 1978; Pravdo et al. 1979; Li et al. 1980; these constraints are summarized in Figure 5 of Li et al. 1980 and Figure 2 of Joss 1980). With these orbital constraints, reasonable mass functions can be found only for very short orbital periods ($P_{orb} \lesssim 0.3$ d) or very long orbital periods ($P_{orb} \gtrsim 200$ d). The lack of a giant or Be stellar companion tends strongly to rule against long orbital periods because a smaller star, which would be far from filling its critical potential lobe and which should have a relatively low mass-loss rate, would very probably be unable to supply sufficient matter to the neutron star.

The existence of $\sim 10^3$-s, quasi-periodic oscillations in the X-ray intensity was first reported by Joss, Avni and Rappaport (1978). A recent two-day pointed observation of 4U1626-67 with SAS-3 (Li et al. 1980) again revealed the presence of such oscillations. The X-ray intensity during one 18-hour segment of the latter observation is shown in Figure 12. The count rate is dominated by rather symmetric, sharply peaked, flare-like events which have a characteristic recurrence time of $\sim 10^3$ s. The power density spectrum of these data, also shown in Figure 12, demonstrates that there is a preferred time scale of $\sim 10^3$ s but that the flare-like events are not strictly periodic. The corresponding autocorrelation function indicates that the characteristic width of a flare is ~200 s (FWHM).

The $\sim 10^3$ s oscillations could be the result of quasi-periodic obscuration of X-rays from matter orbiting near the outer edge of an accretion disk. We note that the $\sim 10^3$ s time scale is the natural period for orbital motion in the outer portions of an accretion disk if the source is in a highly compact binary system with an orbital period of $\sim(2\text{-}3) \times 10^3$ s (Li et al. 1980; see also Joss, Avni and Rappaport 1978).

Optical photometric pulsations at the X-ray pulse period were discovered by Ilovaisky, Motch and Chevalier (1978) and confirmed by McClintock et al. (1979) and Peterson et al. (1980). Recent optical photometric observations by Middleditch et al. (1980) have revealed the presence of two pulsation periods, one at the frequency of the X-ray pulsations and the other downshifted in frequency by ~0.4 millihertz. Portions of the power density spectra for several of the data runs obtained by Middleditch et al. are shown in Figure 13. The prominent peak at 130.26 millihertz corresponds to the frequency of the X-ray pulsations and is interpreted as reprocessed X-radiation from an accretion disk near the X-ray star, while the downshifted peak apparently corresponds to X-radiation reprocessed on or near the surface of the companion star. The difference in pulse period can be understood as the difference between the synodic and sidereal rotation period of the neutron star as seen from the companion, provided that the orbital period is ~2500 s (Middleditch et al. 1980). (Since the frequency of the pulses from the vicinity of the companion star is downshifted, the sense of rotation of the neutron star should be the same as that of the orbit.) The pulses that are downshifted in frequency also suffer a phase delay of ~1 s relative to the unshifted

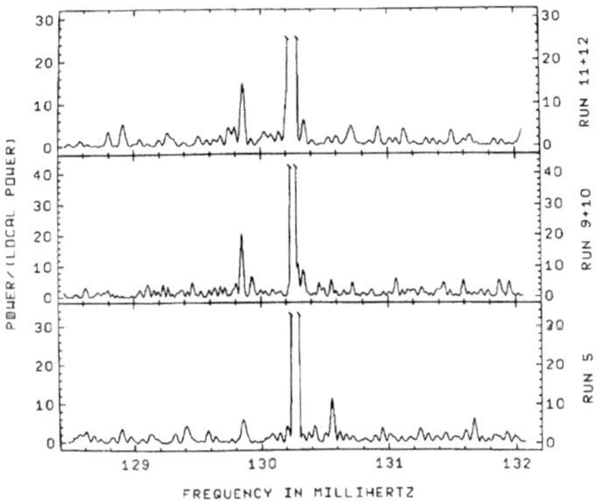

Figure 13: Optical power density spectra for 4U1626-67 (from Middleditch et al. 1980) near the nominal X-ray pulsation frequency of 130.26 mHz for runs 5, 9 + 10 and 11 + 12 by Middleditch et al. The peaks due to the direct optical pulsations at the same frequency are all more than ten times higher than the vertical axes and are truncated in the figure. The lower sidelobe at 129.86 mHz is identified with optical pulsations from the binary companion of the X-ray star, apparently produced by reprocessing of pulsed X-rays that strike the star (see text and Middleditch et al. 1980).

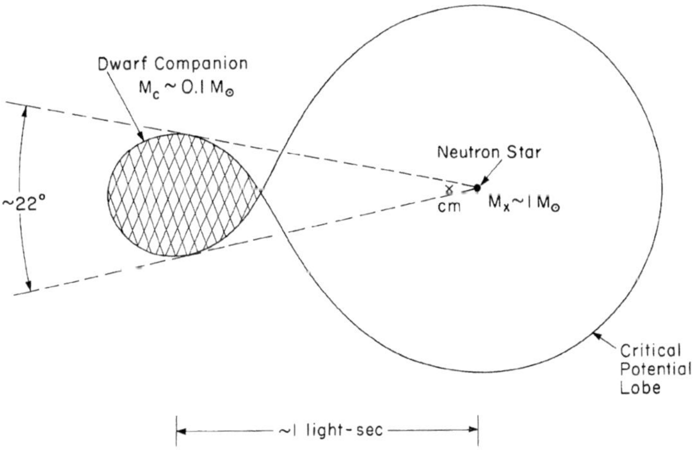

Figure 14: Highly-compact-binary model for 4U1626-67 (adapted from Joss and Rappaport 1979). The figure is drawn to scale for the indicated illustrative values of M_c and M_x.

pulses (Middleditch et al. 1980). If the reprocessing timescale for conversion of X-rays into optical light is short (<< 1 s) or independent of location in the binary system, this result implies that the binary stellar separation is ~1 lt-sec. Other information derived from the optical observations, such as constraints on the masses of the neutron star and companion star and the orbital inclination angle, will be available shortly (Middleditch et al. 1980).

With these observational properties of 4U1626-67, we can present a plausible model for the system that explains many of its observed properties (see Joss, Avni and Rappaport 1978; Li et al. 1980). Consider a main-sequence dwarf with $M_c \simeq 0.1\ M_\odot$ in orbit about a neutron star with $M_x \simeq 1\ M_\odot$ (Figure 14). This type of system would be consistent with the stringent observational limits on $a_x \sin i$ (see Figure 2 of Joss 1980). Such a system may have evolved from a cataclysmic variable, wherein the matter accreting onto the degenerate dwarf eventually drives it over the Chandrasekhar limiting mass, forming a neutron star. This evolutionary scenario has been discussed by Gursky (1976), van den Heuvel (1977), and Joss and Rappaport (1979).

Mass transfer in the binary system could be driven by the decay of the orbit due to gravitational radiation. The rate of extraction of orbital angular momentum can be calculated from the Einstein formula for gravitational quadrupole radiation (see, e.g., Paczyński 1967). Under the assumptions that no mass is lost from the system during mass transfer and that the main-sequence dwarf rotates synchronously and continues to fill its Roche lobe as the orbit decays, the resultant mass transfer rate can then be computed analytically (Paczyński 1967; Faulkner 1971). For a main-sequence dwarf that follows the mass-radius relation $M_c/M_\odot \simeq R_c/R_\odot$, the accretion rate is

$$\dot{M}_x \simeq 7 \times 10^{15}\ (1-\mu)^2\ (4-7\mu)^{-1}\ \mu^{-2/3}\ \mathrm{g\ s}^{-1} \quad , \qquad (9)$$

where $\mu \equiv M_c/(M_x + M_c)$. For the illustrative system parameters sketched in Figure 14, $\mu \simeq 0.09$ and $\dot{M}_x \simeq 10^{16}\ \mathrm{g\ s}^{-1}$, which corresponds to an X-ray luminosity of $\sim 10^{36}$ ergs s^{-1} and is thus marginally consistent with the estimated luminosity of $\sim 7 \times 10^{36}$ ergs s^{-1} (Joss, Avni and Rappaport 1978).

Because the gravitational timescale in this system ($\sim 10^8$ yr) is of the order of or smaller than the Kelvin time of a low-mass main-sequence dwarf ($\sim 10^9$ yr), the assumed mass-radius relation should cease to be appropriate as matter is stripped from the star (Faulkner 1979; Webbink 1979), and hence equation (9) should be substantially modified. It is also possible that the mass transfer is driven or enhanced by a self-excited stellar wind (Faulkner 1974; Basko et al. 1977) and/or by the evolution of the companion (Faulkner 1974).

As an illustrative example, Li et al. (1980) determined the evolution of the system under the assumption that equation (9) continues to be appropriate for small values of M_c. The term $(1-\mu)^2 (4-7\mu)^{-1}$ is nearly constant when $\mu \leq 0.2$ (i.e., when $M_c \leq 0.3\ M_\odot$), so that equation (9) can be integrated trivially to yield

$$\dot{M}_x(t) \simeq \frac{\dot{M}(0)}{(1-t/\tau)^{2/5}} \quad . \tag{10}$$

Here $\dot{M}(0)$ is the mass transfer rate at time $t = 0$, and τ is a time constant given by $\tau = 2 \times 10^{10} [(M_c + M_x)/M_\odot]^{-2/3} [M_c(0)/M_\odot]^{5/3}$ yr. For the illustrative system parameters indicated in Figure 14, $\tau \simeq 5 \times 10^8$ yr. Thus, relatively high mass transfer rates may be achieved as t approaches τ.

If the companion star is a degenerate dwarf rather than a main-sequence dwarf, an even larger mass transfer rate can be driven by the extraction of angular momentum from the orbit due to gravitational radiation. It is uncertain whether a close binary system comprising a neutron star and degenerate dwarf can form. It is possible, however, that the degenerate dwarf could evolve from a main-sequence dwarf after most of its mass has been transferred to the neutron star, nuclear burning has ceased, and collapse ensues. The cessation of steady-state (i.e., main-sequence) hydrogen burning will occur when M_c drops below a critical value of $\sim 0.08\ M_\odot$ (Kumar 1963). The mass-radius relation for chemically homogeneous low-mass degenerate dwarfs is given by $R/R_\odot \simeq 0.013\ (1+X)^{5/3} (M/M_\odot)^{-1/3}$ (see, e.g., Paczyński 1967), where X is the fractional hydrogen content by mass. With this relation, the mass transfer rate can again be computed analytically:

$$\dot{M}_x = 2.8 \times 10^{23}\ (1+X)^{-20/3} \left(\frac{M_x+M_c}{M_\odot}\right)^{16/3} (1-\mu)^2\ (2-5\mu)^{-1}\ \mu^{14/3}\ \mathrm{g\ s^{-1}}. \tag{11}$$

For $\mu \simeq 0.05$ and $X \lesssim 0.1$, the mass transfer rate corresponds to a luminosity of $\sim 2 \times 10^{37}$ ergs s^{-1}.

For small μ, equation (11) can be easily integrated to yield

$$\dot{M}_x(t) \simeq \frac{\dot{M}(0)}{(1+t/\tau)^{14/11}} \quad , \tag{12}$$

where $\tau \simeq 1.2 \times 10^2\ (1+X)^{20/3} [(M_c+M_x)/M_\odot]^{-2/3} [M_c(0)/M_\odot]^{-11/3}$ yr (Li *et al*. 1980). For $M_c = 0.05\ M_\odot$, $M_x = 1\ M_\odot$, and $X = 0$, we obtain $\tau \simeq 10^7$ yr. Here, the evolution is very different from that described by equation (10); the peak mass transfer rate is at $t = 0$ (corresponding to the time of formation of the degenerate dwarf), and the transfer rate decays rather rapidly thereafter. This decay is associated with an *increase* in the orbital separation; the difference between this case and that described by equation (10) results from the inverted mass-radius relation for degenerate dwarfs.

We are presently carrying out detailed numerical calculations of the evolution of highly compact binary X-ray sources. These calculations incorporate the fact that the companion star is not, in general, in thermal equilibrium as the mass transfer takes place. The results of these studies will be published elsewhere.

We note that for a specified mass-radius relation for the companion star and under the assumption that the companion fills its Roche lobe, its mass is uniquely defined by the orbital period (Faulkner 1971). In particular, for a period of ~2500 s (Middleditch et al. 1980), a main-sequence or degenerate dwarf companion would have a mass of ~0.08 or ~0.02 $M_\odot$, respectively (see Fig. 2 of Joss 1980).

C. Observations of X-ray Pulsars with Einstein

Extensive searches for X-ray pulsars have been carried out with a number of satellites, including SAS-3 (Rappaport and Li 1980). For the most part, these searches have been restricted to sources brighter than a few Uhuru flux units (UFU). For example, a 5-hour pointed observation of LMC X-4 with SAS-3 yielded only a rather high upper limit of ~25% on the pulsed fraction (Li, Rappaport and Epstein 1978).

The low background counting rate of the Einstein observatory (Giacconi et al. 1979) makes it an ideal tool to search for new X-ray pulsars. In practice, only relatively short observations (~5000 s) are required for sources brighter than ~0.5 UFU because the detection of only ~ 2×10^3 photons, with little or no background, is sufficient to detect pulsations in sources that are pulsed at the ~10% level. Even very faint objects ($\lesssim$ 0.01 UFU), such as binary X-ray sources in nearby galaxies like M31, which yield several hundred counts in ~10^5 s with the imaging proportional counter (IPC) aboard Einstein, can be subjected to systematic searches for X-ray pulsations. Hence, the brighter X-ray sources in M31 should be detectable as X-ray pulsars if they have pulsed fractions similar to those in our own galaxy (see Fig. 1).

As an illustration of the capability of Einstein to detect very faint X-ray pulsars, we note recently reported evidence for an 8.3 s pulsar in M31 (van Speybroeck and Bechtold 1980). The source, which is located within arc minutes of the center of M31, yielded ~200 counts during a 5×10^4 s exposure with the IPC. In another Einstein observation, Lamb et al. (1980) reported the detection of two X-ray pulsars separated by only ~15 arc min and with pulse periods of ~290 s and ~298 s. Both objects had intensities of the order of ~1 UFU. This result apparently resolves the mystery of two nearly equal X-ray periodicities detected from the same region of the sky, which was first reported on the basis of Ariel V data (White et al. 1978).

Searches for faint X-ray pulsars in data taken over long time intervals (i.e., $\gtrsim 10^4$ s) must take into account the possibility that binary orbital motion may significantly alter the pulse arrival times during the course of the observations. The signature of such an effect would be the development in the power density spectrum of sidelobes of the pulsar frequency shifted by multiples of the orbital frequency. In the case of a faint X-ray pulsar, this division of power would render the pulsations more difficult to detect. However, it is easy to think of methods of data analysis by which such difficulties might be overcome.

Finally, we note that pulsating transient X-ray sources may, in fact, emit X-rays at a low but detectable level during their "off states". It would be very useful to measure such emission to (1) determine the mass transfer rate during quiescent intervals, (2) use Einstein's capabilities to measure precise source locations that were previously unavailable, and (3) keep track of theoretically important variations in pulse periods between outbursts.

IV. CONCLUDING REMARKS

During the past decade, the study of binary X-ray pulsars has provided a great stimulus to many areas of astrophysics. Of particular importance are the measurements of the masses of neutron stars (§II.C) and their relation to nuclear and high-energy physics, gravitation theory, and stellar evolution. Detailed information has been obtained on the companion stars (§II.C), and this may in turn enable a coherent picture of binary stellar evolution in these systems to be developed. Binary X-ray pulsars have also provided a significant new probe into the mass transfer processes in binary stellar systems and an impetus to theoretical investigations of accretion flows, accretion disks, and accretion torques (§II.D). Preliminary observations have also been carried out to utilize the X-ray pulsations as probes of the internal structures of both the emitting neutron stars (§II.B) and their companion stars (§III.A).

At the present time, the data banks of the Ariel 5, OSO-8, SAS-3, and HEAO-1 satellites may well contain significant new data on binary X-ray pulsars. These data will require several years for a complete analysis. We are hopeful that the Hakucho and Ariel 6 observatories can be used to pursue new X-ray pulsar observations over the next few years. The Einstein observatory could be extremely useful in discovering new X-ray pulsars (§III.C), but the observing time required for detailed follow-up studies may be unavailable.

In the future, a dedicated X-ray timing satellite would be able to continue and substantially advance the study of X-ray pulsars and other compact X-ray sources that was carried out so successfully during the 1970's.

Acknowledgements

We thank Ms. T. Welch and Ms. M. Davis for their able assistance in the preparation of the manuscript and figures. We are grateful to Dr. J. Sullivan for his valuable advice concerning the development of the computer programs used in the generation of some of the figures. This work was supported in part by the National Aeronautics and Space Administration under contract NAS5-24441 and grant NSG-7643, in part by the National Science Foundation under grant AST78-21993, and in part by the United States-Israel Binational Science Foundation (BSF). P.C.J. is an Alfred P. Sloan Research Fellow.

Notes

[1] Recent theoretical efforts (see Baym and Pethick 1979 for references) have pointed increasingly to the possibility that neutron-rich matter undergoes a transition to a fluid comprised of free quarks at densities not far in excess of nuclear-matter densities ($\rho \simeq 2.5 \times 10^{14}$g cm^{-3}). If this is the case, then much of the matter within a "neutron" star may actually be composed of free quarks, and such objects might be more accurately called "quark stars". However, many of the observable macroscopic properties of quark stars (masses, radii, moments of inertia, and so forth) need not be different from those of neutron stars (Fechner and Joss 1978). For simplicity, in this paper we shall refer to all such objects as "neutron stars".

[2] Note that our definition of β uses the effective radii of the critical potential lobe and the companion star, and does not refer to distances along the line joining the two stars.

REFERENCES

Arnett, W.D., and Bowers, R.L., 1977. Astrophys. J. Suppl. 33, 415.
Arnett, W.D., and Schramm, D.N., 1973. Astrophys. J. 184, L47.
Avni, Y. 1976. Astrophys. J. 209, 574.
Avni, Y. 1977. "In Highlights of Astronomy", Vol. 4, Part I, p. 137.
Avni, Y. 1978. In "Physics and Astrophysics of Neutron Stars and Black Holes", ed. R. Giacconi and R. Ruffini (Amsterdam: North Holland), p. 43.
Avni, Y., and Bahcall, J.N., 1975a. Astrophys. J. 197, 675.
Avni, Y., and Bahcall, J.N., 1975b. Astrophys. J. Lett. 202, L131.
Avni, Y., and Bahcall, J.N., 1976. In "X-Ray Binaries" (NASA SP-389), p. 615.
Baan, W.A., and Treves, A., 1973. Astron. Astrophys. 22, 421.
Bahcall, J. 1978a. In "Physics and Astrophysics of Neutron Stars and Black Holes", ed. R. Giacconi and R. Ruffini (Amsterdam: North Holland), p. 63.
Bahcall, J. 1978b. Ann. Rev. Astron. Astrophys. 16, 241.
Bahcall, J.N., and Chester, T.J., 1977. Astrophys. J. Lett. 215, L21.
Basko, M.M., and Sunyaev, R.A., 1975. Astron. Astrophys. 42, 311.

Basko, M.M., Hatchett, S., McCray, R. and Sunyaev, R.A., 1977. Astrophys. J. 215, 276.
Baym, G., and Pethick, C., 1979. Ann. Rev. Astron. Astrophys. 17, 415.
Becker, R.H., Swank, J.H., Boldt, E.A., Holt, S.S., Pravdo, S.H., Saba, J.R., and Serlemitsos, P.J., 1977. Astrophys. J. Lett. 216, L11.
Becker, R.H., Rothschild, R.E., Boldt, E.A., Holt, S.S., Pravdo, S.H., Serlemitsos, P.J., and Swank, J.H., 1978. Astrophys. J. 221, 912.
Boynton, P., and Deeter, J., 1979. In "Compact Galactic X-Ray Sources" (ed. F. Lamb and D. Pines), p. 168.
Bradt, H.V., Apparao, K.M.V., Dower, R., Doxsey, R.E., Jernigan, J.G. and Markert, T.H., 1977. Nature 269, 496.
Canal, R., and Schatzman, E., 1976. Astron. Astrophys. 71, 217.
Chester, T.J., 1978. Astrophys. J. 222, 652.
Chevalier, C., and Ilovaisky, S.A., 1977. Astron. Astrophys. 59, L9.
Cominsky, L., Clark, G.W., Li, F., Mayer, W., and Rappaport, S., 1978. Nature 273, 367.
Conti, P.S., 1978, Astron. Astrophys. 63, 225.
Cowley, A.P., Crampton, D., and Hutchings, J.B., 1979. Astrophys. J. 231, 539.
Cowling, T.G., 1938. Mon. Not. R. astron. Soc. 98, 734.
Crampton, D., Hutchings, J.B., and Cowley, A.P., 1978. Astrophys. J. Lett. 225, L63.
Davidson, K., and Ostriker, J.P., 1973. Astrophys. J. 179, 585.
Davies, R.E., Fabian, A.C., and Pringle, J.E., 1979. Mon. Not. R. astron. Soc. 186, 779.
Davis, R. Jr., 1978. In "Proc. Brookhaven Solar Neutrino Conference", Vol. 1, p. 1.
Davison, P.J.N., Watson, M.G., and Pye, J.P., 1977. Mon. Not. R. astron. Soc. 181, 73p.
Dower, R.G., Apparao, K.M., Bradt, H.V., Doxsey, R.E., Jernigan, J.G., and Kulik, J., 1978. Nature 273, 364.
Elsner, R.F., and Lamb, F.K., 1976. Nature 262, 356.
Fabbiano, G., and Schreier, E.J., 1977. Astrophys. J. 214, 235.
Faulkner, J., 1971. Astrophys. J. Lett. 170, L99.
Faulkner, J., 1974, in "IAU Symp. No. 66", ed. R.J. Tayler (Dordrecht: Reidel), p. 155.
Faulkner, J., 1979. Private communication.
Fechner, W.B., and Joss, P.C., 1977. Astrophys. J. Lett. 213, L57.
Fechner, W.B., and Joss, P.C., 1978. Nature 274, 347.
Forman, W., Jones, C., Tananbaum, H., Gursky, H., Kellogg, E., and Giacconi, R., 1973. Astrophys. J. Lett. 182, L103.
Ghosh, P., Lamb, F.K., and Pethick, C.J., 1977. Astrophys. J. 217, 578.
Ghosh, P., and Lamb, F.K., 1978. Astrophys. J. Lett. 223, L83.
Ghosh, P., and Lamb, F.K., 1979. Astrophys. J. 234, 296.
Giacconi, R., 1974. In "Astrophysics and Gravitation, Proceedings of the 16th International Solvay Congress", (Brussels: Universite de Bruxelles), p. 27.

Giacconi, R., Murray, S., Gursky, H., Kellogg, E., Schreier, E., and Tananbaum, H., 1972. Astrophys. J. **178**, 281.
Giacconi, R. et al., 1979. Astrophys. J. **230**, 540.
Gnedin, Yu.N. and Sunyaev, R.A., 1973. Astron. Astrophys. **25**, 233.
Gursky, H. 1976. In "IAU Symp. No. 73", ed. P. Eggleton, S. Mitton, and J. Whelan (Dordrecht: Reidel), p. 19.
Gursky, H., and Schreier, E., 1975. In "Neutron Stars, Black Holes and Binary X-Ray Sources", ed. H. Gursky and R. Ruffini (Dordrecht; Reidel), p. 175.
Hammerschlag-Hensberge, G., Zuiderwijk, E.J., van den Heuvel, E.P.J., and Hensberge, H., 1976. Astron. Astrophys. **49**, 321.
Hutchings, J.B., 1976. In "X-Ray Binaries" (NASA SP-389), p. 531.
Hutchings, J.B., 1980. Private communication.
Hutchings, J.B., Crampton, D., Cowley, A.P., and Osmer, P.S., 1977. Astrophys. J. **217**, 186.
Hutchings, J.B., Crampton, D., and Cowley, A.P., 1978. Astrophys. J. **225**, 548.
Hutchings, J.B., Cowley, A.P., Crampton, D., van Paradijs, J., and White, N.E., 1979. Astrophys. J. **229**, 1079.
Iben, I., 1974. Ann. Rev. Astron. Astrophys. **12**, 215.
Ilovaisky, S.A., Motch, Ch., and Chevalier, C. 1978. Astron. Astrophys. **70**, L19.
Ives, J.C., Sanford, P.W., and Bell Burnell, S.J., 1975. Nature **254**, 578.
Johns, M., Koski, A., Canizares, C., and McClintock, J., 1978. IAU Circ., No. 3171.
Johnston, M., Bradt, H., Doxsey, R., Gursky, H., Schwartz, D., and Schwarz, J., 1978. Astrophys. J. Lett. **223**, L71.
Joss, P.C., 1980. This volume.
Joss, P.C., and Rappaport, S., 1976. Nature **264**, 219.
Joss, P.C., and Rappaport, S., 1979. Astron. Astrophys. **71**, 217.
Joss, P.C., and Rappaport, S., 1980. Rev. Mod. Phys., in preparation.
Joss, P.C., Avni, Y., and Rappaport, S., 1978. Astrophys. J. **221**, 645.
Joss, P.C., Fechner, W.B., Forman, W., and Jones, C., 1978. Astrophys. J. **225**, 994.
Joss, P.C., Li, F., Nelson, J., and Middleditch, J., 1980. Astrophys J. **235**, 592.
Kelley, R., Rappaport, S., and Petre, R., 1980. Astrophys. J., in press.
Kelley, R., Apparao, K., Doxsey, R., Jernigan, G., Naranan, S., and Rappaport, S., 1980a. Submitted to Astrophys. J.
Kelley, R., Rappaport, S., Brodheim, M. and Cominsky, L., 1980b. In preparation.
Kopal, Z., 1965. Adv. Astron. Astrophys. **3**, 89.
Kumar, S.S., 1963. Astrophys. J. **137**, 1121.
Lamb, F.K., Pethick, C.J., and Pines, D., 1973. Astrophys. J. **184**, 271.
Lamb, F.K., Pines, D., and Shaham, J., 1978. Astrophys. J. **224**, 969.
Lamb, F.K., and Boynton, P., 1980. Private communication.
Lamb, R.C., Markert, T.H., Hartman, R.C., Thompson, D.J., and Bignami, G.F., 1980. Astrophys. J., in press.

Landau, L.D. and Lifshitz, E.M., 1962. The Classical Theory of Fields (Oxford: Pergamon), p. 374.
Lecar, M., Wheeler, J.C., and McKee, C.F., 1976. Astrophys. J. 205, 556.
Li, F., Rappaport, S., Clark, G.W., and Jernigan, J.G., 1979. Astrophys. J. 228, 893.
Li, F., Rappaport, S. and Epstein, A., 1978. Nature 271, 37.
Li, F., Joss, P.C., McClintock, J.E., Rappaport, S., and Wright, E., 1980. Astrophys. J. 240, in press.
Manchester, R.N. and Taylor, J.H., 1977. "Pulsars" (San Francisco: Freeman), p. 18.
Mason, K.O., 1977. Mon. Not. R. astron. Soc. 178, 81p.
McClintock, J.E., Canizares, C.R., Bradt, H.V., Doxsey, R.E., Jernigan, J.G., and Hiltner, W.A., 1977. Nature 270, 320.
McClintock, J.E., Canizares, C.R., Li, F.K., and Grindlay, J.E., 1980. Astrophys. J. Lett. 235, L81.
Middleditch, J., and Nelson, J., 1976. Astrophys. J. 208, 567.
Middleditch, J., Mason, K.O., Nelson, J., and White, N., 1980. Preprint.
Nandy, K., and Schmidt, E.G., 1975. Astrophys. J. 198, 119.
Ögelman, H., Beuermann, K.P., Kanbach, G., Mayer-Hasselwander, H.A., Capozzi, D., Fiordilino, E., and Molteni, D., 1977. Astron. Astrophys. 58, 385.
Okuda, T., and Sakashita, S., 1977. Astrophys. Space Sci. 47, 385.
Paczyński, B., 1967. Acta Astronomica 17, 287.
Parkes, G.E., Mason, K.O., Murdin, P.G., and Culhane, J.L., 1980. Mon. Not. R. astron. Soc. 191, 547.
Peterson, B.A., Wallace, P., Elliot, K.H., Hill, R.W., and Manchester, R.N., 1980. Mon. Not. R. astron. Soc. 190, 33P.
Petterson, J.A., 1978. Astrophys. J. 224, 625.
Plavec, M., 1968. Adv. Astron. Astrophys. 6, 201.
Pounds, K.A., Cooke, B.A., Ricketts, M.J., Turner, M.J., and Elvis, M., 1975. Mon. Not. R. astron. Soc. 172, 473.
Pounds, K.A. 1976. Talk presented at the Meeting of the High Energy Astrophysics Division, American Astronomical Society, Cambridge, Massachusetts, January 1976.
Pravdo, S.H., White, N.E., Boldt, E.A., Holt, S.S., Serlemitsos, P.J., Swank, J.H., Szymkowiak, A.E., Tuohy, I., and Garmire, G., 1979. Astrophys. J. 231, 912.
Pravdo, S.H., and Bussard, R.W., 1980. Submitted to Astrophys. J. Lett.
Primini, F., Rappaport, S., Joss, P.C., Clark, G.W., Lewin, W., Li, F., Mayer, W., and McClintock, J., 1976. Astrophys. J. Lett. 210, L71.
Primini, F., Rappaport, S., and Joss, P.C., 1977. Astrophys. J. 217, 543.
Pringle, J.E., and Rees, M.J., 1972. Astron. Astrophys. 21, 1.
Rappaport, S., 1979. In "Proc. NATO Advanced Study Institute on Compact X-Ray Sources", Cape Sounion, Greece, June 1979.
Rappaport, S., Joss, P.C., and McClintock, J.E., 1976. Astrophys. J. Lett. 206, L103.
Rappaport, S., and Joss, P.C., 1977a. Nature 226, 123.

Rappaport, S., and Joss, P.C., 1977b. Nature 266, 683.
Rappaport, S., and Joss, P.C., 1977c. Ann. N.Y. Acad. Sci. 302, 460.
Rappaport, S., Markert, T., Li, F.K., Clark, G.W., Jernigan, J.G.; and McClintock, J.E., 1977. Astrophys. J. Lett. 217, L29.
Rappaport, S., Clark, G.W., Cominsky, L., Joss, P.C., and Li, F.K., 1978. Astrophys. J. Lett. 224, L1.
Rappaport, S., and Li, F.K., 1979. Unpublished SAS-3 data.
Rappaport, S., Joss, P.C., and Stothers, R., 1980. Astrophys. J. 235, 570.
Russell, H.N., 1928. Mon. Not. R. astron. Soc. 88, 641.
Schreier, E.J., Levinson, R., Gursky, H., Kellogg, E., Tananbaum, H., and Giacconi, R., 1972a. Astrophys. J. Lett. 172, L79.
Schreier, E.J., Giacconi, R., Gursky, H., Kellogg, E., and Tananbaum, H., 1972b. Astrophys. J. Lett. 178, L71.
Schreier, E.J., and Fabbiano, G., 1976. "In X-Ray Binaries" (NASA SP-389), p. 197.
Schwarzschild, M., 1958. "Structure and Evolution of the Stars" (Princeton: Princeton University Press).
Shapiro, S.L., and Lightman, A.P., 1976. Astrophys. J. 204, 555.
Sterne, T.E., 1939. Mon. Not. R. astron. Soc. 99, 451.
Stothers, R., 1974. Astrophys. J. 194, 651.
Tananbaum, H., Gursky, H., Kellogg, E.M., Levinson, R., Schreier, E., and Giacconi, R., 1972. Astrophys. J. Lett. 174, L143.
Taylor, J., Fowler, L.A., and McCulloch, P.M., 1979. Nature 277, 437.
Tsuruta, S., 1975. Ann. N.Y. Acad. Sci. 262, 391.
Ulmer, M.P., 1976. Astrophys. J. 204, 548.
van den Heuvel, E.P.J., 1975. Astrophys. J. Lett. 198, L109.
van den Heuvel, E.P.J., 1976. In "Structure and Evolution of Close Binary Systems", ed. P. Eggleton (Dordrecht: Reidel), p. 35.
van den Heuvel, E.P.J., 1977. Ann. N.Y. Acad. Sci. 302, 14.
van Paradijs, J., Zuiderwijk, E.J., Takens, R.J., Hammerschlag-Hensberge, G., van den Heuvel, E.P.J., and de Loore, C., 1977. Astron. Astrophys. Suppl. 30, 195.
van Paradijs, J.A., Hammerschlag-Hensberge, G., and Zuiderwijk, E.J., 1978. Astron. Astrophys. Suppl. 31, 189.
van Speybroeck, L., and Bechtold, J., 1980. Private communication.
Vidal, N.V., 1973. Astrophys. J. Lett. 186, L81.
Wang, Y.-M., 1975. Nature 253, 249.
Watson, M.G., and Griffiths, R.E. , 1977. Mon. Not. R. astron. Soc. 178, 513.
Webbink, R., 1979. Private communication.
White, N.E., 1979. Private communication.
White, N.E., Mason, K.O., Huckle, H.E., Charles, P.A., and Sanford, P.W., 1976. Astrophys. J. Lett. 209, L119.
White, N.E., Mason, K.O., and Sanford, P.W., 1977. Nature 267, 229.
White, N.E., Mason, K.O., and Sanford, P.W., 1978. Mon. Not. R. astron. Soc. 184, 67p.
White, N.E., Parkes, G.E., Sanford, P.W., Mason, K.O., and Murdin, P.G., 1978. Nature 274, 665.
White, N.E., and Pravdo, S.H., 1979. Astrophys. J. Lett. 233, L121.
Yahel, R.Z., 1979. Astrophys. J. Lett. 229, L73.

X-RAY EMISSION FROM NORMAL GALAXIES

L. Van Speybroeck and J. Bechtold
Harvard/Smithsonian Center for Astrophysics

This talk will be restricted to a summary of results obtained with the Einstein Observatory. There are two general categories of normal galaxy investigation being pursued -- detailed studies of nearby galaxies where individual sources can be detected and possibly correlated with galactic morphology, and shorter observations of many more distant objects to determine the total luminosity distribution of normal galaxies. The principal examples of the first type are the CFA study of M31 and the Columbia study of the Large Magellanic Cloud. The Columbia normal galaxy survey is the principal example of the second type, although there also are smaller CFA programs concentrating on early galaxies and peculiar galaxies, and MIT has observed some members of the local group. Studies such as these were not possible before the Einstein Observatory because of limited sensitivity and angular resolution, and even with the Einstein Observatory, there are very few normal galaxies which can be studied in any detail. So, whereas there is no question that X-ray observations provide a useful description of the galactic cluster media, which in turn affects the evolution of galaxies in that environment, the usefulness of X-ray data for studying normal galaxies, as distinct from the study of the X-ray sources themselves, remains to be established.

M31 is the only large galaxy which is close enough for detailed study of sources with intensities comparable to those in our own galaxy. We have obtained three IPC fields which span most of the visible galaxy; these fields are shown superimposed in Figure 1. The exposure times are approximately 3×10^4 seconds, and the typical limiting sensitivity corresponds to about 10^{37} erg s^{-1} in M31. The sensitivity is somewhat poorer than usually achieved with exposures of this duration because the IPC was operated with anomalously high gain during these observations. The appearance of the SW field is different from the other two because there are no adequate guide stars to enable stable pointing in this direction, and so the vehicle was allowed to drift between ground commanded repositionings and the data later were corrected for the instantaneous pointing direction. This also results in a somewhat lower sensitivity in the SW field. In

R. Giacconi (ed.), X-Ray Astronomy with the Einstein Satellite, 153–172.

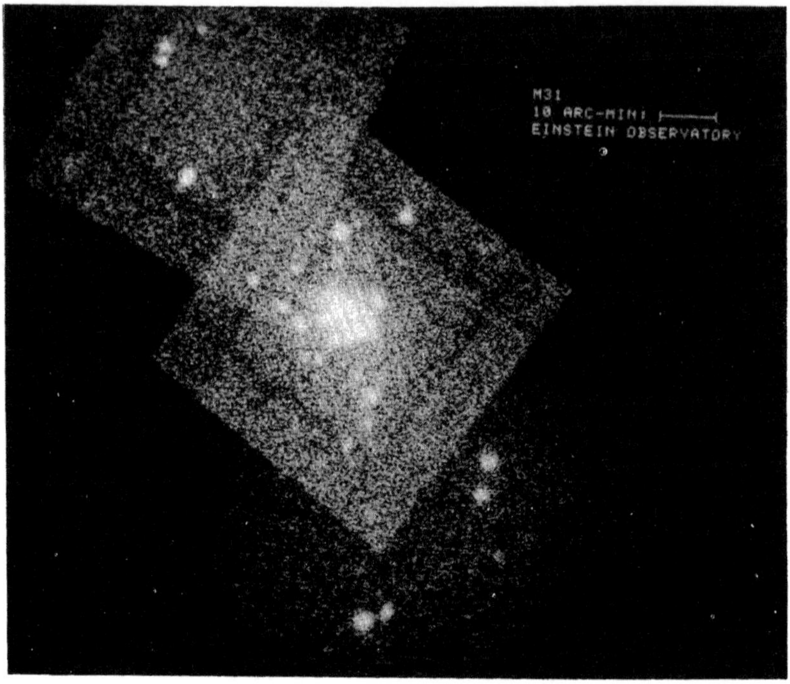

Figure 1. Composite of 3 IPC observations of M31. North is at the top, East is at the right in this and all subsequent illustrations.

exposures of this sensitivity, we expect to obtain of order one serendipitous source--typically a star or QSO--but a small number of such background objects won't affect the principal conclusions to be inferred from the data. Similarly, the figures were prepared while data were being reduced, and there are small discrepancies present as some source identifications were made after the figures were prepared, but there are no important conclusions affected by this.

The overexposed region near the center of the figure is unresolved with the IPC, but contains numerous strong sources which are associated with the bulge of M31 and are resolved with the HRI. The remaining sources are distributed over the disk of the galaxy. The approximate range of luminosities is seen in the NE field. The faintest of the three sources at the extreme northeast has a luminosity (in this exposure) of 1.2×10^{37} erg s^{-1}, which is near the sensitivity limit. The bright isolated source in the NE field has a luminosity of 2.4×10^{38} erg s^{-1} and is one of the brightest of the objects detected. All luminosities in this paper refer to the 0.5 to 4.5 keV band unless otherwise stated. Typical uncertainties are ±50% due to systematic effects, but relative values are much less uncertain.

The source positions are overlaid upon a visible light photograph of the galaxy in Figure 2. These positions include some HRI detections

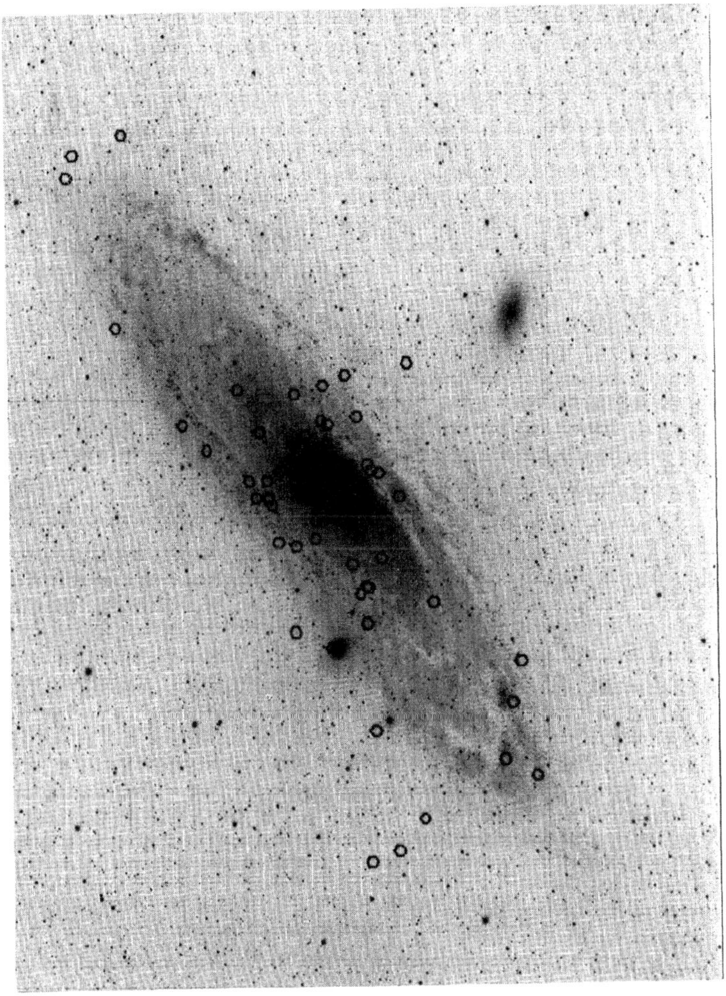

Figure 2. X-ray source locations superimposed upon the visible image of M31.

which are not apparent in Figure 1. There is a large group of sources associated with the bulge and a general association with the spiral arms of the galaxy. There also are two sources which may be associated with M32 and a few sources which do not seem to be associated with the galaxy. The apparently isolated source to the NW, approximately half way between M31 and NGC205, actually is a globular cluster of M31. The sources to the extreme SW are not located sufficiently accurately to secure identifications, but one also may be associated with an M31 globular cluster. The three sources to the extreme NE seem to be associated with a rather faint portion of the disk. The comparison of X-ray source positions and the H_I map of EMERSON (1976) in Figure 3 is more illustrative; here, the three NE sources are seen to be associated with a pronounced H_I feature in that region, and there is a general correlation between the non-bulge X-ray sources and the H_I. This is not surprising because we expect X-ray sources to be associated with young, massive stars, and in M31 the indicators of Population I activity, such as OB associations, H_I, H_{II}, CO, and dust are centered

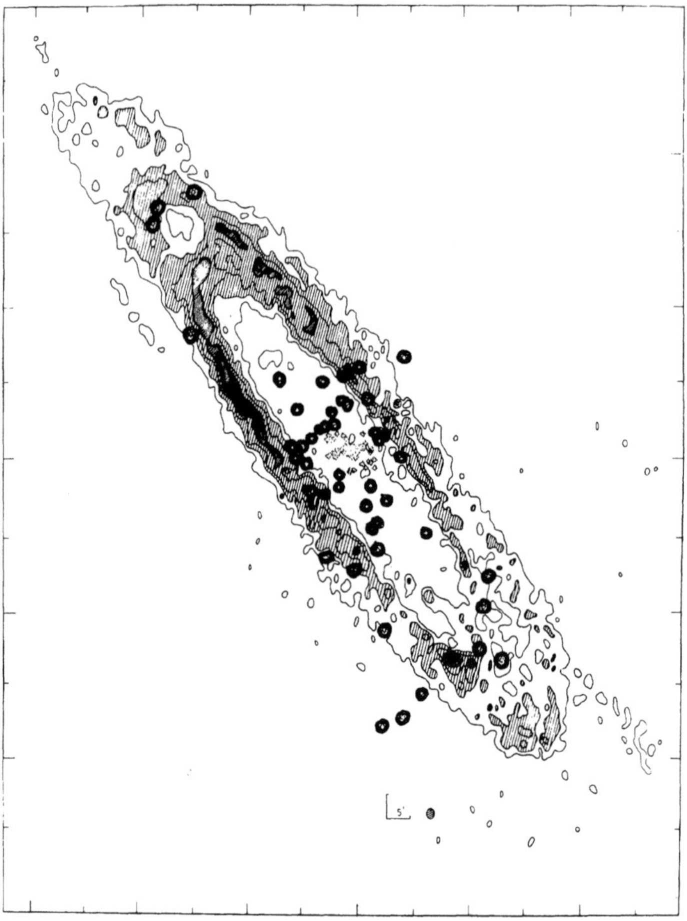

Figure 3. HRI exposure of the center of M31. The nucleus is the bright source approximately 60 arcsec east of the brightest source in the field.

in an arm about 9 kpc from the nucleus (EMERSON 1974, 1976, 1978; Baade and Arp 1964, Berkhuijsen 1977, Combes et al. 1976a,b).

We are beginning to identify individual objects in M31. There are, for example, two possible coincidences with the supernova remnant list of D'Odorico et al. (1979); these have X-ray luminosities of about 2×10^{37} ergs s^{-1} and will be studied further at a later date. The supernova of 1885 is not detected at the 10^{37} erg s^{-1} level.

The largest number of source identifications are globular clusters. We have detected 17 of the M31 globular clusters listed by Sargent et al. (1977) or Battistini et al. (1979) with the HRI; the position errors in these cases are small enough to be confident of almost all identifications. We also have possible coincidences with 6 additional IPC detections, but in this case, the position errors would result in approximately three spurious associations. The 17 HRI detections are listed in Table 1. The number of detections is about that expected; there are 225 objects on the Sargent list in our fields, and we detect 14; for comparison, there are 9 of 130 detected in our own galaxy, or about the same fraction. The globular clusters detected in M31, however, are more luminous in X-rays; NGC6624 with $L_x = 3.5 \times 10^{37}$ erg s^{-1} is the brightest globular cluster in X-rays in our galaxy, whereas the mean of the 17 globular clusters detected in M31 is 5.2×10^{37} erg s^{-1}. In fact, only one of the globular clusters in our galaxy would have been detected in M31, whereas, if the populations are the same, we would expect ten. We thus are left with two possibilities (or combinations thereof):

1. The globular clusters in M31 are brighter X-ray sources than those in our galaxy; or

2. The probability of a globular cluster being an X-ray source is significantly higher in M31 than in our galaxy, and we are observing the high tail of the X-ray luminosity distribution in M31.

The spatial distribution of X-ray globular clusters in our galaxy is more concentrated towards the disk than the total population; the average Z for the two sets being 2.2 and 4.5 kpc respectively. We have noted a similar trend in M31. The 225 globular clusters of Sargent et al. (1977) which fall in our observation fields can be divided into three groups as in Table 2. The three categories are relatively rich in objects located within the bulge, disk, and halo respectively, the OB associations of Baade and Arp (1964) being used as tracers of the disk population. The probability of a globular cluster containing a strong X-ray source is significantly larger if the globular cluster is located within the denser regions of the galaxy. We do not have a satisfactory explanation for this effect. Two possibilities which have been suggested, assuming that the X-ray source consists of a collapsed object in a binary system, are tightening of the binary orbit due to collisions with objects outside the cluster, and exchange collisions

Table 1. M31 X-ray Globular Clusters

Globular Cluster Sargent et al. 1977	Globular Cluster Battistini et al. 1979	Position Error (arc seconds)	X-ray Luminosity ($\times 10^{37}$)
198	143	4.9	1.7
200	148	4.1	2.3
	144	3.2	2.2
	146	1.8	2.0
	153**	2.6	4.5
148	86	4.2	3.1
213	158	3.0	3.2
144	82	4.5	2.9
52	5	4.6	9.4
322		5.3	4.4
307		3.0	26.4
280	225	4.4	5.6
235	185	5.0	1.5
108	45	3.3	5.7
192	135	6.1	10.7
199	147	6.0	1.0
264	213	7.5	1.9

(Less than one chance coincidence is expected based on a 10" error radius.)

**Globular cluster identification supported by spectroscopic data of Liller, Forman and Jones.

between putative but unobserved globular cluster binaries and single external objects (to provide either the collapsed object or a new source of gas).

Table 2

Correlation, M31 Globular Clusters and OB Associations[1]

Globular Cluster Group	Globular Clusters[2]	X-ray Globular Clusters	Fraction
1. R 10' from nucleus	42	4	.095
2. R 10' from nucleus and more than 5 OB associations within 5'	116	9	.078
3. R 10' from nucleus and 5 or fewer OB associations within 5'	67	1	.015

Note: Probability of obtaining 0 or 1 of 14 objects in (67/225) of the sample is .049.

[1]Baade and Arp 1964.

[2]Sargent et al. 1977.

We have observed the central region of M31 with the HRI (∿3" resolution) on three occasions; the first of these is shown in Figure 4. The central region of the galaxy is very dense indeed; there are 19 resolved sources plus 10 diffuse or confused regions with $L_x > 10^{37}$ within two arc minutes of the nucleus of M31. The sources near the center of the galaxy also appear to be elongated along the EW axis. The positions of these inner sources are shown superimposed upon an optical photograph in Figure 5. The density in the inner bulge region is quite apparent.

The luminosity distribution of the X-ray sources in M31 is shown in Figure 6. The mean luminosity for all detected sources is 3.6×10^{37} erg s^{-1}; the lower cutoff corresponds to our detection threshold and is not real, although the contribution of sources below our threshold must be relatively small because the sum of the individual sources we have detected is comparable to previous measurements of the total X-ray luminosity of M31. Luminosity distributions by source category also are shown in Figure 6. The globular cluster categorization is based

Figure 4. HRI exposure of the center of M31. The nucleus is the bright source approximately 60 arc sec east of the brightest source in the field.

Figure 5. X-ray source locations near the center of M31.

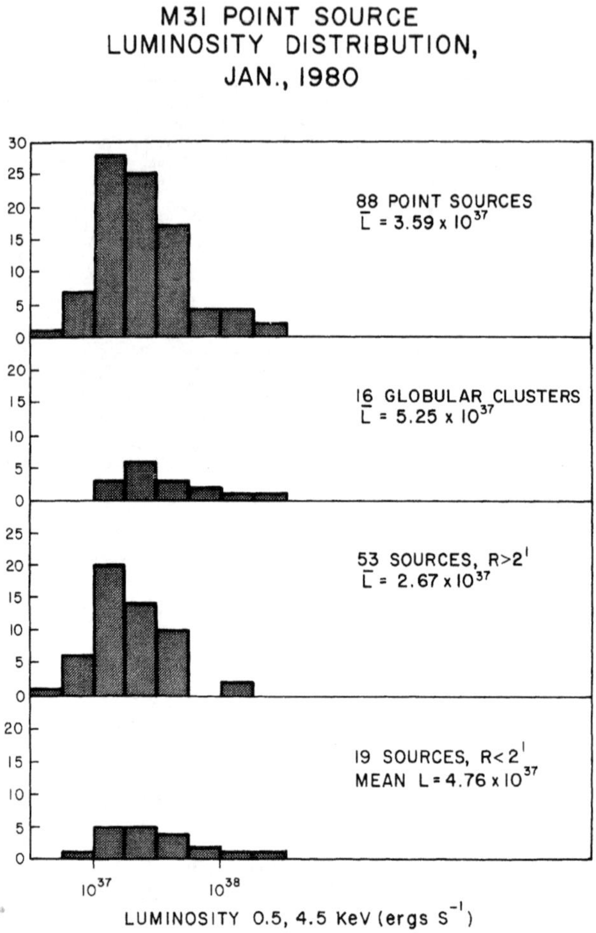

Figure 6. M31 X-ray source luminosity distributions.

upon optical identifications. The two remaining categories are an attempt to separate disk and bulge sources, for, in our own galaxy, the bulge sources are more luminous and softer than the disk sources and presumably are a different population (Jones 1977, Markert et al. 1977, Clark et al. 1978). We originally attempted to base this separation upon optical morphology--calling sources associated with spiral arm features, dust lanes, etc. Pop. I, but in practice our division was one of radius, and it seems better to use this more objective criterion. We find that, similar to our own galaxy, the inner bulge sources in M31 are brighter than the outer sources, as indicated in Figure 6. These inner bulge X-ray sources are more concentrated than the optical bulge of M31--Morton, Anderech, and Bernard (1977) quote an optical bulge radius of approximately 5 arc minutes. The 16 X-ray sources between 2' and 5' from the bulge, however, have a mean luminosity of 2.2×10^{37} erg s^{-1}, or more similar to the outer sources (mean $L_x = 2.7 \times 10^{37}$) than to the inner bulge group (mean $L_X = 4.8 \times 10^{37}$). The inner sources thus appear to comprise a different population from the outer disk group and have X-ray luminosities comparable to the M31 globular cluster sources. The extent and location of this group of sources is similar to the 400 pc region in which Rubin and Ford (1971) find evidence of gas expansion velocities of order 100 km s^{-1} superimposed upon a rapid rotation. This inner region thus may constitute a distinct structure within M31.

The sources in this inner region appear to lie approximately along an EW axis, but it is not certain if the sources actually form a dense, quasi spherical population or, if instead, the sources are located in the plane of the disk. If they are a spherical population, then the source density is much greater than near the nucleus of our own galaxy. A 2 arc minute radius in M31 corresponds to approximately 2.3° at the center of our galaxy; there is only one persistent source (and three known transients) within 2.3° of the nucleus of our own galaxy which could have been detected in M31, whereas there are 19 point sources and 10 diffuse or confused regions within 2' of the nucleus of M31. If the M31 sources form a spherical population, then they constitute a structure very different from any found in our own galaxy.

Alternatively, the X-ray sources near the center of M31 may not be a spherical population, but instead may lie largely within the plane of the galaxy. This geometry is suggested by the bulge sources in our own galaxy, which certainly lie in the plane and do not form a spherical population. If we assume this in M31, then we can deproject the X-ray source positions to obtain the appearance of the galaxy as it would be seen if it were viewed face on. This is done in Figure 7, in which the Baade and Arp OB associations and the X-ray source locations are deprojected. The association of the X-ray sources with the spiral arms is apparent. The inner group of X-ray sources now appear as a bar in the galaxy. This does not prove that such a structure exists, but if it is assumed, then M31 and our own galaxy become very similar in their X-ray characteristics--the principal difference being a larger number of X-ray sources in M31 corresponding approximately to the ratio of masses of the two galaxies. In this view, the "bulge" sources would delineate bars in both galaxies.

The Columbia group has performed a detailed study of the Large Magellanic Cloud. The LMC has approximately 1/10 the mass of M31, but is a much younger and generally more active galaxy. Its smaller distance also permits less luminous objects to be detected. The Columbia group has detected at least 130 sources in the LMC, and possibly as many as 190 depending upon evaluations of the expected distribution of statistical fluctuations which are still in process. The luminosity

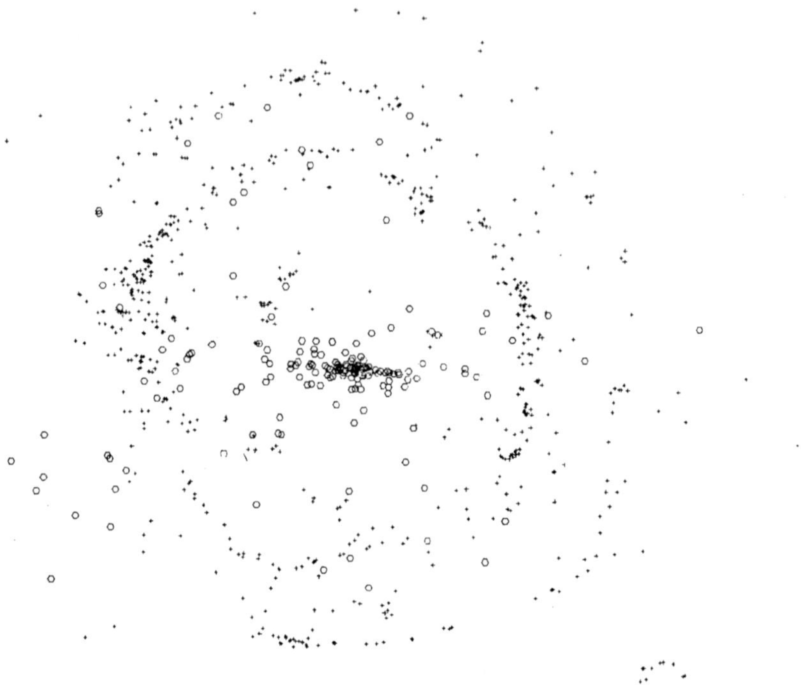

Figure 7. M31 X-ray source locations (larger symbols) and the Baade and Arp (1964) OB associations as they would appear if the galaxy were viewed face on, assuming that both populations exist only in the plane of the galaxy.

distribution of these sources is shown in Figure 8. The source list includes 25 supernova remnants which will be discussed by David Helfand later in this conference. They find ages ranging from 1,000 to 20,000 years and initial energies between 0.8 x 10^{50} and 3.4 x 10^{50} ergs.

Helfand and Long (1979) have used the measured luminosity, size, and spectra of the SNR to study the LMC interstellar media. They find densities between .05 and 10 cm^{-3}, with densities highly correlated with SNR diameters. Simple blast waves propagating into a homogeneous media are not consistent with the data; some higher density clouds in the interstellar median are required. The number of bright SNR in the LMC seems larger than would be expected from our own galaxy; note that there are four with $L_x > 10^{37}$ erg s^{-1}. This, however, is partly an

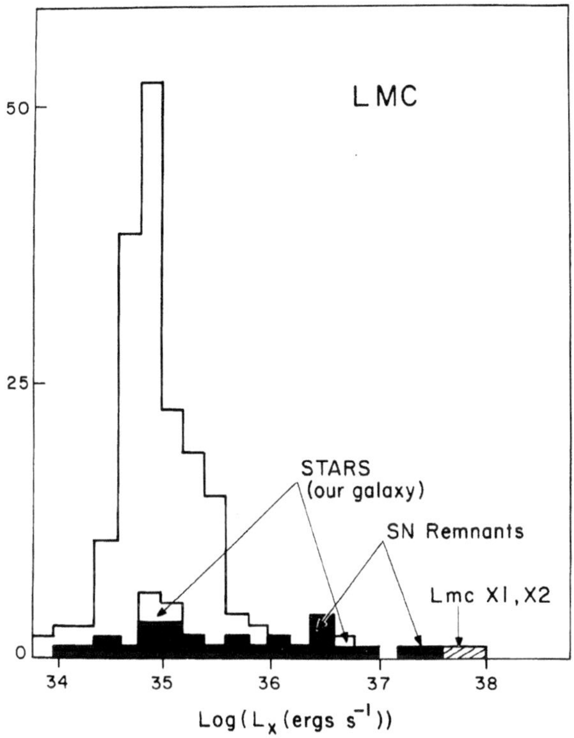

Figure 8.
LMC X-ray source luminosity distribution (from K. Long 1980).

observational effect due to the small absorption in the direction of the LMC; we can detect these very soft sources only within a small portion of our own galaxy. The other surprising result of the LMC survey is the large number of sources with luminosities of order 10^{35} erg s^{-1}; there is no equivalent population known in our own galaxy. The locations of the X-ray sources in the LMC are shown in Figure 9. There is a large concentration in the very active 30 Doradus region, and a general association with emission nebulae and the rather weak spiral arm. The bar of the LMC is not densely populated with X-ray sources--perhaps arguing against our interpretation of M31, or perhaps indicating that the M31 type of sources have not yet evolved to the X-ray emitting phase in the younger LMC.

A number of other nearly normal galaxies have been observed and are being analyzed. Fred Seward and Melanie Mitchell and also various guest observers are performing a similar study of the Small Magellanic Cloud; so far, 26 sources have been detected, including six SNR. These numbers are reasonably consistent with the mass ratios of the two Magellanic Clouds. The SMC sources are in regions of HI enhancement, but are not well correlated with the peak of the emission. The Columbia group has located four sources in M101, which seems to be an under-

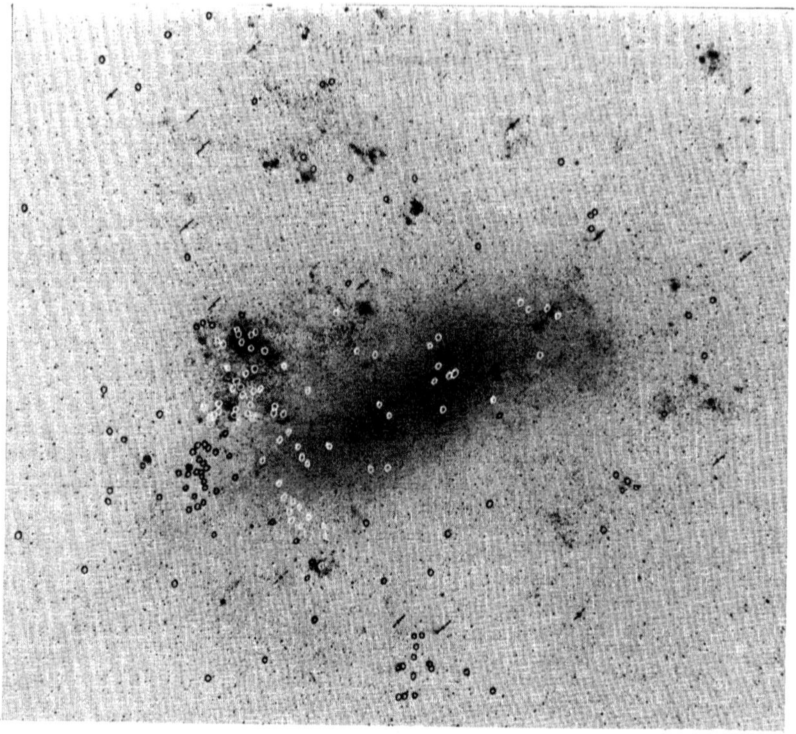

Figure 9. LMC X-ray source positions (from K. Long, 1980).

luminous galaxy in X-rays. Columbia and MIT are studying M33 jointly, and have found at least nine sources in this galaxy. We at CFA have observed M83 and find approximately ten sources. The central region of M83 has a total X-ray luminosity of about 2.5 x 10^{39} erg s^{-1} in the Einstein band--or about twice that of the bulge region of M31. The X-ray sources are located primarily in the spiral arms of these three galaxies; the nucleus of M33 is a strong source which will be discussed below.

There are few normal galaxies which are near enough to be studied in the same detail as M31 or the Magellanic Clouds, but the total X-ray emission of more distant objects can be detected and compared with other properties, such as type of galaxy, optical luminosity, etc. Long and Ku (1979) of the Columbia group are conducting such a program; they have observed 43 galaxies and detected 31. They find that most galaxies emit 10^{-3} to 10^{-4} of their total luminosity in X-rays, although a few fall significantly below these limits. Their measured X-ray fluxes and detection limits are compared to apparent visual magnitude in

Figure 10, and the correspondence between X-ray and visible luminosities for the X-ray detected galaxies is shown in Figure 11. The scatter in these plots is quite small--M101 being the only object to fall significantly below $L_x/L_{opt} = 10^{-4}$. The brightest galaxies observed are ellipticals in the Virgo cluster, and perhaps have a slightly larger value of L_x/L_{opt} than the mean observed value. It is remarkable, however, that the same range of L_x/L_v describes both small spirals and irregular galaxies and the large ellipticals, even though the stellar contents of these extremes are quite different. Fabbiano et al. (1979) have performed a similar study of galaxies selected for "peculiar" characteristics, thought to be interacting, etc.; these galaxies appear to have slightly higher ratios of L_x/L_{opt} than the more normal galaxies in the Columbia survey.

M82 and NGC253 are two nearby galaxies which may be classified as moderately active, with evidence of significant recent star formation. Rieke et al. (1980) have reviewed the extensive radio, infrared, and optical observations of these galaxies. The central regions of both galaxies emit approximately 3 x 10^{10} $L_\odot$ in the far infrared; this is

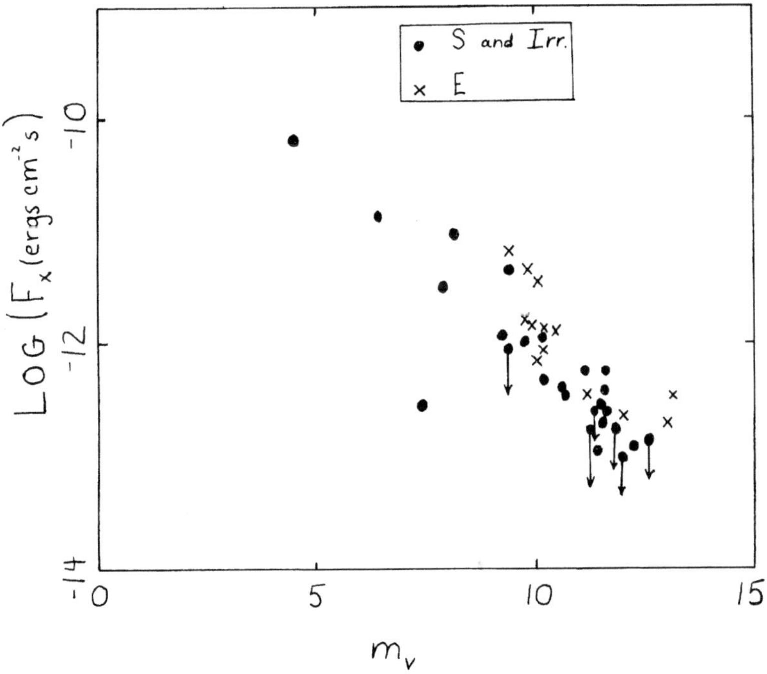

Figure 10. Normal galaxy X-ray flux vs. apparent visual magnitude (from K. Long 1980).

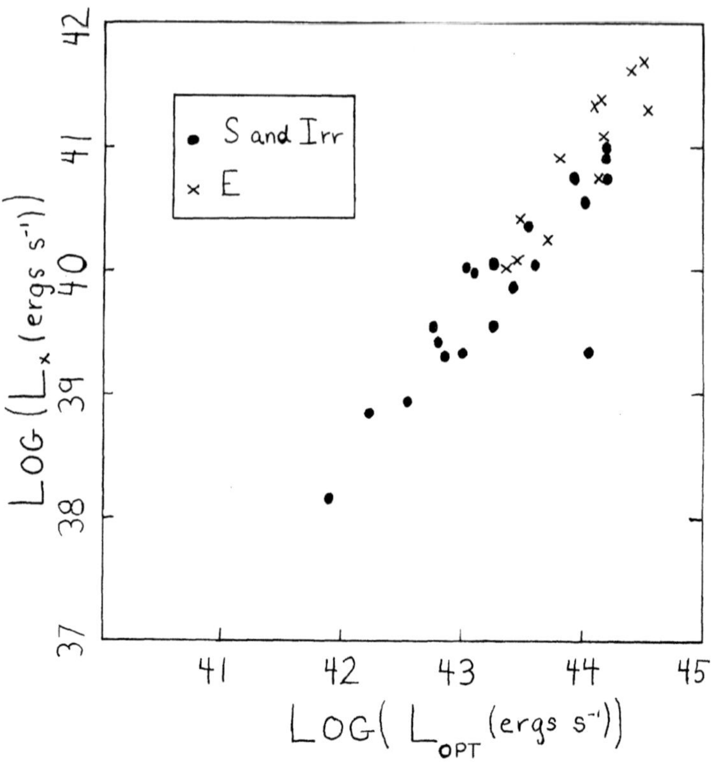

Figure 11. X-ray vs. visual luminosity for normal galaxies (from K. Long 1980).

thought to be radiation from dust heated by numerous young stars. Both galaxies have non-thermal radio sources, that in M82 being of order 20" x 35" in size, and that in N253 having a characteristic size of order 20 arc seconds. M82 also contains a 0.0015 arc second non-thermal nucleus (Geldzahler et al. 1977). Ulrich (1978) has found evidence of ionized gas flow to the SE of the nucleus of N253, and Gottesman and Weliachew (1977) have found that M82 is connected to M81 (36' to the south) by a tenuous gas envelope and is tidally interacting with that galaxy. M82 was detected in X-rays with the UHURU, Ariel, and HEAO-A instruments; Griffiths et al. (1979) obtained $L_x = 2.3 \pm 0.6 \times 10^4$ erg s^{-1} in the 2.6 - 5.4 keV band; they did not detect the source in the 0.9 - 2.6 keV band, or at higher energies. NGC253 had not been detected in X-rays prior to the Einstein observations. CFA and the University of Leicester are studying the X-ray emission from M82, and CFA and Columbia are studying NGC253. Contours of the X-ray data for these two galaxies are shown superimposed upon optical photographs in Figures 12 and 13. Both of these galaxies include significant diffuse

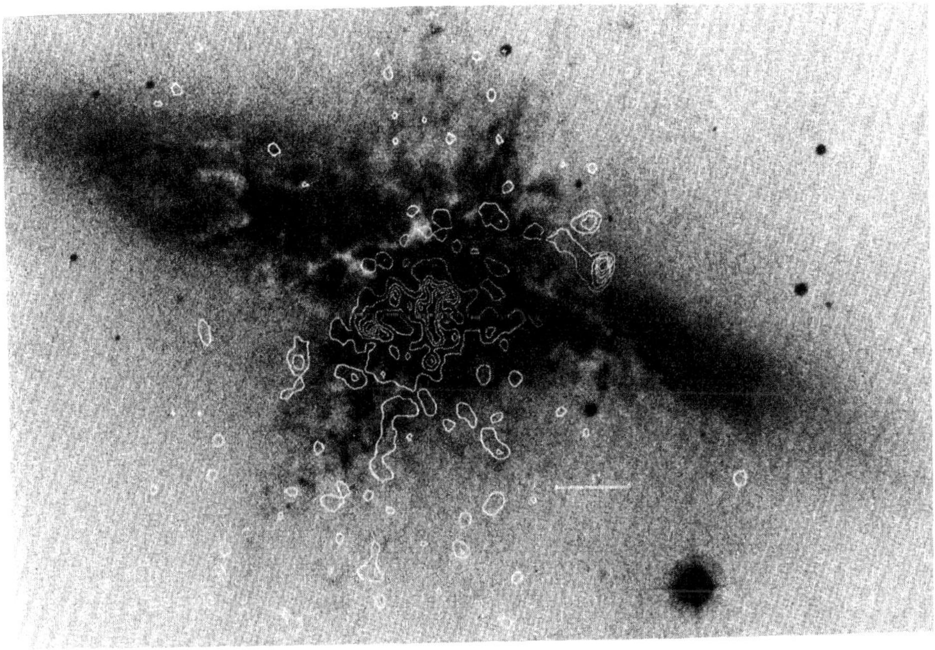

Figure 12. X-ray contour of M82 superimposed upon an H_α image.

emission in addition to point sources. This diffuse component is about 8×10^{39} erg s^{-1} (0.5 - 4.5 keV) in M82, and approximately 4×10^{39} erg s^{-1} in NGC253; in both cases, more than the total emission from M31 and far more than the possible diffuse emission from that galaxy. The M82 X-ray emission is offset towards the south, or in the direction of M81. The NGC253 X-ray emission is in approximately the same region in which Ulrich (1978) found emission from ionized gas. The interpretation of this diffuse X-ray emission is not complete; Rieke et al. (1980) have shown that the X-ray emission could be produced by the expected number of SNR in these galaxies, but inverse Compton interactions between the electrons producing the non-thermal radio emission and the IR photons also may be a viable model.

The nuclei of normal galaxies can be luminous in X-rays. We have observed the galactic center with the IPC and find a 5×10^{35} erg s^{-1} source consistent with the galactic nucleus, but this is a crowded region and a weak source such as this needn't be associated with the galactic nucleus at all, and in any event, is not a significant part of its emission. The case of M31 is more interesting; we have observed the center of M31 with the HRI on three occasions separated by approximately 6 month intervals. The first and third data sets include a point

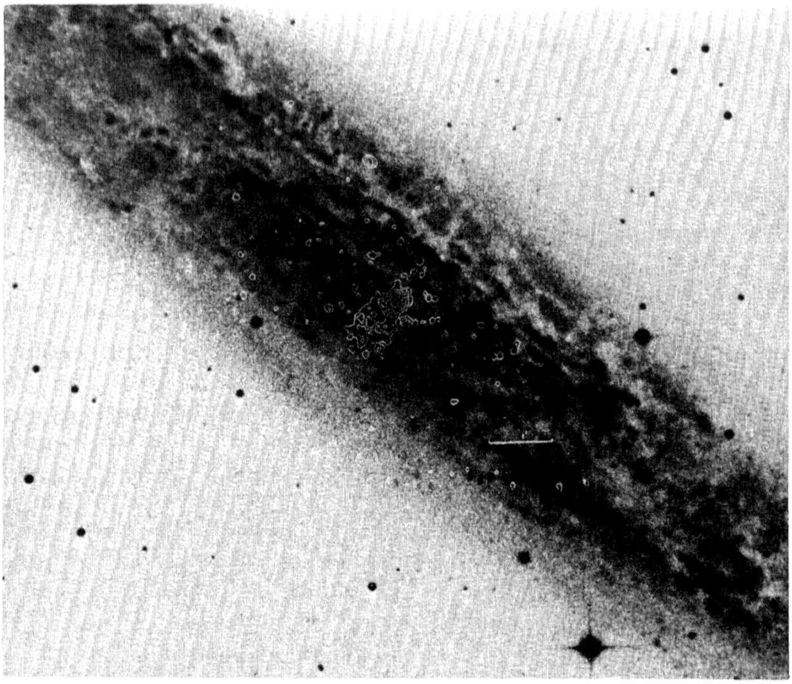

Figure 13. X-ray contours of the center of NGC253 superimposed upon an optical image.

source within our measurement errors of the nucleus; the source density is such that the probability of a chance coincidence between an ordinary X-ray source and the nucleus is about 0.1. The luminosity observed was 10^{38} erg s^{-1} during the first observation, 5 x 10^{37} erg s^{-1} during the third (a year later), and less than 10^{37} erg s^{-1} during the second observation. The first and second observations are shown in Figure 14, which also provides other examples of source variability. The nucleus of M81 appears to be an X-ray source with L_X = 2 x 10^{39} erg s^{-1} and Palumbo et al. (1980) find that the nucleus of M100 has an X-ray luminosity of 1.5 x 10^{40} erg s^{-1}. All of these, however, are major systems in which the possibility of a chance association is not negligible. Perhaps the most certain identification is the nucleus of M33, which is being studied jointly by MIT and Columbia. Markert et al. (1979) found the luminosity of the M33 nucleus to be approximately 10^{39} erg s^{-1} on one occasion and about 6 x 10^{38} erg s^{-1} six months later. M33 is much smaller than M31, M81, or M100, does not have a prominent bulge, and the probability of a chance coincidence between an ordinary X-ray source and the nucleus of this system is correspondingly less. Gallagher (1979) has studied the nucleus of M33 optically and finds it to be stellar with $M_V \propto -10$. The spectrum is like that of cool stars, or a composite population; there are no emission lines. The velocity

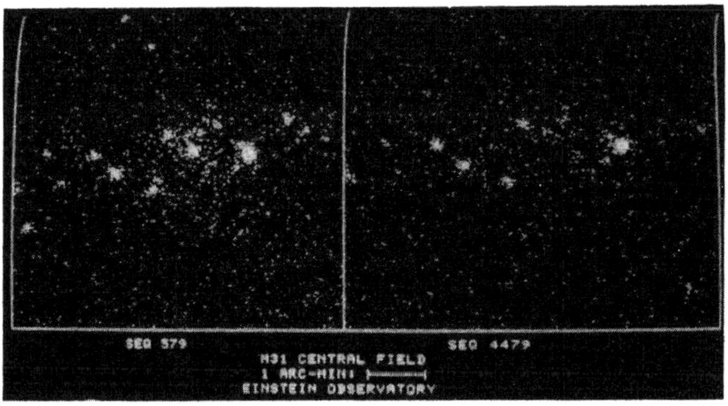

Figure 14. Two HRI exposures taken 6 months apart showing the variability of a source coincident with the nucleus of M31. The nucleus is approximately 1 arc minute east of the brightest source in the left exposure. There also are other examples of source variability present.

dispersion is less than 30 km s^{-1}, and so no extremely heavy object is present, and yet this is the site of an extremely strong X-ray source.

We thus have several examples of the nuclei of ordinary galaxies being variable X-ray sources with $L_X \sim 10^{39}$ erg s^{-1}, and it is likely that this is a common phenomenon; there may well be a continuity of objects typified by the nuclei of normal galaxies at one extreme, and Seyferts and QSOs at the other.

Acknowledgements: We have benefitted greatly from conversations with J. Gallagher, L. Smarr, C. Jones, and W. Forman in the preparation of this material.

REFERENCES

Baade, W. and Arp. H. 1964, Ap.J., 139, 1027.
Berkhuijsen, E.M. 1977, Astr. Ap., 57, 9.
Clark, G., Doxsey, R., Li, F., Jernigan, F.G., and van Paradijs, J. 1978, Ap.J. (Letters), 221, L37.
Coombes, F., Encrenaz, P.J., Lucas, R., and Weliachew, L. 1977a, Astr. Ap., 55, 311.
Coombes, F., Encrenaz, P.J., Lucas, R., and Weliachew, L. 1977b, Astr. Ap., 61, L7.
D'Odorico, S., Dopita, M.A., and Benvenuti, P. 1979, Astr. Astrophys. Supp., 40, 67.
Emerson, D.T. 1974, MNRAS, 169, 607.
Emerson, D.T. 1976, MNRAS, 176, 321.

Emerson, D.T. 1978, Astr. Ap., 63, L29.
Fabbiano, G., Feigelson, E., Elvis, M., Griffiths, R.E., and Schreier, E.J. 1979, Bull. Amer. Astron. Soc., 11, 796.
Geldzahler, B.J., Kellermann, K.I., Shaffer, D.B., and Clark, B.G. 1977, Ap.J. (Letters), 215, L5
Gottesman, S.J. and Weliachew, L. 1977, Ap.J., 211, 47.
Griffiths, R.E., Doxsey, R.E., Johnston, D.A., Schwartz, D.A., Schwarz, J., and Blades, J.C. 1979, Ap.J. (Letters), 230, L21.
Helfand, D.J. and Long, K.S. 1979, Bull. Amer. Astron. Soc., 11, 796.
Jones, C. 1977, Ap.J., 214, 856.
Kronberg, P.P. and Wilkinson, P.N. 1975, Ap.J., 200, 430.
Long, K.S. 1980, presented at 156th Meeting of the American Astronomical Society.
Long, K.S. and Ku, W.H.M. 1979, Bull. Amer. Astron. Soc., 11, 797.
Markert, T.H., Canizares, C.R., Clark, G.W., Hearn, D.R., Li, F.K., Sproti, F., and Winkler, P.F. 1977, Ap.J., 218, 801.
Markert, T.H., Kriss, G.A., Canizares, C.R., McClintock, J.E. and Winkler, P.F. 1979, Bull. Amer. Astro. Soc., 11, 633.
Palumbo, G.G.C., Maccacaro, T., Zamorani, G., Panagia, N. and Vettolani, G. 1980, submitted to Ap.J.
Rieke, G.H., Lebofsky, M.J., Thompson, R.I., Low, F.J., and Tokunaga, A.T. 1980, Ap.J., 238, 24.
Rubin, V. and Ford, W.K., Jr. 1971, Ap.J., 170, 25.
Ulrich, Marie-Helene. 1978, Ap.J., 219, 424.

SPECTROSCOPY OF COMPACT EXTRAGALACTIC X-RAY SOURCES

Stephen S. Holt
Laboratory for High Energy Astrophysics
Goddard Space Flight Center
Greenbelt, MD 20771

Abstract

The X-ray spectra of compact extragalactic sources obtained from the HEAO-1 A-2 experiment and the Solid-State Spectrometer onboard HEAO-2 (the Einstein Observatory) are reviewed. Seyfert spectra are remarkably consistent, with characteristic power-law spectra of energy index α=0.7+.1 over a dynamic range of almost 100 in both luminosity for the whole sample, and energy for individual members. Radio-quiet quasars have similar spectra, perhaps slightly steeper, for the limited sample available. New solid-state spectrometer results for NGC 4151 yield a consistent picture for the geometry of the broad-line clouds in both these related radio-quiet classes of galactic nuclei. Radio-loud objects, especially BL Lacs, are considerably more variable in spectrum as well as in luminosity. Direct synchrotron and synchrotron-self-Compton components are consistent with what we observe from these objects. Finally, the role of spectroscopy in addressing the extent to which compact extragalactic nuclei might contribute to the diffuse X-ray background is discussed.

The Einstein Observatory has been an important step forward in our ability to observe the whole catalog of celestial X-ray-emitting objects. Its imaging experiments are capable of interrogating the luminosity of individual sources in the .5-5 keV band with a sensitivity almost 1000 times better than that afforded by previous instrumentation, so that the sample of compact extragalactic sources which has become available for study has increased dramatically.

Spectroscopically, our capabilities have increased significantly although not quite as measurably as has the raw sensitivity of the imaging detectors. The resolving power of the dispersive Einstein spectrometers, the Focal Plane Crystal Spectrometer (FPCS) and Objective Grating Spectrometer (OGS), is largely negated by their inherently low efficiency for sources which

R. Giacconi (ed.), X-Ray Astronomy with the Einstein Satellite, 173–185.

give photon fluxes as low as do the compact extragalactic sources; these techniques must exhibit their power in the interrogation of much brighter sources contained within our galaxy. The Solid-State Spectrometer (SSS), a non-dispersive device, has the capability to do useful spectroscopy down to a level of less than 10^{-4} of the flux from the Crab nebula, with a resolving power of about 20 at energies below the Einstein telescope cutoff at 5 keV. This nicely complements the proportional counter spectra available from the HEAO-1 A-2 experiment in the energy range 2-60 keV, which have a typical resolving power about a factor of 3 worse (the resolving powers of both instruments are energy dependent: in the region of overlap, the resolving power of the SSS is about 5 times better).

The large fraction of this review will be devoted to the spectra of compact extragalactic sources obtained with both of these instruments, the A-2 and the SSS, and the consistencies which the increased dynamic range of the two, together, has allowed us to infer. I shall discuss only the most luminous classes of galactic nuclei, i.e., Seyferts, BL Lacs and quasars, and concentrate on the general spectral properties of the classes rather than the peculiarities of individual objects. Finally, I shall address the spectroscopic evidence which relates to the role that active galactic nuclei might play in explaining the diffuse X-ray background (XRB).

Seyfert galaxies are distinguished by their broad optical emission lines (of order 10^4 km/sec). The lack of broad forbidden lines indicates that the lines arise in clouds or filaments with densities in excess of 10^8 cm^{-3}, while the line species argue that the density must be less than about 10^{10} cm^{-3}. The prevailing opinion is that the broad lines originate in condensed material at a density of about 10^9 cm^{-3} (and a temperature 10^4 K) which is suspended in a much hotter medium and irradiated from a central continuum source. Davidson and Netzer (1979) have recently reviewed all the relevant pre-Einstein theoretical and experimental evidence for this consensus picture. The condensations are found at a distance from the central source which scales like the square root of the continuum luminosity (with typical value ~0.1 pc), while the narrow line cores and narrow forbidden lines are created at distances of the order of 1 kpc. Invariably, Seyferts are spiral galaxies. Since quasars have virtually the same defining properties, save for the fact that they are "quasi-stellar" rather than "spiral", it would appear that the difference between the two may be more operational than real (at least with respect to quasars which are radio quiet, as Seyferts almost always are). Quasars which are radio loud, on the other hand, are much more variable, and distinguishable from BL Lac objects only in that the latter exhibit no emission lines at all.

The first correlations of Seyfert X-ray emission with optical characteristics were performed by Elvis, et al. (1978) from Ariel-5 data. While the X-ray emission did not correlate very well with optical continuum emission, it correlated extremely well with the breadth of the emission lines and the H_α intensity. The latter correlation, in particular, implied a strong connection between the X-ray emission and the central continuum source. The central power plant has long been assumed to be the manifestation of gravitational potential energy converted from infall to a black hole of mass 10^{6-8} $M_\odot$, but the details of the process and geometry are rather obscure. Recent reviews of the likely mechanisms for X-ray production illucidate the difficulties associated with attempting to infer their nature from the X-ray output spectra alone. Whether direct bremsstrahlung or synchrotron radiation, or some variety of singly or multiply Compton-scattered radiation, almost all the possible scenarios result in smooth, featureless power-law spectra over the limited dynamic range available to the X-ray detectors. In general, the individual measured spectra are as uninteresting as the qualifications above predict them to be. Nevertheless, the measurement of broader-band spectra and/or the identification of consistent similarities or differences among the classifications of compact extragalactic sources may prove to be fruitful exercises.

X-ray measurements have provided an upper limit to the size of the central source in the sense that temporal variations are assumed to be relatable to the light travel time across the entire central source. Some controversy presently exists over the reproducability of individual reports of very short timescale variations, but several active galactic nuclei have exhibited marked variability on timescales as short as about one day. The first report of such variation which was comfortably outside the realm of possible systematic uncertainties or statistical flukes was from the nearby Seyfert NGC 4151 (Mushotzky, Holt and Serlemitsos, 1978). The implications of short timescale variability have been discussed by Cavallo and Rees (1978) insofar as they can be related to the efficiency with which X-radiation can be produced. A dimension for the central source of the order of a light-day implies an efficiency for X-ray production of order 10^{-2} for the Seyfert sources.

Data from the A-2 experiment provided a surprising consistency in the spectra of Seyfert galaxies. Mushotzky, et al. (1980) found that a sample of seven, ranging over two orders of magnitude in luminosity, could all be well-fit with the same power-law index (and better fit with a power law than with a hot isothermal spectrum). The sample has since been extended to about two dozen, with a best-fit α=0.7+.1 (where the error is the rms deviation of all the individual best fits, and is smaller than the typical individual error). Data from a sample of eleven Seyferts measured with the SSS are generally consistent with the same indices, as indicated in Figure 1, but the dispersion in indices is clearly larger. All but NGC 4151 (see below) were fit with a simple power-law-plus-uniform-absorber.

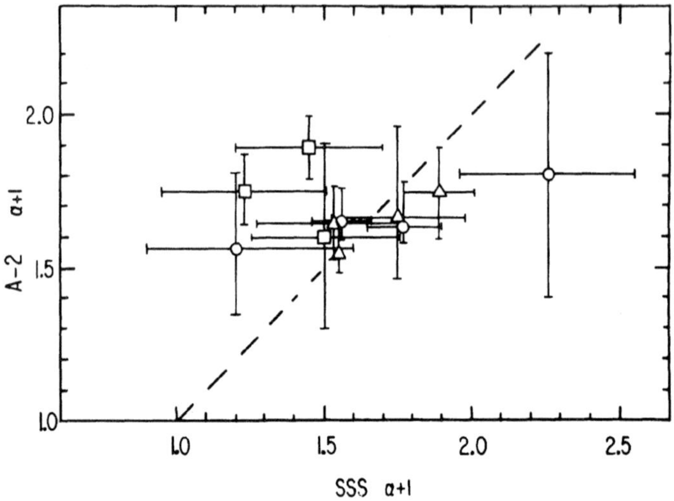

Figure 1. Best-fit power-law indices to Seyfert galaxies measured with the HEAO-1 A-2 experiment, and with the SSS. The circles are the 4 highest luminosity objects in the sample, the triangles the 4 with lower luminosity that have traditionally been labelled Seyfert I (including NGC 4151), and the squares even lower luminosity sources usually classified as emission line galaxies, but which have recently been observed to exhibit broadened wings in permitted optical lines (J. Huchra, private communication).

The typical X-ray power law index is approximately one unit flatter than the typical optical index for Seyferts, and the bright Seyfert ESO 141 was the only Seyfert measured with HEAO-1 to exhibit (occasionally) a steepening below 5 keV which might be plausibly related to an extrapolation of the optical index into the soft X-ray band. The right-most point in Figure 1 is the same source (where the HEAO-1 index on the vertical scale is that measured above 5 keV for the source). The remainder of the significant dispersion for the SSS data is due to the lowest luminosity sources, which seem to have flatter SSS than HEAO-1 spectra. This may be "explainable" in terms of the detailed interpretation of the high precision SSS data from NGC 4151.

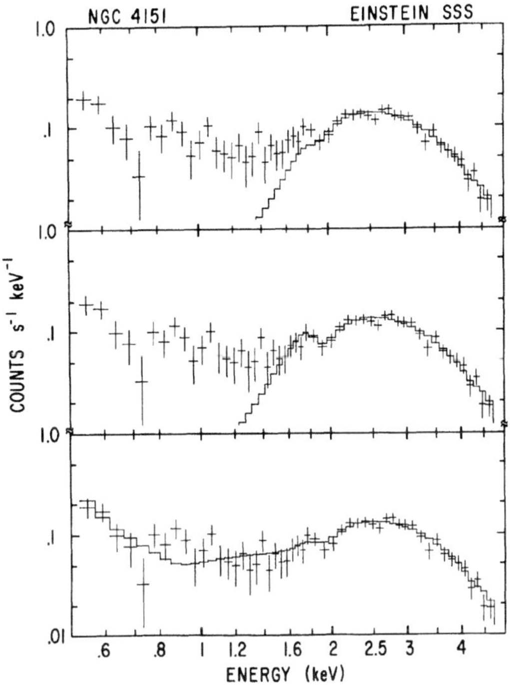

Figure 2. Raw (background-subtracted) SSS data for NGC 4151 fit with three trial spectra. The top trace represents a single power law with a uniform cold absorber in the line-of-sight. The second trace has the silicon abundance in the absorber increased by an order of magnitude, and oxygen completely ionized. The bottom trace, which is the only one of the three fits which is acceptable, has almost the same parameters as the top trace except for the fact that the absorber has "holes" over ~10% of the source.

The three traces of Figure 2 all represent the same raw (background-subtracted) data, fit with three different models. The top trace is the "standard" power-law-plus-uniform-absorber, which matches the data well above ~2 kev, but fails to reproduce the detection at lower energies. Since the higher A-2 energy range is less sensitive to absorption, and since the power-law index is apparently independent of instantaneous intensity (Mushotsky, Holt and Serlemitsos, 1978), we have fixed the index at α=0.55 in the SSS analysis, i.e., the best-fit A-2 index when Fe K-absorption and fluorescent re-emission are considered (consistent with this approach, the NGC 4151 point in Figure 1 at α=.55 has no horizontal error bars). The second trace is an unsuccessful attempt to reproduce the experimental data by allowing the uniform absorber to display peculiar elemental abundance and ionization-state parameters. Finally, the bottom trace is a successful fit to the data which arises from the assumption that the absorber has "leaks" at the 10% level. As described in detail by Holt et al. (1980), the spectroscopic analysis of NGC 4151 allows us to infer several characteristics of the source which imaging alone cannot: the approximate spherical symmetry of the broad-line region (from the ratio of fluorescent Fe K-emission to Fe K-absorption measured with the A-2 experiment), the approximate 90% "covering factor" of the broad-line clouds, and the consistency with a scenario in which the clouds are smaller than the source. Interestingly, this last point naturally explains the observed variation in apparent absorption of NGC 4151 data >2 keV taken at different times (e.g. Barr, et al. 1977) as just the Poisson statistical variations in a total of ~ 100 clouds which cover the source along our line of sight, with an average thickness of ~ 2 clouds. Remembering that the distance at which the clouds are found scales like the square root of the central source luminosity, it would not be surprising if the more luminous Seyferts (and quasars) give no hint of absorption when fit to the "standard" model, because the covered fraction would be low, while the low luminosity Seyferts would appear to have somewhat flatter spectra at low energies, because the absorption would be only partial and would not easily be recognized as such. It is conceivable, therefore, that the most straightforward extrapolation of the NGC 4151 picture revealed from the SSS data gives a natural explanation for the lack of absorption in the high luminosity sources, and the apparent flattening of the SSS spectra of the low luminosity sources. We should always bear in mind, however, the danger of generalizing too broadly from the most easily-studied class members to infer class properties. In some cases the prototypes are the most easily studied simply because they are the most closeby; in others, such as the Crab nebula as a supernova prototype, their high X-ray fluxes arise from their anomalous rather than their class properties.

In contrast to the relative uniformity in the Seyfert spectra, the BL Lac sources form, at first glance, a fairly diverse sample. Spectral components with $\alpha > 3$ and $\alpha \sim 0$ have been reported, sometimes from the same source at the same time over different energy bands. Much more variable in intensity as well as spectrum than the Seyferts, the 3 keV differential luminosity of BL Lacs can easily change by more than an order of magnitude, while a Seyfert variation of more than a factor of two is the exception rather than the rule. A detailed study of the spectra of the five BL Lacs with intensity greater than about 10^{-3} of the Crab nebula, measured with both the A-2 experiment and the SSS, has revealed that the spectra below ~ 3-5 keV can always be fit with a steep ($\alpha \geq 2$) spectrum. Occasionally, in approximately 1/5 of the measurements of all BL Lac spectra we have, a harder spectrum is measured at high energies. This suggests a synchrotron origin for the low energy component, and a self-Compton origin for the more variable high energy component. A variety of consistency checks for the source PKS 2155, including an examination of the overall spectrum at all frequencies as well as the high efficiency implied by the short-term temporal variations, have been found by Urry, et al (1980) to be consistent with such a picture. The radio loudness and the violent variability of BL Lacs are quite independent reasons for expecting that a synchrotron source is more likely to be associated with BL Lacs than with Seyferts.

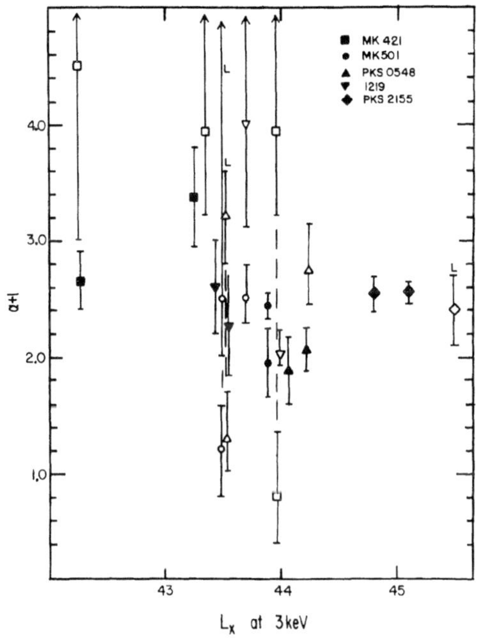

Figure 3. Best-fit spectral indices of BL Lac sources, as a function of luminosity (log base 10) in a 1 keV band at 3 keV. Each shape represents a different source, with the filled shapes representing SSS data points. The open shapes are data taken with the HEAO-1 A-2 experiment (L representing data taken with the low-energy LED detectors). The three points with $\alpha+1 \leq 1.4$ were all obtained as high energy hard components, in addition to steeper spectra measured below 5 keV concurrently, as indicated by the dashed lines.

The quasars for which we have spectral information constitute a considerably more meager sample of the detected total than do either the Seyferts or BL Lacs. More than 100 quasars have now been observed with the Einstein imaging experiments (Zamorani, et al. 1980), but most are much too weak for spectroscopic interrogation. 3C 273 has the same α=0.4 index in the SSS energy range as it does at higher energies (Worrall, et al. 1979) and QSO 0241 has an SSS index consistent with the α=1 measured by A-2 (Worrall, et al. 1980), although the heavy obscuration through the galactic plane makes the SSS spectral sensitivity less constraining for this source. Other quasars (or extreme Seyfert I or N galaxies which have sometimes been labeled quasars) for which SSS spectra have been obtained tend to be slightly steeper than the Seyfert I "universal" α=0.7 (e.g. Fairall 9, IC 4329A, III ZW2), but the total sample is too small to attempt any strong conclusions.

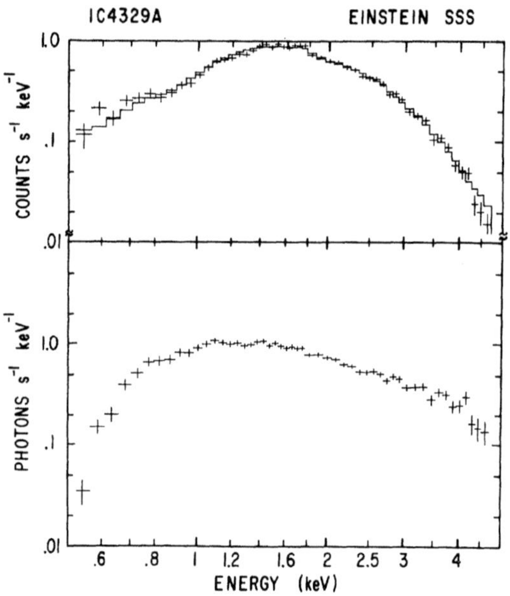

Figure 4. SSS data from the Seyfert/quasar IC 4329A. The top trace is the raw data fit with a power law of α=.86 masked by precisely the amount of cold gas (mostly in its own galaxy) inferred from the reddening. The bottom trace is the raw data of the top trace formally inverted through the SSS response function to display the incident spectrum at the detector.

Additional data which may help to suggest some initial categorization scheme is the recent result from the Einstein imaging data that the radio-loud quasars seem to exhibit an effective spectral index α_{OX} from point measurements in the optical and soft X-rays of approximately 1.3, while the radio-quiet quasars exhibit a considerably steeper index, i.e. α_{OX}~1.7. The latter represent more than 90% of quasars, and it is interesting that this α_{OX} is consistent with the average optical index for Seyferts (i.e., one unit steeper than the X-ray index). In view of the many similarities previously noted between Seyfert 1 and radio-quiet quasars, one might even make the tentative suggestion that the steep low energy component observed in ESO 141, one of the very brightest of the Seyfert I galaxies in X-rays, is a manifestation of the same effect responsible for the fact that radio-quiet QSOs seem to exhibit spectra slightly steeper than do the Seyfert I galaxies. The detailed modeling of X-ray transport out of radio-quiet quasars and Seyferts may have to accommodate this apparent slight softening in spectrum with luminosity.

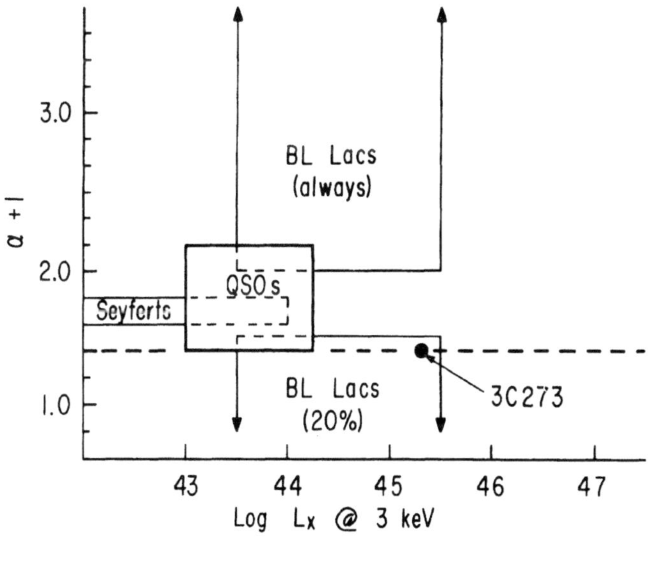

Figure 5. Summary of spectral index measurements reported here for the three classes of compact extragalactic sources.

Conversely (or similarly, depending upon your point of view), the radio-loud QSOs may be associable with the BL Lacs. In this case, the relatively flat 3C 273 spectrum may represent the Compton-scattered component, while many radio-loud quasar

X-ray spectra may be of the much steeper spectral form suggested above to arise from synchrotron emission. It should be remembered, however, that the smaller α_{ox} for the radio quasars suggests that it is their Compton components which are being observed. Further, the higher redshifts of the fainter (in apparent X-ray intensity) objects implies an X-ray sampling at higher energies at the source, which slightly complicates the interpretation given above for the radio-quiet quasar spectra. Nevertheless, the associations made here between the Seyferts and radio-quiet quasars, and the BL Lacs and radio-loud quasars, remain plausible, if not completely justified in all respects.

Finally, the extent to which the diffuse X-ray background (XRB) may be connected to X-ray emission from quasars must be addressed. There are several relevant constraints to consider, including spatial fluctuations, spectra and luminosity functions. The fluctuation analyses performed to date depend to some extent upon the evolution of the quasar population, and will not be addressed here. The optical luminosity function indicated, even before Einstein, that the XRB might easily be explainable from the superposition of individual quasars. Setti and Woltjer (1979), for example, suggested that if 3C 273 and Fairall 9 were representative quasar X-ray emitters, current estimates of the density evolution of the quasar population at high redshift would assure that the XRB would be entirely explained well before 23rd magnitude. The early Einstein imaging results seemed to bear this out, as approximately 25% of the ~1 keV XRB is reconciled with quasars detectable with Einstein sensitivity (Giacconi, et al. 1979).

The spectral results may represent the most formidable obstacle to complete reconciliation. The XRB spectrum has been measured with high precision by Marshall, et al. (1980) between 3 keV and 60 keV. Below 3 keV the effects of galactic contamination (both in absorption and emission) do not allow a useful measure of the XRB spectrum, and the counter efficiency of the A-2 detectors is too low above 60 keV. The data in this energy range are remarkably well-fit by a bremsstrahlung temperature of about 40 KeV, and definitely cannot be fit with a single power-law. It is important to note several different points about this XRB spectrum. First, it indicates that the large fraction of the energy density in the XRB is >10 keV. Second, the aforementioned effects of the galaxy render a measure of the spectrum below 3 keV impossible, so that any attempt to reconcile a potential source with the XRB below 3 keV is faced with the necessity of assuming an extrapolation to lower energies, in addition to performing the consistency check far from where the energy density peaks. Third, measurements made at still higher energies (>100 keV) with other instruments indicate that the XRB measured here exceeds the extrapolation of the

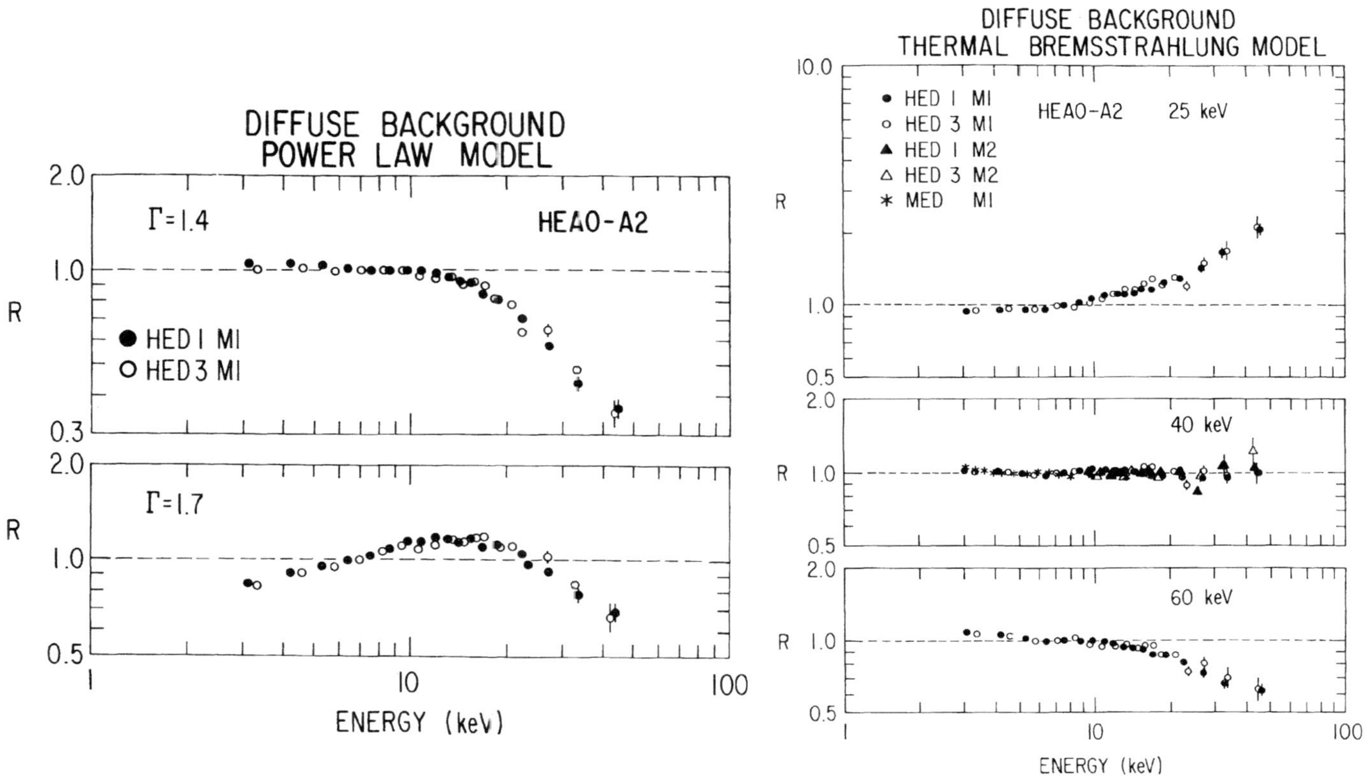

Figure 6. The ratio of the XRB spectrum measured with the A-2 experiment to trial idealized spectra. Different symbols represent independent measures with different detectors. The power-law index $\Gamma = 1 + \alpha$.

thermal spectrum. This suggests that the apparent thermal component is only one of a potentially large set of contributors to the total XRB, but that it deserves special consideration because the energy density is highest there.

The ~25% contribution from quasars near 1 keV is based upon an extrapolation of this thermal spectrum to lower energies. If this fraction (or a higher one) is correct, then the reality of a true substantial thermal component to the XRB is in considerable doubt, since a large non-thermal spectral component subtracted from a thermal one will not yield a net thermal spectrum. It is important to appreciate, however, that the actual composite XRB intensity at 1 keV is not measurable, and may far exceed the extrapolation of the thermal spectrum if the quasar spectra are steep. Holt (1980) has carefully considered the effects of synthesizing an XRB spectrum from Seyfert, quasar and cluster spectra similar to those measured from individual members of these classes to determine the extent to which an apparent thermal component may still contribute; surprisingly, approximately 80% of the XRB intensity in the peak energy density range can still arise from this component. Even though the total XRB substantially exceeds it at both higher and lower energies owing to the contributons from active galactic nuclei, the apparent thermal component can still be preserved even after a large quasar contribution is removed.

This analysis suggests that the XRB near 1 keV is truly dominated by quasars, but that the XRB in the range where its energy density peaks may have a quasar contribution of no more than ~10%. The origin of the apparent thermal component responsible for the large fraction of the XRB is not specified, however. Nor is the necessity of its thermal form: this analysis is meant only to demonstrate that all of the available data are not inconsistent with an apparent thermal component which can supply more than 3/4 of the XRB intensity.

It is unlikely that this component arises in a hot intergalactic medium, for theoretical reasons discussed by many authors. It is difficult to reconcile this component with quasars, based upon the tentative conclusions regarding the spectra of active galactic nuclei discussed above. Only 3C 273, a radio-loud quasar, can match the XRB spectrum, if it is redshifted to $z > 1$ (and then the consistency check can be performed over a dynamic range of only about 1/4 of the 3-60 keV interval). Figure 6 illustrates the futility of attempting to find a universal power-law component with which the XRB can be fit. More realistically, Cavaliere, et al. (1979) have attempted to determine what combination of near-power-law spectra might work. Their analysis suggests not only that $\alpha \sim 0.4$ components must sharply cutoff < 100 keV if they dominate > 10 keV, but also that the softer spectra reviewed here must considerably steepen the net XRB spectrum at lower energies. Unless quasars at large redshift have spectra very different than the active galactic nuclei which we can interrogate locally (e.g. if they exhibit near-bremsstrahlung spectra of 100-200 keV at the source), it will be exceedingly difficult to convincingly synthesize the XRB spectrum from them.

The postulation of quasars with X-ray spectra very different at early epochs than those of local active galactic nuclei may be the ultimate resolution of the difficulty, but a more natural explanation has recently been suggested by Bookbinder, et al. (1980). If very young galaxies are rich in supernovae and binary X-ray sources, they should give rise to intense galactic winds at temperatures of the order of 100-200 keV. Because this explanation meets the spectral constraints of the XRB with an origin which naturally arises at $z > 1$, it is a very attractive alternative to the quasar hypothesis. It will remain for the next generation of X-ray astronomical instrumentation to settle the question unambiguously.

REFERENCES

Barr, P., White, N.E., Sanford, P.W., and Ives, J.C. 1977, M.N.R.A.S. 181, pp. 43

Cavaliere, A., Danese, L., DeZotti, G., and Franceschini, A. 1979, Astron. and Ap. 79, pp. 169

Cavallo, G. and Rees, M.H. 1978, M.N.R.A.S. 183, pp. 359

Davidson, K. and Netzer, H. 1979, Rev. Mod. Phys. 51, pp. 715

Elvis, M., Maccacaro, T., Wilson, A.S., Ward, M.J., Penston, M.V., Fosbury, R.A.E., and Perola, G.C. 1978, M.N.R.A.S. 183, pp. 129

Fabian, A.C. 1979, Proc. R. Soc. London A. 366, pp. 449

Giacconi, R. et al. 1979, Ap. J. (Letters) 234, pp. L1

Holt, S.S. 1980, in X-Ray Astronomy, Giacconi, R. and Setti, G., eds, D. Reidel, Dordrecht-Holland, p. 35, 89, 327.

Holt, S.S., Mushotzky, R.F., Becker, R.H., Boldt, E.A., Serlemitsos, P.J., Szymkowiak, A.E., and White, N.E. 1980, Ap. J. (Letters), in press

Marshall, F.E. et al. 1980, Ap. J. 234, pp. 4

Mushotzky, R.F., Holt, S.S., and Serlemitsos, P.J. 1980, Ap. J. (Letters) 225, pp. L115

Mushotzky, R.F., Marshall, F.E., Boldt, E.A., Holt, S.S. and Serlemitsos, P.J. 1980, Ap. J. 235, pp. 377

Setti, G. and Woltjer, L. 1979, Astron. and Ap. 76, pp. L1

Urry, M., Mushotzky, R.F., Boldt, E.A., Holt, S.S., and Serlemitsos, P.J. 1980, preprint

Worrall, D.M., Boldt, E.A., Holt, S.S., and Serlemitsos, P.J. 1980, Ap. J., in press

Worrall, D.M., Mushotzky, R.F., Boldt, E.A., Holt, S.S. and Serlemitsos, P.J., Ap. J. 232, pp. 683

Zamorani, G. et al. 1980, preprint

EINSTEIN IMAGING OBSERVATIONS OF CLUSTERS OF GALAXIES

W. Forman, J. Bechtold, W. Blair, and
C. Jones
Harvard-Smithsonian Center for Astrophysics

X-ray astronomy began to play a major role in the study of clusters of galaxies ten years ago with the discovery of extended X-ray sources associatd with rich clusters (Gursky et.al. 1971, Kellogg et.al. 1972, and Forman et.al. 1972). Initially, two competing models were proposed to account for this emission. One model relied on inverse compton scattering of the microwave background photons on energetic electrons whose existence was inferred from radio observations of extended synchrotron sources in clusters (Harris and Romanishin 1974). The second model was thermal bremsstrahlung from a hot gas of 100 million degrees (see for example Solinger and Tucker 1973). The gas was presumed primordial, having been left over from the formation of the galaxies and then heated as the cluster collapsed.

High quality spectral observations have shown that the bulk of the emission is indeed due to thermal bremsstrahlung from a hot gas. These observations showed evidence of iron line emission in the cluster spectra with an approximately solar abundance (Mitchell et al. 1976 and Serlemitsos et.al. 1977). Thus, the intergalactic gas was not primordial but must have undergone heavy element synthesis before injection into the intracluster medium.

These X-ray observations have stimulated intensive work in other fields. Theoretically, the mechanism for removing the gas from the galaxies was discussed (Gunn and Gott 1972) as was the final distribution of the gas in the cluster potential (e.g. Bahcall and Sarazin 1978). Theoretical models for the appearance of X-ray emission in clusters at large redshifts were developed (Perrenod 1978). Radio astronomers used the X-ray observations to compute

R. Giacconi (ed.), X-Ray Astronomy with the Einstein Satellite, 187–213.

the gas density around head-tail radio sources in clusters since it was the X-ray emitting gas that was responsible for producing the observed radio structure. The imaging capabilities of the Einstein Observatory now permit us to readdress through the X-ray observations some of the problems which have been developed in relation to clusters by theoretical, radio, and optical as well as X-ray astronomers.

Many seemingly varied questions can be discussed in the context of the dynamical evolution of clusters as has been described by Jones et.al. (1979). This scheme of dynamical evolution can be used to classify the clusters based on their X-ray images coupled with optical properties. The X-ray observations suggest a broad division between clusters having a centrally located, X-ray dominant galaxy characterized by centrally peaked X-ray emission and clusters having a broader X-ray profile. Within these two broad categories the evolutionary indicators such as X-ray temperatures, velocity dispersions, spiral fractions and central galaxy densities may indicate relative stages of evolution. We will describe in some detail the X-ray imaging observations of the Virgo cluster, other dynamically unevolved clusters, "double" clusters which may be in an intermediate evolutionary stage, and dynamically evolved clusters in the context of this classification scheme. Cowie (1980) suggests that the difference between clusters with and without centrally peaked X-ray emission results from cooling in those clusters having a dominant galaxy at the cluster center. At this time, however, the experimental evidence in favor of this possibility is not definitive and we have chosen to carry out our analysis with the assumptions of an isothermal X-ray gas which is gravitationally bound by the galaxies and the clusters.

THE VIRGO CLUSTER

The Virgo cloud is the nearest rich cluster of galaxies and we have planned an extensive program of observations. Figure 1 shows five fields in the Virgo cluster which have been surveyed with the Einstein Observatory superposed on a reproduction from the Zwicky catalog. The individual galaxies are indicated by squares, circles and triangles. The cluster is typical of a dynamically unevolved system with a low velocity dispersion and a high fraction of spiral galaxies (Bahcall 1977a).

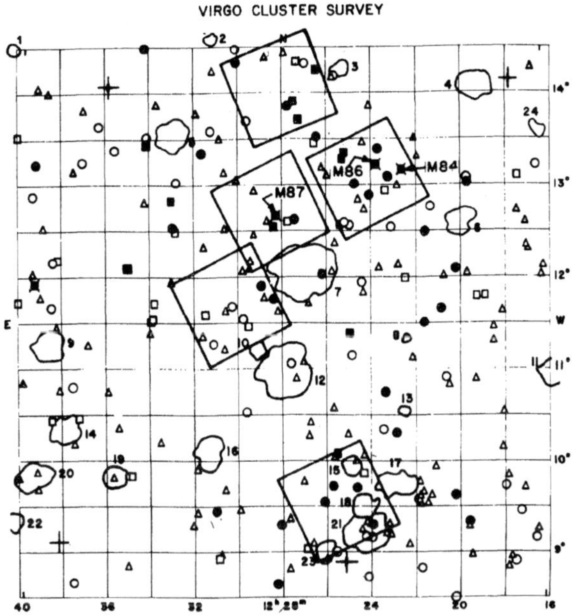

Figure 1 The fields in the Virgo cluster first surveyed with the IPC (squares) superposed on the Zwicky Catalog. Galaxies are indicated by small symbols and background clusters are outlined in solid lines. The galaxies M86, M84 and M87 are labelled.

Figure 2 shows an observation of the M86-M84 field made with the large field of view, moderate resolution IPC. The analysis of the observations of this field are taken from Forman et.al. (1979). Five sources are detected and all but one (lower right) are identified with galaxies in the Virgo cluster. The emission from one source is clearly extended and partially obscured by the detector window support structure. Figure 3 shows the X-ray contours superposed on an optical photograph and establishes the coincidence of the X-ray sources with this interesting group of galaxies which include NGC 4388 (strong narrow emission lines), NGC 4438 (tidally disrupted) and M84 (3C radio source). The brightest X-ray source in this field is extended with a core radius of 15 kiloparsecs and is associated with the galaxy M86. The spectrum of this emission is consistent with that of hot gas at a temperature of a few kilovolts. The key to the model for

Figure 2 The 0.5-3.0 keV X-ray image of the field containing M86 (bright extended source at upper center), M84 (at right edge), N4388 (center right) and N4438 (upper left).

this source was the high velocity of approach of M86 - nearly 1500 km/sec - with respect to the mean of the Virgo cluster. We suggested that the gas now associated with M86 was accumulated during the intervals between the periodic passages of M86 through the Virgo core. In the core there is a relatively high density of 10 keV intracluster gas similar to that in other clusters. The evidence for this hotter gas includes the multi-component spectra found by Mushotsky (1978) and Davison (1978) as well as the increased surface brightness observed in this field away from the galaxies. This intracluster gas with a density of $\sim 5 \times 10^{-4}$ particles per cm^3 in the core region can ram pressure strip the outer parts of the gaseous envelope of M86. However, the high velocity of M86 permits it to spend sufficient time far from the cluster core and thereby replenish its gas halo. M86 spends about 5×10^9 years between the time it exits the core, rises to its maximum height above the core, and then falls back to the cluster center. Assuming a mass loss rate of 10^{-12} $M_\odot$/year per solar mass, and a galactic mass of 10^{12} $M_\odot$, then we would expect to find about ~ $5 \times 10^9\ M_\odot$ of gas surrounding M86. This is the total mass of gas inferred from an isothermal sphere model of the X-ray emission around M86.

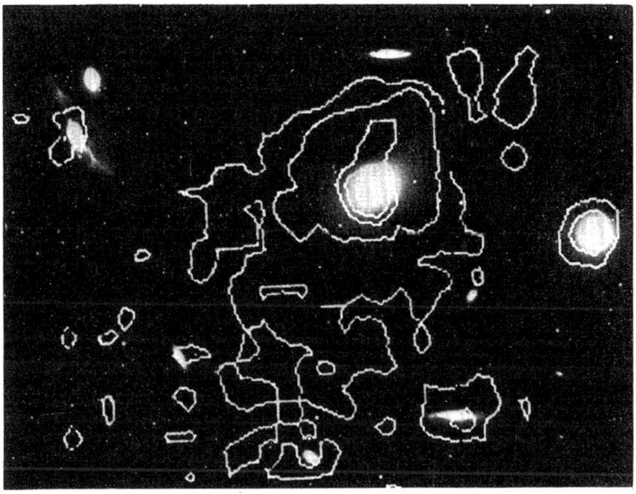

Figure 3 Iso-intensity contours of Figure 2 superposed on an optical photograph. The four galaxies are clearly visible. The extended emission around M86 is apparent.

This scenario would predict that other galaxies in the cluster should show similar gaseous halos. As already noted, the peculiar properties of some of the galaxies detected created considerable uncertainty as to the nature of their x-ray emission mechanism. For example, M84 contains the radio source - 3C272.1 - and it was possible that the x-ray emission was associated with a compact nuclear source with the radio emission as an indicator of this unusual nuclear activity. Therefore, we obtained a high resolution image of M84 to determine whether the emission from the galaxy was nuclear or extended. Figure 4, which shows the surface brightness distribution of M84 observed with the high resolution imager compared to a point source, demonstrates the extended nature of the x-ray emission. If we assume that this emission is also the result of thermal bremsstrahlung, we find a central gas density of about 0.01 cm^{-3} which is approximately that of M86. The temperature was taken to be $\sim$1 keV since the x-ray spectrum of M84 is similar to that of M86.

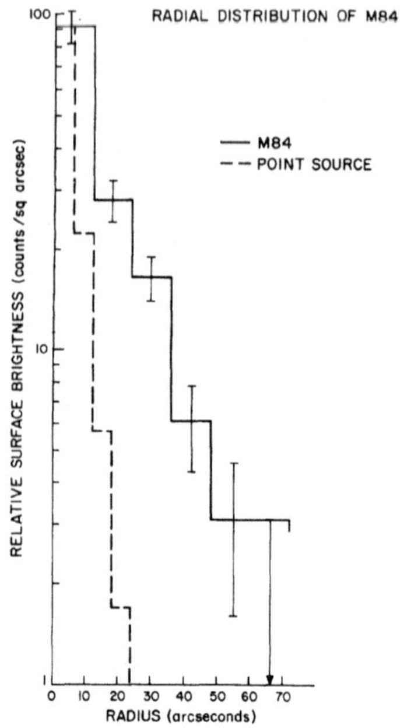

Figure 4 The surface brightness distribution of M84 compared to a point source. These observations are made with the high resolution imager. The extended nature of M84 is evident.

Why is the gas distribution about M84 so much smaller than that surrounding M86? M84 has a velocity in the line of sight which is nearly equal to that of the mean of the Virgo cluster. Therefore, if M84 is a permanent resident of the cluster core, it is continually exposed to the intracluster gas which can strip its outer gaseous envelope. We then observe only the core of the gas distribution which remains bound to the galaxy and cannot be stripped by the ram pressure of the intracluster gas.

In order to study M86 in more detail, we observed it in the center of the IPC field of view. Figure 5 shows the moderate resolution image of M86. In addition to the extended emission centered on the galaxy there is a second region of semi-detached emission. M86 also has been observed with the high resolution detector which shows this emission to be diffuse rather than a few unresolved point sources. We have used the moderate spectral resolution of the IPC to determine the energy spectra of M86 and the detached region or plume. The pulse height spectra are very similar and can be characterized by thermal bremsstrahlung from gas with a temperature of a few keV.

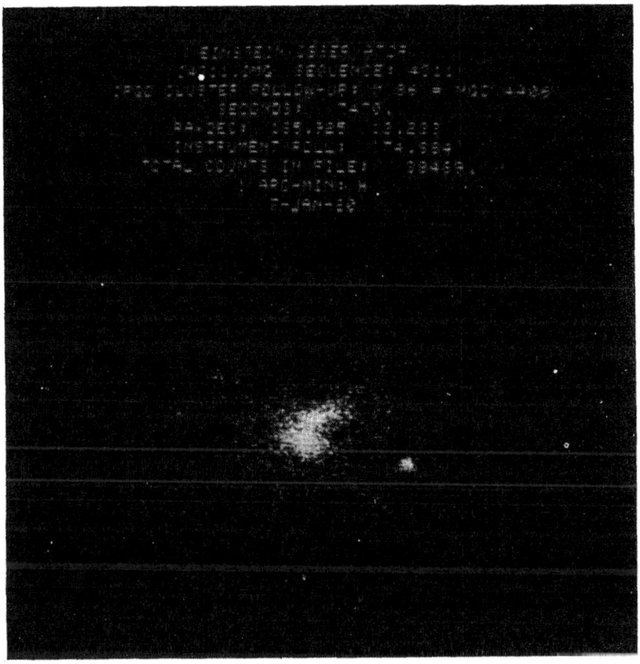

Figure 5 The moderate resolution image (IPC) of M86 showing the extended emission around the galaxy as well as a plume extending to the northwest. M84 is at the right.

Since we have not seen extended emission like this in other Virgo fields, of which we have studied about 10, or in other non-Virgo fields, it is reasonable in light of their similar x-ray spectra, to associate the plume with M86. Therefore, we have modified the original suggestion for the location of M86 relative to the cluster core. Rather than being on the far side of the core, M86 appears to be well into the core region and the plume represents gas that has been ram pressure stripped from M86 but not yet heated and incorporated into the 10 keV intracluster medium.

Figure 6 pictorially summarizes the scenario for the Virgo cluster. We are viewing the cluster from above so that the earth is in the plane of the figure. M86 has entered the hot, low density intracluster core and is being stripped of its outer halo of gas which it accumulated when it was far from the core, in the cluster outskirts. The plume or tail would be elongated opposite to the galaxy's direction of motion. Supporting this view is the small 20 arcsecond gaseous halo surrounding M84 which we presume to be a permanent resident of the denser cluster core.

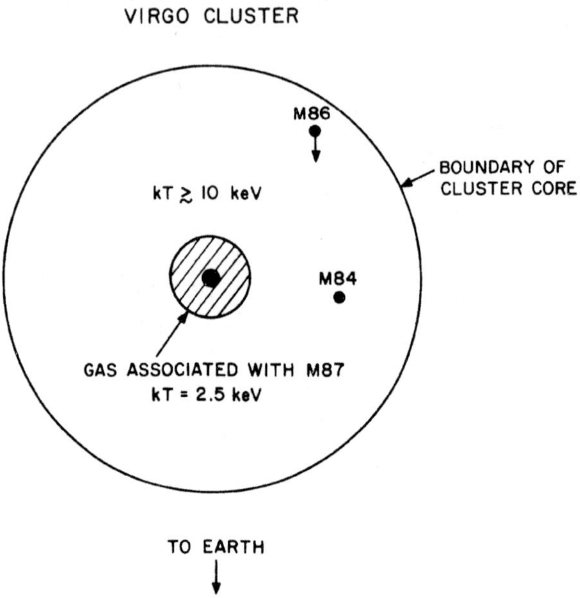

Figure 6 Schematic of the Virgo cluster showing M87 lying at the cluster center, M86 entering the cluster core region, and M84 embedded in the high density gas contained there.

These observations provide evidence that at least some of the hot, x-ray emitting gas in clusters of galaxies is produced by ram pressure stripping of gas from the member galaxies of the cluster.

The schematic representation of the Virgo cluster shows M87 occupying a unique position at the center of the cluster's core. M87's line of sight velocity is consistent with such a position. In addition to M87's unusual radio and optical jet, it is also remarkable in x-rays since it is 100 times more luminous than any other galaxy in the cluster.

Fabricant, Gorenstein, and Lecar (1980) have used the IPC observations of M87 to trace the gravitational potential of the galaxy and thereby determine the mass of the galactic halo needed to bind the x-ray emitting gas. Their analysis is similar to that carried out by Bahcall and Sarazin (1978), and by Mathews (1978). They argue that the gas surrounding the galaxy is in hydrostatic equilibrium since the cooling time of the gas is much longer than the free fall time. They also find that over a large range of radii the temperature of the x-ray gas is nearly constant.

Figure 7 shows their results for the density distributions. The visible galaxy dominates the inner few arc minutes while the dark halo dominates everywhere else. The x-ray emitting gas, which is very luminous and serves to trace the gravitational potential and therefore the mass distribution of the dark halo, represents only a small fraction of the total inferred mass.

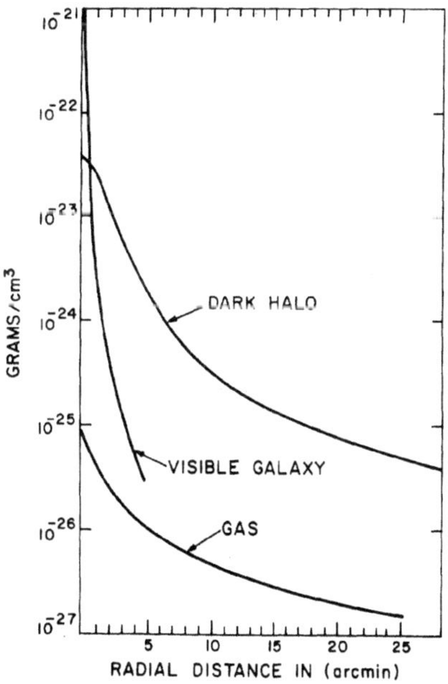

Figure 7 The radial distributions of the visible material, the X-ray gas and the derived distribution of the dark material inferred from the X-ray surface brightness distribution of M87.

Fabricant et al (1980) have traced the M87 x-ray emission out to a distance of 230 kpc before it can no longer be detected above the hot cluster gas and where the pressure effects of this cluster gas are believed to become important. Out to that radius, the galaxy mass is computed to be ~ 2-4×10^{13} $M_{\odot}$. Thus the x-ray observations demonstrate the existence of a halo at least ten times more massive than either the optical galaxy or the x-ray emitting gas. Similar analyses have been performed for clusters as a whole (see below) whose mass estimates agree with their Virial masses and suggest that massive halos may be a general characteristic associated with all galaxies.

DYNAMICALLY UNEVOLVED CLUSTERS

While M87 is a unique object in the Virgo cluster, other centrally located galaxies in other clusters are qualitatively similar. Figure 8 shows the x-ray iso-intensity contours superimposed on the optical image of A262.

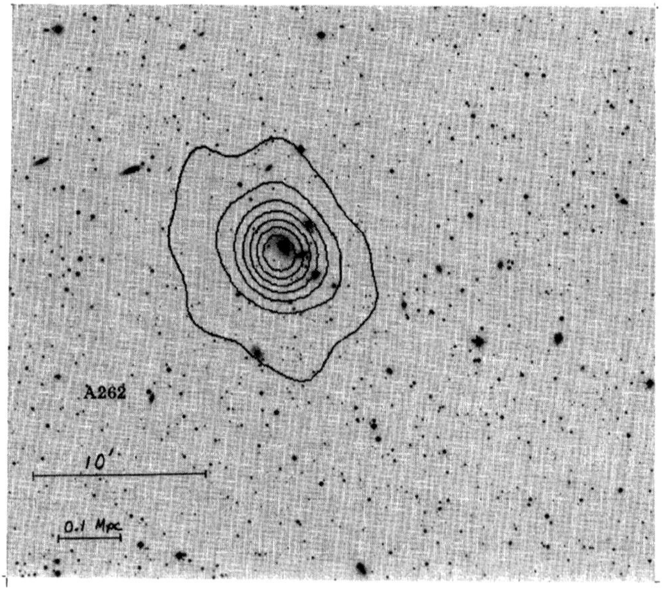

Figure 8 The smoothed (Weiner-filter) contour map of A262 superposed on an optical photograph. The X-ray centroid and the bright elliptical galaxy NGC708 coincide within the one arc minute positional uncertainty of the IPC.

As with M87, the temperature of the x-ray emitting gas is cool, ~ 2 keV (Mushotsky 1980) and centered on a bright elliptical galaxy. A similar image has been obtained for A2199 which also has a centrally located, bright galaxy. Although the x-ray emission from all three galaxies is dominated by cool gas, both Virgo and A2199 are known to contain some hot gas at about 10 keV (Mushotsky 1980). This finding of cool gas in the cluster A2199 is quite surprising since it is dominated by a classical cD galaxy (Rood and Sastry 1971) and yet it appears closer in its x-ray properties to clusters like A262 and Virgo than to other x-ray emitting cD clusters.

Clusters like Virgo and A262 are prototypes of dynamically unevolved clusters containing dominant x-ray galaxies. Both contain a high percentage of spiral galaxies and their galactic distributions are not centrally condensed (Bahcall 1977a and Rood and Sastry 1971). These clusters have two temperature component x-ray spectra with the cool gas associated with individual galaxies presumably embedded in a hot intergalactic medium.

In Virgo and A262, a substantial faction of the total x-ray emission originates from cool gas surrounding a single, prominent galaxy in the cluster core. There is some evidence from both optical and x-ray observations for an analogous class of unevolved clusters having no dominant central galaxy. This class is best typified by A1367 whose x-ray image is shown in Figure 9. The projection of the x-ray observations onto the horizontal and vertical axes shows the broad nature of the emission. Its irregular shape is also apparent. The brightest x-ray source in the cluster is the radio galaxy 3C264, at the lower left edge of the cluster emission. Its luminosity is 10^{42} ergs/sec which is only a few percent that of the total cluster emission. In A1367, only a small percentage of the x-ray emission is associated with individual bright galaxies compared to Virgo where 50% of the emission is concentrated around M87. The optical evidence for the early evolutionary stage of A1367 is its low velocity dispersion, high spiral fraction, and late Bautz-Morgan classification. The cool x-ray temperature and irregular x-ray emission can be considered as x-ray indicators of the dynamical state corresponding respectively to velocity dispersion (depth of the cluster potential) and Bautz-Morgan class (mass distribution).

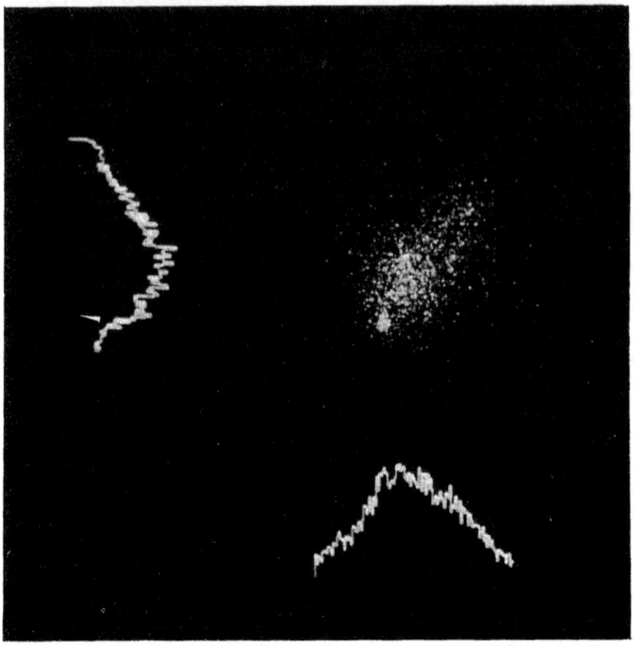

Figure 9 The X-ray image of A1367 made with the Imaging Proportional Counter. The source at the lower left coincides with 3C264 =NGC3862 and produces a few percent of the total cluster emission. The horizontal and vertical projections emphasize the irregular shape of the cluster.

For the X-ray dominant galaxies in unevolved clusters (Virgo, Centaurus, A262, and A2199), we have derived the total mass necessary to produce the observed x-ray distribution following Fabricant et.al. (1980) under the assumptions of hydrostatic equilibrium and an isothermal gas. These masses range from 1-2 x 10^{13} $M_{\odot}$ within 100 kpc. The mass of the x-ray emitting gas itself is less than 10% of the total mass.

For the Abell 1367 type clusters their x-ray emission is not symmetrical and therefore they cannot be characterized by a spherical distribution nor can we compute the cluster mass since they may not be relaxed systems.

A common characteristic of all these dynamically young systems is that they show two component x-ray spectra.

Mushotsky (1980) has recently summarized the spectral data for the brighter clusters. These dynamically young clusters exhibit a 2 keV component - primarily that of gas presumably associated with individual galaxies - and generally a 10 keV component which is the gas no longer bound to any single galaxy but which can only be contained by the entire cluster potential.

"DOUBLE" CLUSTERS

As part of a survey of nearly fifty clusters with the Imaging Proportional Counter on the Einstein Observatory, we have observed the clusters A98, A115, A1750 and SC0627-54 whose x-ray surface brightness distributions exhibit two enhancements. Figure 10 shows the x-ray images along with a one dimensional projection of the surface brightness distributions which emphasizes the double structure of each cluster.

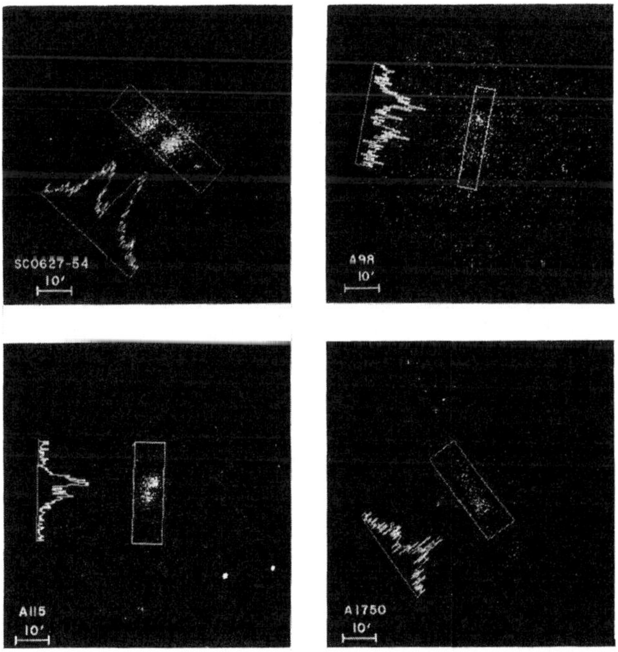

Figure 10 The X-ray images of SC0627-54, A98, A115, and A1750 are shown as observed by the IPC in the energy band 0.5-3.0 keV. The one dimensional projections include only those photons inside the rectangular boxes superposed on each image.

In Figure 11 the x-ray iso-intensity contours are superposed on prints from the Palomar Sky Survey or the European Southern Observatory Sky Survey. Each subcluster is centered on one of the bright galaxies in the cluster to within our one arc minute positional uncertainty. In addition to the relatively small angular separation of the pairs, the available redshifts are consistent with the doubles being physically associated. Spectra of the central galaxies in seven of the eight subclusters have been obtained, the exception being the southern component of A115. In the other three subclusters the velocities differ by less than 1500 km/sec, a value comparable to the line of sight velocity dispersions of rich clusters.

We have calculated in several different ways the probability that the double clusters are chance superpositions. We have used the separation of the pairs in both velocity and angle and assumed that our population of objects was detectable out to a distance corresponding to z = 0.2 (the redshift of the most distant cluster in our sample). Using the Smirnov-Cramer-Von Mises test (Eadie et. al. 1971), we computed a probability of less than 10^{-3} that the double clusters consist of a coincidence between a known cluster with another uniformly distributed population. We have made a second calculation and compared the volumes containing each cluster pair to the volume within which we could have detected such a pair (within z = 0.2 and the central 34 x 34 arc minutes of the IPC). In this test the probability that all four are chance coincidences and lie within a fraction of space equal to the largest double cluster volume is 2×10^{-5}. Therefore, we conclude that these double clusters can not all be random coincidences and that they represent a class of clusters with physically associated subclusters.

In order to compare the properties of the components of the double clusters to those of previously studied objects, the radial distributions of both components of each cluster have been fitted to an isothermal sphere model with the x-ray surface brightness distribution given by the King approximation (Lea et. al. 1973) convolved with the instrument response of the IPC (approximately a Gaussian of width 0.6 arc minutes). None of the eight cluster components is consistent with a single point source. The core radii vary from 0.53 to 0.80 Mpc with uncertainties of up to 30%. These sizes are similar to the 0.6 Mpc core radii found for evolved clusters with no dominant galaxy.

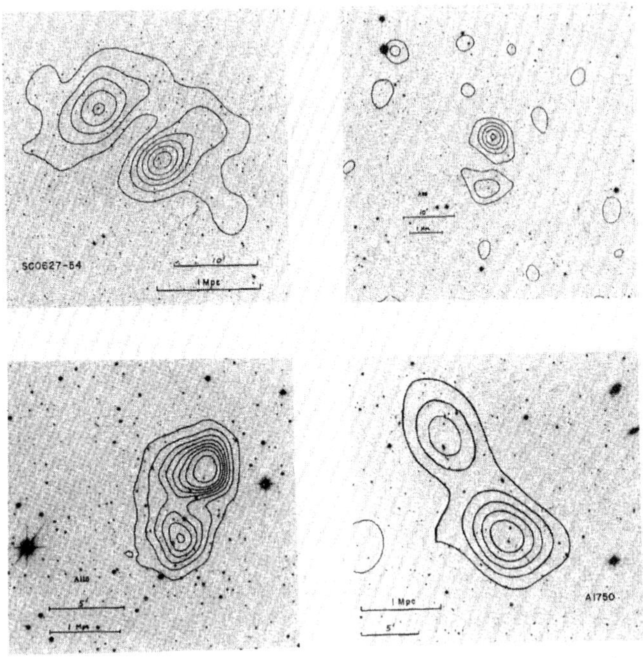

Figure 11 The X-ray iso-intensity contours are shown superposed on the Palomar Sky Survey Prints (A98, A115, A1750) and on the European Southern Observatory Pring (SC0627-54). The X-ray contours have been generated by deconvolving the image data with a Weiner filter. The optical photographs show galaxy sub-clustering which matches the X-ray observations in the northeast portion of SC0627-54, both parts of A98 (see also Faber and Dressler, 1977 plate Ia), and the southern parts of A115 and A1750. The contour levels are given below as the number of counts in each 64 x 64 square arcsecond bin:

SC0627-54 - 12.0, 21.4, 30.2, 39.1, 48.5, 57.3
A98 - 5.2, 7.0, 8.8, 10.8, 12.6
A115 - 5.3, 7.7, 10.5, 13.3, 16.1, 18.9, 22.1, 27.3
A1750 - 3.9, 6.8, 9.6, 12.6, 15.5

The luminosities of the subclusters range from 0.8 to 3.2 x 10^{44} ergs/sec which are typical of rich clusters. The central gas densities, assuming isothermal gas distributions with temperatures from 2 to 10 keV, are approximately 10^{-3} particles/cm^3 and are similar to those found for evolved clusters without dominant galaxies.

Only one of the double clusters - A98 - has been optically studied in some detail (Faber and Dressler 1977 and Dressler 1976). Henry et al. (1980) have compared the x-ray and optical observations and find that the optical galaxy counts also show a bimodal distribution similar to that observed in the x-ray.

Unfortunately there is no precise information available for many of the dynamical tracers - spiral fractions, accurate velocity dispersions, or x-ray temperatures - which would independently imply an evolutionary classification for these clusters. However, numerical simulations do suggest a classification. Hierarchical gravitational clustering has been analyzed through numerical simulations by Peebles (1970), Aarseth (1969), and White (1976) among others. These numerical experiments have shown that an initially expanding cloud of galaxies will collapse and eventually reach equilibrium with an extended halo surrounding a denser core. White (1976) has shown that during this evolution, subclustering occurs about the more massive galaxies. Subclusters continue to merge into a few large concentrations before their final coalescence. This evolution is shown in Figure 12 which is taken from White (1976). We suggest that the x-ray double clusters represent an intermediate evolutionary stage (Figure 11c) before the final merger of the subclusters into a relaxed Coma-type cluster. The observed subcluster separations of 1 Mpc (a minimum separation since projection effects cannot be removed) as well as the 0.6 Mpc core radii agree with those expected from White's (1976) model for an intermediate dynamical phase. White assumed that the initial mass function is derived from a Schechter luminosity function with a constant mass to light ratio and his analysis did not allow for galactic cannibalism. Therefore, his simulation is most appropriate to those clusters with no dominant galaxy. The interval during which clusters exhibit a bimodal distribution is substantial since free fall times for subclusters with 1 Mpc separations are 10^9 years. Gingold and Perrenod (1979) have studied the properties of the intra-cluster medium during the final subcluster merger. They calculated that the x-ray emitting gas should have the same bimodal distribution exhibited by the galaxies.

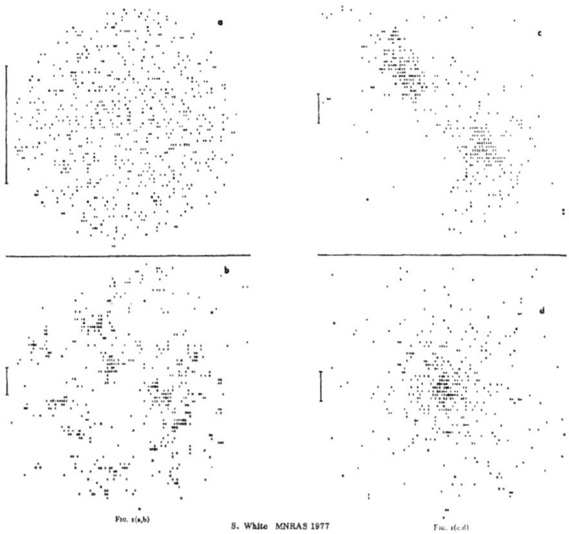

Figure 12 Four phases of a numerical simulation of cluster dynamics taken from White (1976). The model shows an expanding system of galaxies (upper left) which begins clustering on the smallest scales producing at first groups of galaxies (lower left). As this clustering continues, the next phase (upper right) shows the cluster dominated by two major condensations and finally, a relaxed cluster is seen at the bottom right.

EVOLVED CLUSTERS

The class of evolved clusters with a dominant central galaxy is exemplified by A85 whose x-ray image is shown in Figure 13.

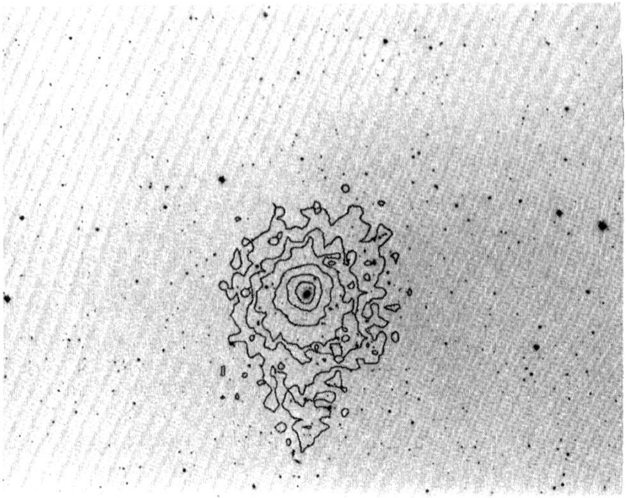

Figure 14 The iso-intensity contours from Figure 13 are superposed on an optical photograph. The dominant galaxy lies at the center of the gravitational potential as defined by the X-ray emitting gas.

We find the extents for the 11 cD clusters studied to be quite constant. This excludes A2199 which was mentioned earlier because of its cool temperature. The cD clusters as a class, have temperatures around 7 keV. Assuming hydrostatic equilibrium and a constant temperature profile, we have used the x-ray distributions to derive total masses of about 10^{14} $M_{\odot}$ within 250 kpc. This compares well for the three clusters (A85, A1795, and A2063) whose velocity dispersions have been measured and for which virial masses can be computed. In these clusters the x-ray gas accounts for only 4-15% of the total mass.

The second class of evolved clusters comprises those which, like Coma (Ku et al. 1980), have no dominant central galaxy. One such cluster, A2256, is shown in Figure 15. Again x-ray contours are overlayed on a Sky Survey print. The emission of the Coma-type cluster is similar to the other evolved class of clusters, the cD's since both are characterized by temperatures of about 7 keV. As would be expected for gas at those temperatures, no emission is seen associated with individual galaxies in Coma, the best studied example (Ku et al. 1980).

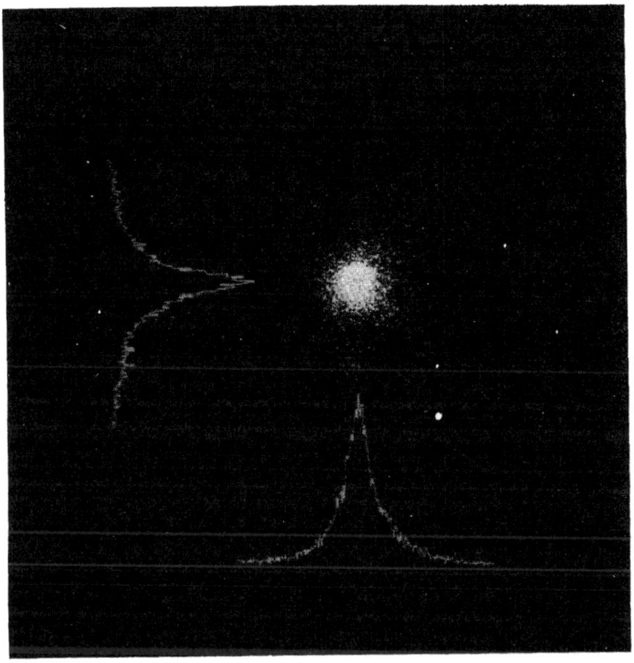

Figure 13 The IPC image of A85 in the energy range 0.5-3.0 keV. The horizontal and vertical projections show both the symmetry and strong central concentration of this cluster.

The projections onto the two axes show the centrally peaked distribution. Figure 14 shows the x-ray contours superposed on an optical photograph. The peak of the emission coincides with the bright cD galaxy which dominates the cluster and resides at the bottom of the cluster's gravitational potential well. The x-ray emitting gas in such clusters is hot - about 7 keV and must be associated with the cluster as a whole, not with individual galaxies. The x-ray observations are consistent with this since no emission is seen which is associated with a single galaxy as was observed in the Virgo cluster. The x-ray distributions are quite symmetrical in the central regions of these clusters suggesting a relaxed system.

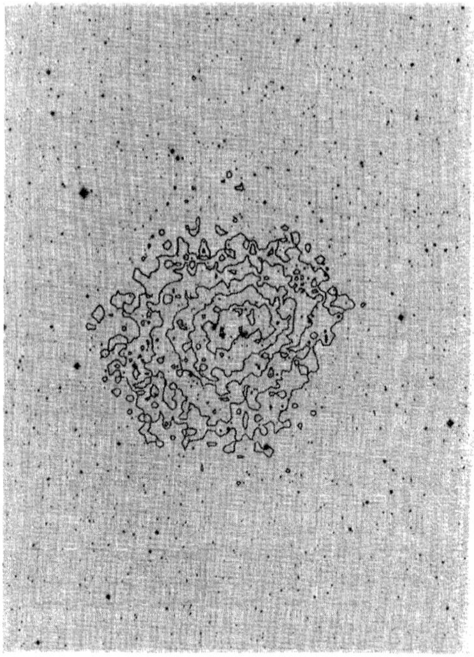

Figure 15 The X-ray iso-intensity contours of the moderate resolution (IPC) image of A2256 in the energy range 0.5-3.0 keV. There is a slight east-west elongation.

Figure 16 shows the radial distributions for four Coma-type clusters and A85, a cD, for comparison. The distributions are normalized to the same peak value and are plotted on the same linear scale. Qualitatively the Coma-type clusters are more extended than the cDs.

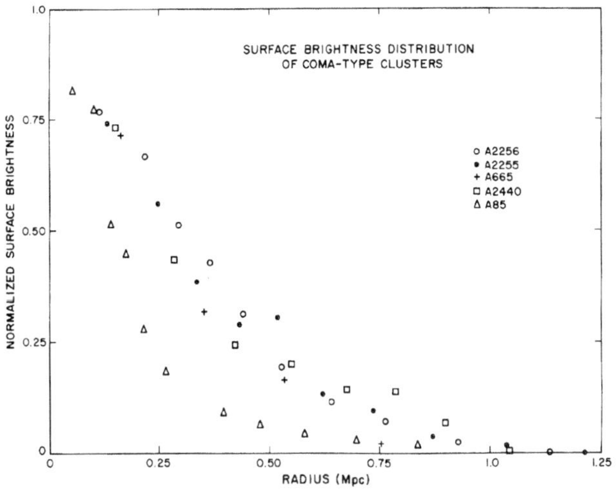

Figure 16 Surface brightness distributions of four Coma-type clusters are shown along with that of the cD-cluster A85. All the distributions are normalized to the same peak value.

The two evolved types of clusters differ in the shape of their surface brightness profiles. The Coma-type clusters can be fit very well by isothermal sphere distributions. The cD clusters, however, are better described by power laws over the range from ~ 30 kpc to 500 kpc. This behavior is similar to that found for M87 by Fabricant et al. (1980) and for other dominant galaxies like that in A262. Cowie (1980) and Fabian et al. (1980) interpret this departure from an isothermal sphere as evidence for cooling in the denser central regions of some clusters. However, Fabricant et al. (1980) have shown for M87 that the X-ray surface brightness distribution cannot solely be produced by cooling but that a massive dark halo is required. If this interpretation is correct for the cD clusters as well, then the cD clusters may be similar in type, but at later evolutionary stages than clusters like Virgo and A262. However, more detailed spectral analyses are necessary to definitively resolve this problem. In the following section we discuss a possible classification system which assumes that cooling is a significant process only in the dense central regions of clusters and that the outer regions can be modelled by isothermal, gravitationally bound gas distributions.

IMPLICATIONS FOR CLUSTER CLASSIFICATION

Optical classification systems have been developed for clusters of galaxies partially with the goal of providing dynamical information. For example, Hausman and Ostriker (1978) suggested that the Bautz-Morgan system is directly related to the evolution of clusters with those of type I (systems with optically dominant galaxies) being the most highly evolved. Bahcall (1977b) has reviewed other classification systems and their relation to the dynamical evolution of clusters. Several optical properties of cluster member galaxies have been suggested as indicators of dynamical evolution. For example, evolved clusters tend to be associated with lower fractions of spiral galaxies, higher velocity dispersions, and larger central densities (see Bahcall 1977b and references therein).

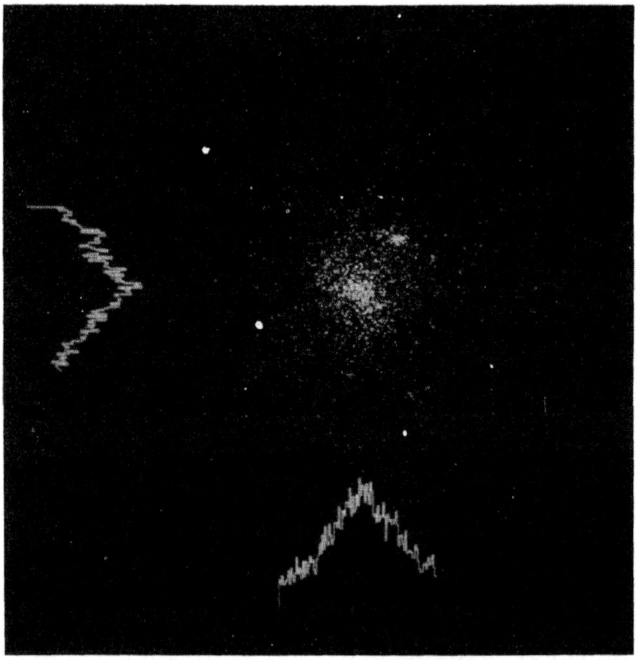

Figure 17 The moderate resolution image of A2634 is shown along with the horizontal and vertical projections. This cluster shows no strong central concentration as does A85 (Figure 13).

However, these dynamical indicators do not always correlate with the Rood-Sastry or Bautz-Morgan classification system. In particular, rich clusters like Coma and A2256 (Figure 15) appear in both their x-ray and optical properties to be relaxed systems, yet they do not contain single dominant galaxies. Alternatively, the cluster A2634 has been classified as a cD type cluster (Rood and Sastry 1971) although the x-ray appearance exhibits little or no central condensation indicating the absence of a truly massive central galaxy. Figure 17 shows the x-ray image of A2634 with projections about the two axes. This figure should be compared to that of A85 (Figure 13) which shows the marked difference in the depth of the gravitational potential as traced by the x-ray gas. The x-ray contours are shown superposed on the optical image for A2634 in Figure 18. The galaxy which gives A2634 its cD classification is playing a very different role than the corresponding galaxy in A85 (Figure 14). The x-ray observations of these clusters show the inadequacy of using only the optically defined morphological systems to determine the dynamical state of a cluster.

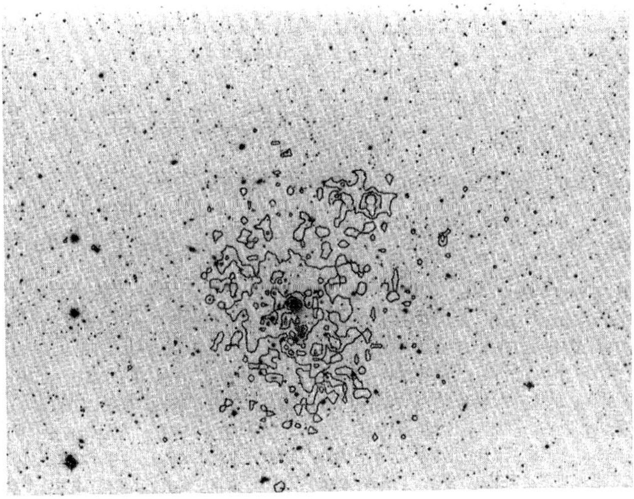

Figure 18 The iso-intensity contours of Figure 17 are superposed on an optical photograph. There is an absence of central concentration about the central dominant galaxy.

Einstein x-ray observations have begun to help resolve the problems noted above by suggesting a modified classification system for clusters of galaxies. The efficacy of the x-ray images of clusters in classification studies derives from their sensitivity in mapping the mass distribution within clusters. Supplemented with x-ray temperatures for the nearby clusters from previous x-ray satellites, and the other optical indicators, we can suggest a tentative classification system which may change as the analysis of the x-ray observations continues.

The proposed classification system divides clusters into two groups - those with dominant centrally located galaxies characterized by centrally peaked X-ray surface brightness distributions and those having no strong central X-ray concentration. This division is made in terms of the x-ray observations which easily distinquish between A85 and A2634 which are both classified optically as cD clusters. These two clusters are not only at different evolutionary states but also A85 belongs to that group of clusters exhibiting centrally peaked X-ray emission while A2634 does not. Within these two major divisions, the optical and x-ray indicators suggest the relative evolutionary stages for different clusters. Table 1 shows the main features of this classification system with evolutionary indicators and examples of the different types of clusters. For example, the low x-ray temperature of A1367, its large spiral fraction, and its low velocity dispersion all indicate a relatively unevolved system. Similarly, Virgo and A262 are in early evolutionary stages. In contrast Coma and A85 with their low spiral fractions, and high x-ray temperatures are apparently more evolved systems.

The images of clusters show that the x-ray observations are useful in determining a cluster's dynamical state within the two major classes. Further detailed optical analyses are necessary to understand the processes that occur as evolution proceeds. In particular, studies of spiral fractions, galaxy colors, and cluster velocity dispersions will aid in understanding the ram pressure stripping of galaxies and the evolutionary effects of the cluster environment on individual galaxies. In addition optical galaxy counts for the double clusters should confirm the bimodal mass distributions as has been done for A98 (Henry et al. 1980). Numerical experimentation could better define the importance of the initial conditions upon the final state of the cluster. Future X-ray observations over a broader energy range may also provide evolutionary information through the X-ray temperature profiles.

Table 1

CLASS I XD CLUSTERS Clusters containing an X-ray Dominant, Centrally-located Galaxy	CLASS II NXD CLUSTERS Clusters with no X-ray Dominant Galaxy	Evolutionary Indicators
Virgo (M87), A262, A2199	A1367, A2634	Low velocity dispersions High fraction of spiral galaxies Cool X-ray Temperatures ($\sim$2 keV) Irregular galaxy distribution
A85, A1795	Coma, A2256	High velocity dispersion Small fraction of spiral galaxies Hot X-ray Temperatures ($\sim$8 keV) Regular galaxy distribution

ACKNOWLEDGMENTS

We wish to thank D. Fabricant, M. Lecar and P. Gorenstein for providing their M87 observations prior to publication. J. Schwarz, L. Van Speybroeck, R. Giacconi and A. Fabian participated in the analysis of M86, M84, and the double clusters. We think K. Gilleece for preparing this manuscript. This work was supported by NASA under contract NAS8-30751.

REFERENCES

Aarseth, S. 1969, MNRAS, 144,537.

Bahcall, J. and Sarazin, C.L. 1978, Ap.J., 219, 781.

Bahcall, N. 1977a, Ap.J.(Letters), 218, L93.

Bahcall, N. 1977b, Ann. Rev. of Astr. and Astrophysics, 15, 505.

Cowie, L. 1980, this volume.

Davison, P.J.N. 1978, MNRAS, 183, 39p.

Dressler, A. 1976, Ph.D. thesis, University of California, Santa Cruz.

Eadie, W.T., Drijard, D., James, F.E., Roos, M., and Sadoulet, B. 1971, in Statistical Methods in Experimental Physics (North Holland: London) p.268.

Faber, S., and Dressler, A. 1977, A.J., 82, 187.

Fabian, A.C. and Grindlay, J. 1980, preprint.

Fabricant, Gorenstein, and Lecar (1980) submitted to Ap.J.

Forman, W., Kellogg, E., Gursky, H., Tananbaum, H., Giacconi, R. 1972, Ap.J.(Letters), 178, 309.

Forman, W., Schwarz, J., Jones, C., Liller, W., and Fabian, A.C. 1979, Ap.J.(Letters), 234, L27.

Gingold, R.and Perrenod, S. 1979, MNRAS, 187, 371.

Gunn, J.E., and Gott, J.R. 1972, Ap.J., 176, 1.

Gursky, H., Kellogg, E., Murray, S., Leong, C., Tananbaum, H., and Giacconi, R. 1971, Ap.J.(Letters), 167, L81.

Harris, D.E. and Romanishin, W. 1974, Ap.J., 188, 209.

Hausman, M.A. and Ostriker, J.P. 1978, Ap.J., 224, 320.

Henry, J.P., Henriksen, M.J., Charles, P.A., and Thorstensen, J.R. 1980, Ap.J.(Letters), in press.

Jones, C., Mandel, E., Schwarz, J., Forman, W., Murray, S.S., Harnden, F.R., Jr. 1979, Ap.J.(Letters), 234, L21.

Kellogg, E., Tananbaum, H., Giacconi, R., and Pounds, K. 1972, Ap.J.(Letters), 174, L65.

Ku, W.H.M. and Abramopoulus, F. 1980, San Francisco AAS Meeting.

Lea, S.M., Silk, J., Kellogg, E., and Murray, S. 1973, Ap.J.(Letters), 184, L105.

Mathews, W.G. 1978, Ap.J., 219, 413.

Mitchell, R.J., Culhane, J.L., Davison, P.J.N., and Ives, J.C. 1976, MNRAS, 176, 29p.

Mushotsky, R.F., Serlemitsos, P.J., Smith, B.W., Boldt, E.A., and Holt, S.S. 1978, Ap.J., 225, 21.

Mushotsky, R.F. 1980 in Proceedings of the NATO Advanced Study Institute, Erice, Italy, July 1979, Eds. R. Giacconi and G. Setti (Reidel: Holland), p. 171.

Peebles, P.J.E. 1970, A.J., 75, 13.

Perrenod, S.C. 1978, Ap.J., 226, 566.

Rood, H.J. and Sastry, G.N. 1971, Pub.As. P., 83, 313.

Salinger, A.B. and Tucker, W.H. 1973, Ap.J.(Letters), 175, L107.

White, S.D.M. 1976, MNRAS, 177, 717.

White, S.D.M. 1978, MNRAS, 179, 33.

X-RAY SPECTRA OF CLUSTERS OF GALAXIES

Claude R. Canizares
Department of Physics and Center for Space Research
Massachusetts Institute of Technology

INTRODUCTION

X-ray spectroscopy has played a crucial role in the study of clusters of galaxies, possibly more so than in any other area. In one stroke, the detection of iron line emission from clusters by Mitchell et al. (1976) and Serlemitsos et al. (1977) showed that the dominant X-ray emission mechanism is thermal bremsstrahlung and that the intracluster gas is enriched with heavy elements -- since the iron line strengths imply near-solar abundances.

There are several comprehensive papers which summarize the extensive spectroscopic surveys carried out with the Ariel 5 and OSO-8 satellites and which describe some of the earlier work (Gursky and Schwartz 1977; Mitchell et al. 1977; Mushotzky et al. 1978; Mitchell et al. 1979, and Smith et al. 1979). I will not review those results here except to note that the picture which has emerged is that of a more or less isothermal cluster gas bound in the cluster potential. The temperature is roughly correlated with the depth of the potential as measured by the cluster velocity dispersion.

Of course, this description can only be true at the simplest level. Numerous individual cases of clusters with spectral complexity have been reported in the literature. I will focus on the spectral complexity of the most recent data from the Einstein Observatory and attempt to explain at least some of it in terms of models of radiatively regulated accretion in the central parts of the clusters and/or onto massive galaxies.

First, I will very briefly review the range of models which have been proposed by the many workers in the field (see also L. Cowie's paper in this volume) with specific emphasis on the predicted temperature distribution. These are illustrated in Fig. 1. The most widely applied models (Lea 1975, Gull and Northover 1975, Cavaliere and Fusco-Femiano 1976, Bahcall and Sarazin 1978) are those which treat the gas as hydrostatic and polytropic (i.e., the density ρ and pressure P

R. Giacconi (ed.), X-Ray Astronomy with the Einstein Satellite, 215–226.

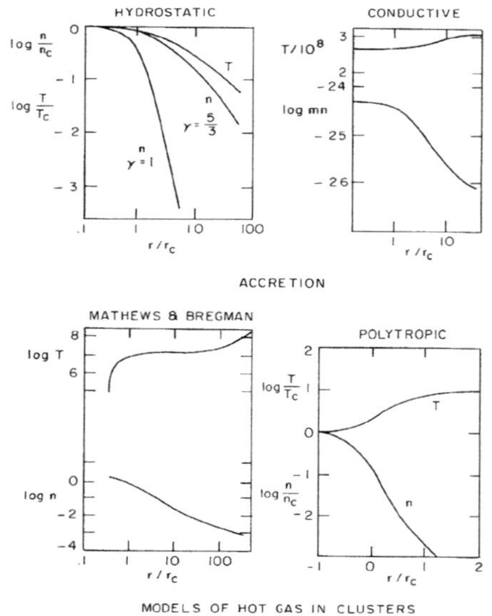

FIG. 1: Schematic temperature (T) and density (n) profiles for various representative cluster models plotted vs. radius (r) in units of a core radius (r_c). References are given in the text.

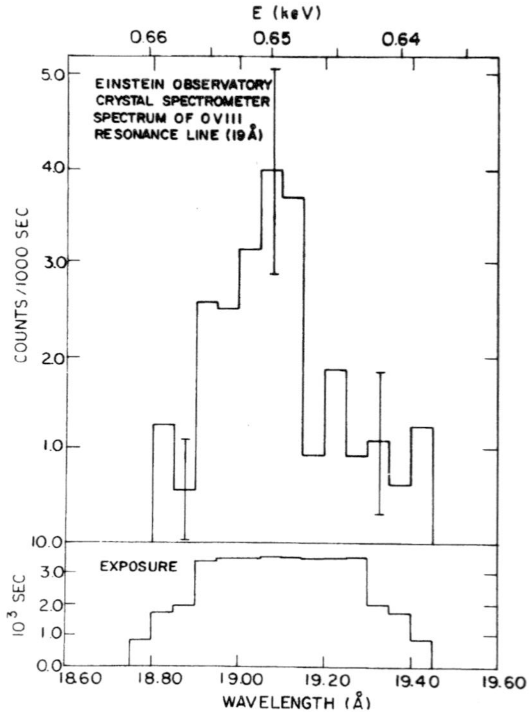

FIG. 2: The O VIII Ly α line from M87 (Canizares et al. 1980)

are related by $(\partial \log P/\partial r) = \gamma(\partial \log \rho/\partial r)$, where r is the radius, with polytropic index between $\gamma=1$ (isothermal) and 5/3 (adiabatic). Note that for all $\gamma>1$ the temperature is a universal function of r and decreases with increasing r. A number of authors have explored the consequences of radiative cooling of the gas in the dense central regions of a cluster or cluster plus central galaxy (Silk 1976, Cowie and Binney 1977, Fabian and Nulsen 1977, Mathews and Bregman 1978). One detailed model of M87 by Mathews and Bregman (1978) is shown in Fig. 1. All accretion models have temperatures which decrease towards the center. One expects that at very large r the temperature will again decrease as in the adiabatic models (Cowie and Binney 1977, Fabian and Nulsen 1977). Accretion can be represented by polytropic models with $\Upsilon<1$. Garrett Jernigan has computed such models with constant Υ ($\Upsilon=0.65$ in the example of Fig. 1), and more recently with Υ varying smoothly from $\gamma<1$ to $\Upsilon=5/3$ in a manner which represents a constant accretion rate. So far there is only one temperature profile for a model describing the conductive heating and expulsion of galactic matter (Takahara and Takahara 1979; see also Cowie and Songaila 1977). They assume a high conductivity and obtain the inverted but nearly isothermal temperature profile shown in Fig. 1.

M87

The X-ray source surrounding the giant elliptical galaxy M87 in the Virgo Cluster has been studied in more detail than any other. It has been recognized for some time that the kT=2-3 keV emission around M87 is associated more with the galaxy than with the cluster as a whole; the latter has its own larger, hotter source (e.g., Gorenstein et al. 1977; Bahcall and Sarazin 1977; Lawrence 1978; Davison 1978; Fabricant et al. 1978; and Lea et al. 1979). On the basis of X-ray surface brightness profiles, Gorenstein et al. (1977) and Mathews (1978) suggested that accretion onto M87 may be taking place.

One of the first observations which we carried out with the high spectral resolution Focal Plane Crystal Spectrometer (FPCS; for a description of the instrument see Canizares et al. 1980; Giacconi et al. 1979) on the Einstein Observatory gave strong support to the accretion picture. Using a 3x30 arc minute aperture (19x190 h_{50}^{-1} kpc where h_{50} = Ho/50 km s^{-1} kpc^{-1}) we detected a strong line from O VIII -- the Ly α line of hydrogenic oxygen at 654 eV as shown in Fig. 2 (Canizares et al. 1979, 1980). This detection implies the presence of material at temperatures less than 1 keV, and the line flux is in good agreement with that predicted by the accretion model of Mathews and Bregman (1978). Mushotzky (1979) showed some preliminary data from the Solid State Spectrometer (SSS) on the Einstein Observatory which suggest a somewhat intermediate temperature of ~2 keV based on the detection of the H and He-like lines of S and Si (see Fig. 3).

We have recently completed a more extensive study with the FPCS of the Fe lines from M87 (Canizares et al. (1980)). Fig 4 shows our data for one of two regions inside the Fe emission complex around 1

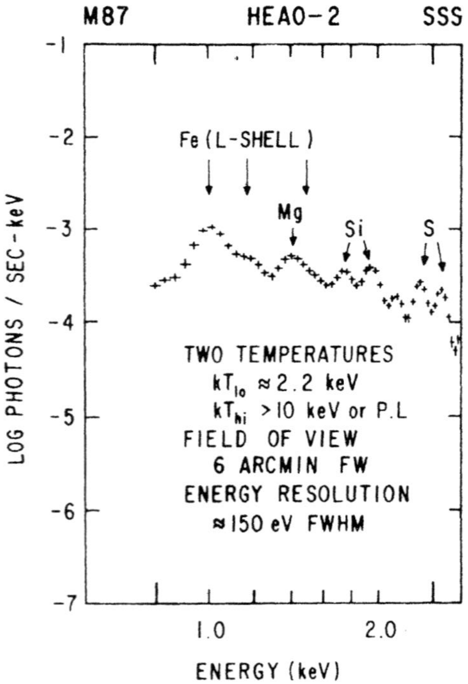

FIG. 3: The SSS spectrum of M87 (Mushotzky 1980)

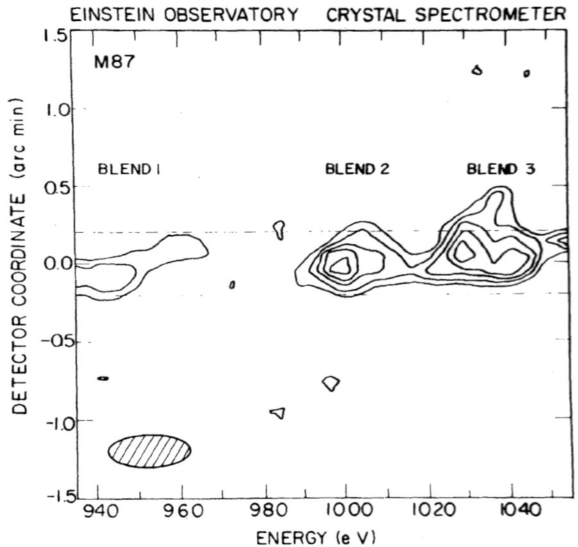

FIG. 4: Count rate contours from an FPCS observation of M87 showing emission from three blends of Fe L lines. The meaning of the ordinate is described in the text.

keV. The plot gives count-rate contours vs. energy on the abscissa. The ordinate is a detector coordinate which effectively divides the plot into a band containing source+background counts (within the dashed lines) and background only (above and below the dashed lines). There are clear regions of enhanced emission which correspond to blends of (predominantly) Fe lines. Table 1 lists the details of all our line/blend detections with the contributing ionic species from Raymond and Smith (1977; in Table 1 we assume a lower absorbing column density than in Canizares et al. 1979 for reasons described in Canizares et al. (1980).

The blends and lines of Table 1 are excited over a range of temperatures. In Fig. 5 I show the absorption-corrected strength of each line divided by the Raymond and Smith (1979) emissivity as a function of temperature (solid curves). This gives the volume emission measure at any single temperature required to account for the line (we plot $\langle n^2 \rangle V/4\pi D^2$ where n is the density, V the volume and D the distance to the source, all in c.g.s. units). The lack of a common intersection point shows that no single temperature plasma can account for the observed lines. (A plasma at logT = 7.2 would give Fe line fluxes close to those observed but would require an implausible overabundance of oxygen by a factor of eight relative to iron [Arnett 1978, DeYoung 1978] and would fail to agree with the SSS data.) The most plausible interpretation is that the emission comes from plasma over a range of temperatures. In that case, the distribution of volume emission measure vs. log T will more or less follow the lower envelope of all the solid curves in Fig. 5. In fact, the simple model of a flat distribution shown in Fig. 5 (filled circles) accounts moderately well for the observed line strengths (a formal chi squared is 13 for 7 degrees of freedom).

As I mentioned above, the models of radiatively regulated accretion predict the existence of matter over a wide range of temperature as we observe. The arbitrarily normalized distribution of emission measure with log T for one of Garrett Jernigan's polytropic accretion models is shown in Fig. 5 (filled squares), and one can see that it is nearly flat and follows the lower envelope of the solid curves rather well. An accretion rate of ~ 10-30 $M_\odot$ yr^{-1} is indicated. I feel that this agreement gives further strong support to the radiative accretion picture for M87.

Of course, the accretion models are not unique; in principle, other models could explain the range of temperatures we observe. We can, however, rule out several alternative pictures. The cool component cannot be a point source ($\lesssim$1 arc min) since such a source would be bright enough to be distinguished easily in the imaging data of Fabricant et al. (1980). Similarly, the cool line-emitting material cannot be the cooler outer halo predicted by the adiabatic hydrostatic models (see Bahcall and Sarazin 1978) -- our aperture subtends $\lesssim$5% of the outer region and its surface brightness is much too small (Fabricant et al. 1980). We can also rule out the detailed model of

TABLE 1 X-RAY LINES FROM M87

Line (1)	Rest Energy (eV) (2)	Corrected Line Flux (10^{-3} cm^{-2} s^{-1}) (3)	Line Luminosity[a,b] ($10^{41} h_{50}^{-2}$ erg s^{-1}) (4)	Dominant Ions (5)
O VIII Lα	654	9.4±2.2	5.7	
Blend 1	939-971	2.1±0.4	1.9	Fe XX
Blend 2	997-1017	1.8±0.4	1.7	Fe XVII, XX, XXI, XXII
Blend 3	1021-1043	1.6±0.4	1.5	Ne X, Fe XXI
Blend 4	1106-1128	0.9±0.4	0.9	Fe XVII, XXIII, XXIV
Blend 5	1134-1180	1.0±0.2	1.1	Fe XIX, XXIV
Blend 6	1182-1210	0.6±0.3	0.7	Fe XX
Fe XVII	826	<1.7	<1.3	Fe XVII, O VIII Lγ, δ

NOTES:

a) Corrected for a column density of 2×10^{20} cm^{-2} (Fabricant _et al_. 1978, Lea _et al_. 1979) and the abundances of Brown and Gould (1969).

b) The distance to M87 is taken to be 22 h_{50}^{-1} Mpc, where h_{50}=Ho/50 kms^{-1} Mpc^{-1}.

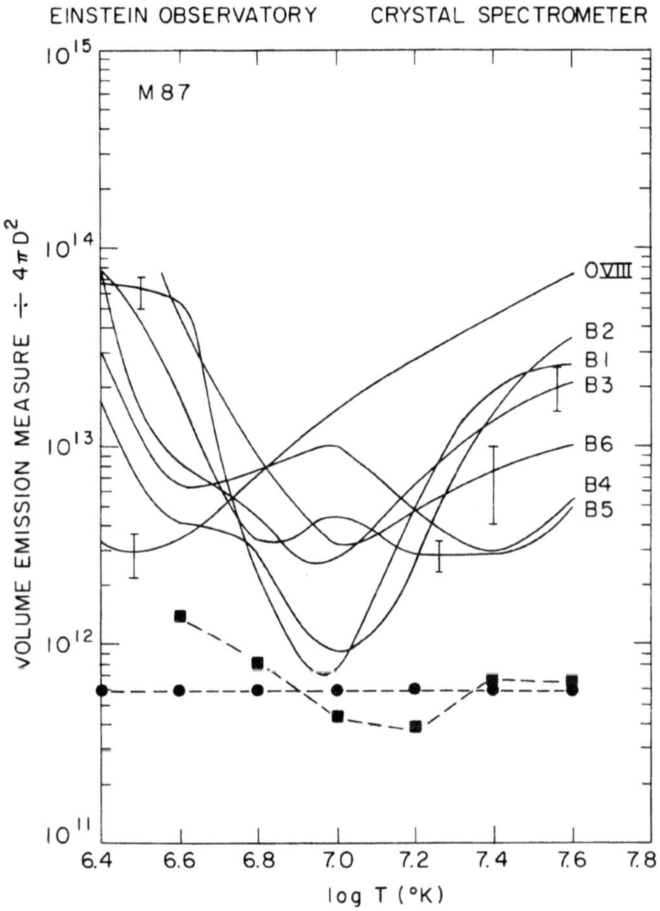

FIG. 5: The volume emission measure required to account for all the flux from a given line at a given log T (solid curves; B refers to "blend"). The units are cm^{-5}. The filled circles show a flat distribution of emission measure which accounts reasonably well for all the lines. The filled squares show an arbitrarily normalized distribution from a polytropic accretion model (see text).

Takahara and Takahara (1979) describing mass outflow due to conductive heating of gas in the galaxy since it predicts only a 20% drop in temperature (see Fig. 1). Future models with lower conductivity might work (L. Cowie and J. Binney, private communication). But recent observations by Andernach et al. (1979) of Faraday rotation suggest that the magnetic field within ~5 arc min of M87 is circumferential; this would effectively suppress thermal conduction to the central part of the galaxy.

NGC1275

There is some new spectral data for NGC1275 in the Perseus Cluster. In this case the galaxy is at the center of the cluster and the overall cluster X-ray temperature is about 7 keV (cf. Mushotzky et al. 1978). Ulmer and Jernigan (1978) found that the cluster appears smaller at lower energies. Both the FPCS and SSS show evidence for cooler material in the vicinity of NGC1275. The preliminary spectrum shown by Mushotzky (1980) includes lines of hydrogenic and helium-like Si and S which are emitted at temperatures less than 2-3 keV (see Fig. 6).

With the FPCS we scanned the region around a prominent Fe XVII line (Canizares et al. 1979 , Jernigan et al. in preparation). The upper part of Fig. 7 shows that we have detected three lines identified as Fe XVII, and O VIII Lyman γ and δ. All three lines are indicative of relatively cool material. The relative strengths of the O and Fe lines can be accounted for with material at a range of temperatures from $\sim 10^7$ to $\sim 10^8$ °K as expected from radiative accretion models. For example, the lower portion of Fig. 7 shows the predicted relative line fluxes for the polytropic accretion model.

As in the case of M87, the data do not uniquely require an accretion model but most of the above arguments against alternative pictures for M87 apply equally well to NGC1275.

OTHER MULTICOMPONENT CLUSTERS

No other cluster sources have been studied with the detail of M87 and NGC1275. But there are other clusters which show two-component spectra. These are A478 (Schnopper et al. 1977), A1060 and Centaurus (Mitchell and Mushotzky 1980, Mushotzky 1980), A2147 (Pravdo et al. 1979, Mushotzky 1980), A2199 and the NGC1129 group (Mushotzky 1980). Analyses of IPC data presented at this meeting show lower temperatures near the center for A576 (White and Silk 1980) as for Perseus (Fabian and Stewart 1980). More circumstantial evidence for spectral complexity in clusters comes from the systematic differences in the cluster temperatures measured by OSO-8 and Ariel-5. These might be explained by the presence of composite spectra in general and of cooler central regions in particular (Mitchell et al. 1979).

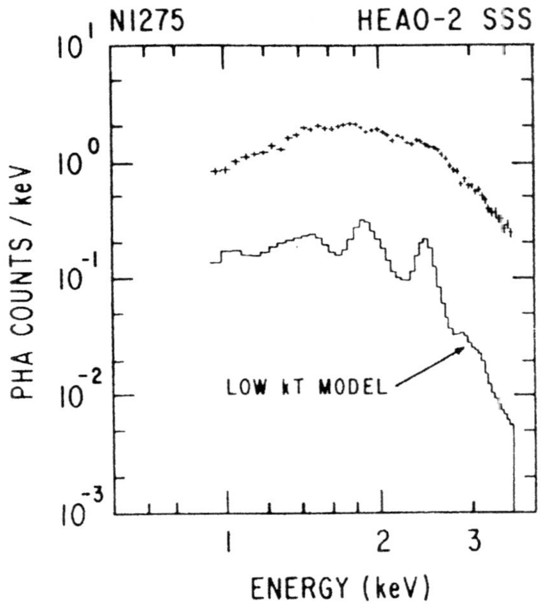

FIG. 6: The SSS spectrum of NGC1275 (Mushotzky _et al_. 1980)

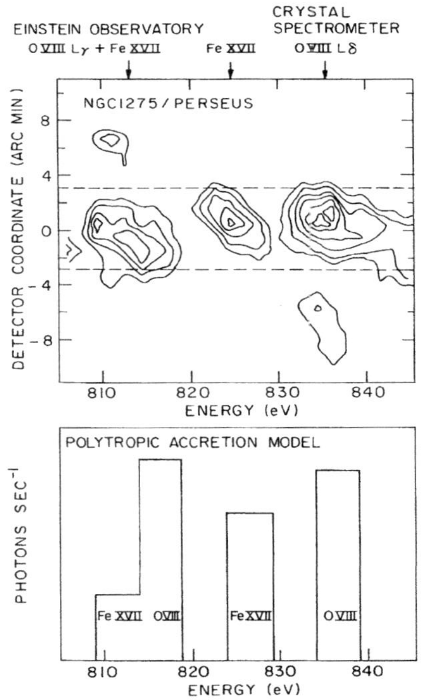

FIG. 7: (Upper) Count rate contours from an FPCS observation of NGC1275 showing three lines as indicated.
(Lower) Relative intensities of lines from a polytropic accretion model and the emissivities of Raymond and Smith (1979).

DISCUSSION

On the basis of the evidence at hand I would like to hypothesize that radiative accretion is a widespread phenomenon in X-ray clusters. The requirement is that the cooling timescale near the center of the cluster be less than a Hubble time. Using Mathews and Bregman's (1978) calculation of the timescale for the growth of thermal instability (τ_{TI}) together with the emissivities of Raymond and Smith (1979) for $10^7 < T < 10^8$ °K gives

$$\tau_{TI} \simeq 6\, t_{50} \left(\frac{T}{10^8\ °K}\right)^{1.3} \left(\frac{n}{10^{-3}}\right)^{-1},$$

where t_{50} is the Hubble time (Ho=50 km s^{-1} Mpc^{-1}), T is temperature and n the gas density in cm^{-3}. The central gas densitites of most clusters are not known. But crude estimates based on emission measures from Mushotzky et al. (1978) and typical X-ray core radii of 200-500 kpc (Jones et al. 1980) show that a significant number of clusters could have sufficiently small values of τ_{TI} to have initiated radiative accretion (see also Schwartz et al. 1979).

As for M87 and NGC1275, low velocity, massive galaxies near the centers of other clusters may serve as the "nucleation sites" for radiative accretion. In order for the galaxy to play more than a passive role it would have to provide a sufficiently deep local gravitational potential to enhance the gas density and give a significantly shorter local value of τ_{TI} (some of the enhanced density might be the result of the galaxy's own trapped material; J. Binney private communication). Early estimates of the characteristics of single galaxies in an X-ray cluster by Gull and Northover (1975) discounted their effect (but cf. Hunt and Sciama 1972). However, many clusters are cooler than their "standard" cluster (cf. Mushotzky et al. 1978, Mushotzky 1980), some galaxies may be more massive than their "standard" galaxy (such as M87, cf. Fabricant, et al. 1980), and even a modest density enhancement may be sufficient to cause nucleation in the many clusters which could have boarderline values of τ_{TI}.

The most logical place to look for the cooling, accreting material is at the centers of dense, compact clusters which contain massive cD galaxies. But often the low temperature emission may be an extremely small fraction of the total luminosity and thus be difficult to detect without high sensitivity X-ray line measurements. Many of the above mentioned clusters which show multicomponent spectra have low luminosities -- this may have facilitated the detection of their cool components which might be masked in higher luminosity clusters. Clusters which are not centrally condensed may reveal less extreme analogs of M87 - massive ellipticals around which radiative accretion has been initiated. It is not known how low the galaxy velocity relative to the intra-cluster gas must be to allow the accretion flow to become established, but De Young et al. (1980) estimate that M87 may have a relative transverse velocity as high as ~100 km s^{-1}. It is at least conceivable that some of the galaxies with enhanced X-ray surface brightness shown by Bill Forman (this volume) are examples of accretion.

ACKNOWLEDGEMENTS

I thank my many colleagues at MIT, CFA and GSFC who contributed to the work presented here, particularly Garrett Jernigan and Richard Mushotzky. I am grateful to John Bahcall, James Binney, Lennox Cowie, William Mathews, Paul Nulsen and Craig Sarazin for useful discussions and comments, and I thank John Raymond and Barry Smith for their unpublished emissivity calculations. This work was supported in part by NASA under contract NAS8-30752.

REFERENCES

Andernach, H., Baker, J.R., von Kap-herr, A., Wielebinski, R. 1979, Astron. Ap. 74, 93.
Arnett, W.D. 1978, Ap.J. 219, 1008.
Bahcall, J.N. and Sarazin, C.L. 1977, Ap.J. (Letters), 213, L99.
Bahcall, J. and Sarazin, C. 1978, Ap.J. 219, 781.
Brown, R.L. and Gould, R.J. 1969, Phys. Rev. D. 1, 2252.
Canizares, C.R., Berg, C., Clark, G.W., Jernigan, J.G., Kriss, G., Markert, T.H., Schattenburg, M. and Winkler, P.F. 1980, presented at IAU, Montreal; to appear in Highlights of Astronomy 5, 657.
Canizares, G.R., Clark, G.W., Markert, T.H., Berg, C., Smedira, M., Bardas, D., Schnopper, H. and Kalata, K. 1979, Ap.J. (Letters) 234, L33.
Cavaliere, A. and Fusco-Femiano, R. 1976, Astron. & Ap. 49, 137.
Cowie, L.L. and Binney, J. 1977, Ap.J. 215, 723.
Cowie, L. and Songaila, A. 1977, Nature 266, 501.
Davison, P.J.N. 1978, M.N.R.A.S. 183, 39P.
DeYoung, D.S. 1978, Ap.J. 223, 47.
DeYoung, D.S., Condon, J.J., and Butcher, H.H. 1980, Bul. Amer. Astron. Soc. 11, 655.
Fabian, A.C. and Nulsen, P.E.J. 1977, M.N.R.A.S. 180, 479.
Fabian, A.C. and Stewart, G.C. 1980, Bul. Amer. Astron. Soc. 4, 785.
Fabricant, D. Topka, K., Harnden, Jr. F.R. and Gorenstein, P. 1978, Ap.J. (Letters) 226, L107.
Fabricant, D., Lecar, M. and Gorenstein, P. 1980, Ap.J. (in press).
Giacconi, R. et al. 1979, Ap.J. 230, 540.
Gorenstein, P., Fabricant, D. Topka, K., Tucker, W. and Harnden, Jr. F.R. 1977, Ap.J. (Letters) 216, L95.
Gull, S.F. and Northover, K.J.E. 1975, M.N.R.A.S. 173, 585.
Gursky, H. and Schwartz, P. 1977 Ann. Rev. Astron. Ap. 15, 541.
Hunt, R. and Sciama, D.W. 1972, Nature 238, 320.
Jones, C., Mandel, E., Schwarz, J., Forman, W., Murray, S.S., Harnden, Jr. F.R. 1979, Ap.J. 234, L21.

Lawrence, A. 1978, M.N.R.A.S. 185, 423.
Lea, S.M. 1975, Ap. J. (Letters) 16, 141.
Lea, S.M., Mason, K.O., Reichert, G., Charles, P.A. and Riegler, G. 1979, Ap.J. (Letters) 227, L67.
Mathews, W.G. 1978, Ap.J. 219, 413.
Mathews, W.G. and Bregman, J.N. 1978, Ap.J. 224, 308.
Mitchell, R.P., Culhane, J.L., Davison, P.J.N., Ives, J.C. 1976, M.N.R.A.S. 175, 29P.
Mitchell, R.J. and Culhane, J.L. 1977, M.N.R.A.S. 178, 75P.
Mitchell, R.J., Ives, J.C. and Culhane, J.L. 1977 M.N.R.A.S. 181, 25P.
Mitchell, R.J., Dickens, R.J., Bell-Burnell, S.J. and Culhane, J.L. 1979 M.N.R.A.S. 189, 329.
Mitchell, R.M. and Mushotzky, R.E. 1980, Ap.J. 236, 730.
Mushotzky, R.E. 1980 in X-Ray Astronomy, Proceedings of the NATO Advanced Study Institute held at Erice, Sicily, July 1979, G. Setti and R. Giacconi, eds., (D. Reidel, Dordrecht), p.171.
Mushotzky, R.F., Serlemitsos, P.J., Smith, B.W., Boldt, E.A. and Holt, S.S. 1978, Ap.J. 225, 21.
Pravdo, S.H., Boldt, E.A., Marshall, F.E., McKee, J., Mushotzky, R.F., Smith, B.W., Reichert, G. 1979, Ap.J. 234, 1.
Raymond, J.C. and Smith, B.W. 1978, Ap.J. Suppl. 35, 419.
Raymond, J.C. and Smith, B.W. 1979, private communication.
Schnopper, H.W., Delvaille, J.P., Epstein, A., Helmken, H., Harris, D.E., Shrom, R.G., Clark, G.W., Jernigan, J.G. 1977, Ap.J. (Letters) 217, L15.
Schwartz, D.A., Schwartz, J., and Tucker, W. 1979, preprint.
Serlemitsos, P.J., Smith, B.W., Boldt, E.A., Holt, S.S. and Swank, J.H. 1977, Ap.J. (Letters) 211, L63.
Silk, J. 1976, Ap.J. 208, 646.
Smith, B.W., Mushotzky, R.F. and Serlemitsos, P.J. 1979, Ap.J. 227, 37.
Takahara, M. and Takahara, F. 1979, Progress of Theoretical Phys. 62, 1253.
White, S.D.M. and Silk, J. 1980, Bul. Amer. Astron. Soc. 4, 783.
Ulmer, M.P. and Jernigan, J.G. 1978, Ap.J. (Letters) 222, L85.

THEORETICAL MODELS OF X-RAY EMISSION FROM RICH CLUSTERS OF GALAXIES

Lennox L. Cowie
Princeton University Observatory

1. INTRODUCTION

At the present time it seems best to divide the X-ray emission from rich clusters into two broad classes according to morphology, which we will call homogeneous and clumped respectively. (Further subdivision does not seem useful as yet.) The homogeneous clusters such as Perseus and Coma show uniform emission with relatively high temperatures ($\gtrsim 10^8$°K) while in clumped clusters, such as Virgo, emission is concentrated around the member galaxies and is relatively cool ($\lesssim 10^7$°K). In theoretical terms the homogeneous case requires us to consider uniform cluster atmospheres while the clumped case is a problem of individual galaxy atmospheres.

Table I. Cluster Morphology

	Homogeneous	Clumped
	Uniform X-ray morphology	Patchy X-ray emission around galaxies
	Spiral deficient ($\lesssim$30%)	Standard spiral fraction ($\sim$ 50%-60%)
	Deficient in nuclear emission lines from ellipticals	Standard elliptical emission line fraction ($\sim$ 15%)
	High X-ray luminosity ($\gtrsim 10^{44}$ ergs)	Low X-ray luminosity ($\lesssim 10^{44}$ erg/s)
Rood-Sastry	CD, B, C, L, F	Ir
Bautz-Morgan	I, II, III	III

In Table I we have attempted to summarise these classifications in terms of their X-ray properties and to show how they relate to other morphological classifications of the clusters. Reviews of cluster morphology have recently been given by N.Bahcall(1979) and Jones et al(1979).

R. Giacconi (ed.), X-Ray Astronomy with the Einstein Satellite, 227–240.

The classification roughly divides the dynamically evolved homogeneous clusters with deep potential wells and the less evolved clumped clusters where the cluster potential is weaker and galaxy potentials correspondingly more important. The hotter denser cluster atmosphere in the homogeneous case appears capable of stripping or evaporating gas from galaxies (as is evidenced by the low spiral fraction in these clusters) and homogenising it, while in the clumped case the cluster atmosphere is either too tenuous or too cool to do this. As the morphology of the underlying clusters evolves, the transition from one case of X-ray morphology to the other may be very rapid as the cluster atmosphere forms by destroying galaxy atmospheres (Norman and Silk 1979, Sarazin 1979).

The determination of the morphology of gas in the X-ray homogeneous rich clusters should in principle be a straightforward problem. We know that the gas must be in quasi-hydrostatic equilibrium within the potential well of the cluster, since sound crossing times given by $t \sim r/$ sound speed $\approx$ 600 kpc/10^8cm/s $\approx 6.10^8$ yrs are short compared with the cosmological timescales over which cluster and gas parameters are evolving. In addition it is probable that cool material cannot coexist with the hot gas (Cowie and Songaila 1977, Bergeron and Gunn 1977) and that the gas will be highly homogeneous in temperature and density on scalelengths shorter than those of the cluster.

Under these conditions it remains only to specify the equation of state of the gas and the shape and depth of the cluster potential to obtain the gas distribution. Conversely if the gas distribution is known we may invert the process to obtain the cluster potential. The processes by which the cluster gas formed appears only in the local equation of state and even this information may be lost if the gas is well mixed subsequently, if thermal conductivity redistributes energy, or if radiative losses become important.

In section 2 we review the theory of cluster gas with polytropic equations of state and discuss what local polytropic indices in the gas distribution are predicted by the various models of the gas origin. In section 3 we review the consequences of modifying the equation of state by thermal cooling in the central regions of the cluster (radiatively regulated flows).

If the equation of state can be understood locally we may attempt to predict the X-ray luminosity of the clusters at distant cosmological epochs and measure the evolution of the cluster's potential by observing the evolution of the X-ray luminosity and surface brightness. Work on this topic is reviewed in section 4.

Comparatively little attention has been paid as yet to the somewhat more complex problem of gas distribution in the clumped clusters where individual galaxy potentials are important owing to the relative coolness ($T \approx 10^7$°K) of the gas. The presence of hotter cluster gas ($T \approx 10^8$°K) is presently ambiguous in these cases. At most its density is a factor 10 less than that in the homogeneous clusters. Most work has been done on the high luminosity X-ray emission around M87 in the Virgo cluster and theories for the more general galaxy emission are now being put forward. In section 5 we discuss these models and attempt to predict what we may hope to learn from observations of the clumped clusters.

2. GAS DISTRIBUTIONS

Lea (1975), Gull and Northover (1975), and Cavaliere and Fusco-Femiano (1976) have pointed out that if the gas is described by a polytropic equation of state $\underline{P} = \kappa\rho^{\gamma}$ then the equation of hydrostatic equilibrium $\nabla\underline{P} = -\rho\nabla\phi(\underline{r})$ and a specified gravitational potential complete the description of the distribution of gas within the cluster. Since the central density (ρ_c) and temperature (T_c) must also be specified, these results form a three parameter family. However the central density is only a normalising factor for the density profile so that the form of the solution is a function only of (T_c, γ). The equation of hydrostatic equilibrium may be directly integrated to give

$$T = T_c\left\{1 - \frac{(\gamma-1)}{\gamma}\frac{(\phi-\phi_c)}{a_c^2}\right\} \qquad \gamma \neq 1 \qquad (1a)$$

$$\rho = \rho_c\left\{1 - \frac{(\gamma-1)}{\gamma}\frac{(\phi-\phi_c)}{a_c^2}\right\}^{\frac{1}{\gamma-1}} \qquad (1b)$$

$$T = T_c \qquad \gamma = 1 \qquad (1c)$$

$$\rho = \rho_c \exp\left\{-\frac{(\phi-\phi_c)}{a_c^2}\right\} \qquad (1d)$$

where a_c is the isothermal sound speed at the central temperature.

In Figure 1 we plot temperature and density as a function of ϕ/ϕ_c for various values of $(\gamma-1)\phi_c/a_c^2$. The upper axis shows this translated into a functional dependence on radius based on the model of King (1972)(a realistic fit to shape of the Coma cluster). Convolving such profiles to surface brightness profiles and fitting to clusters allows the various parameters to be determined. As an example of the application of this procedure Bahcall and Sarazin (1977) derived $\gamma = 1 \rightarrow 5/3$, $T_c = 1.1 \rightarrow 2.8\times10^8\,°K$, $n_c = 1.5 \rightarrow 3.3\times10^{-3}\,°K$ and $M_{gas} = 4\times10^{13} \rightarrow 4.6\times10^{14}M_{\odot}$ for Coma.

Physical processes (mixing: $\gamma = 5/3$; conductivity: $\gamma = 1$) may give the equation of state as a polytrope, with a specified index. Alternatively the mechanism of gas input, either injection of gas from

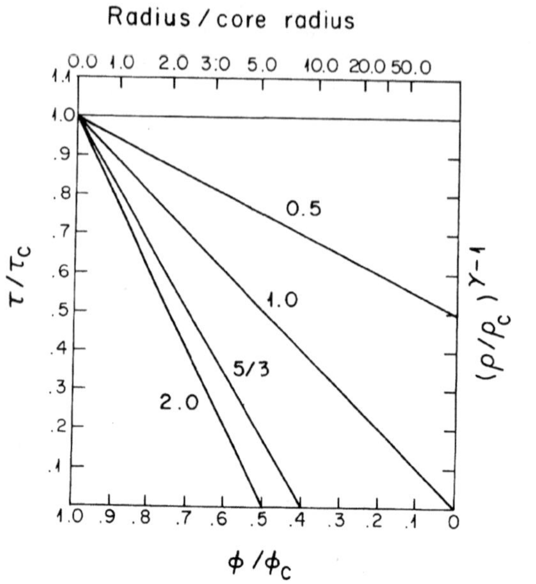

Figure 1 Temperature and density profiles as a function of gravitational potential and of normalized radius.

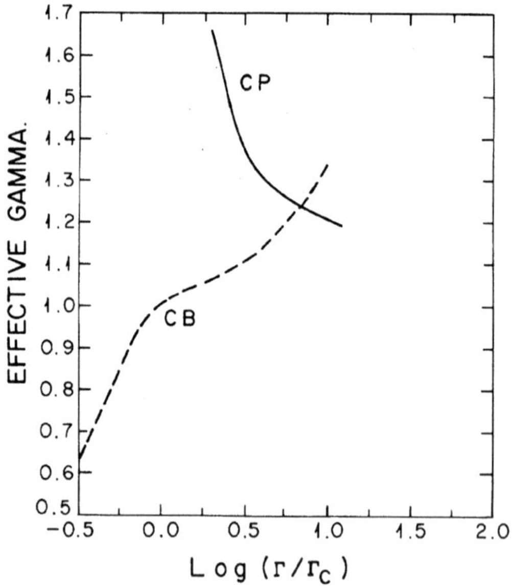

Figure 2 Local indices of polytropic fits to infall models (CP) and galaxy injection models (CB).

galaxies or infall of material into the cluster may set up a particular equation of state (not generally a polytrope). We can parameterise numerical models of these last two cases by a "local polytropic index" $\equiv \gamma(r) \equiv (d\ell nT/d\ell np\rho + 1)$, and this is shown in Figure 2 based on computations by Cowie and Binney 1977 (galaxy injection) and Cowie and Perrenod 1978 (gas infall into clusters). The gas injection models of Cowie and Binney have significant radiative cooling inside the core radius and may not be generally applicable in this region (section 3). Out to a few core radii they are well approximated by an isothermal equation of state ($\gamma = 1$). The infall models, on the other hand, have indices between 4/3 and 5/3 in this region, intermediate between adiabatic and isothermal cases, and are not well fit by a polytrope.

3. RADIATIVELY REGULATED ACCRETION FLOWS

Cooling times of the X-ray emitting gas in the central regions of the luminous homogeneous rich clusters are strikingly close to cosmological timescales. Setting $\Lambda(T) = 2\times10^{-27}\ T^{1/2}$ ergs cm^3 s^{-1} then $t_{cool} \equiv 3kT/n\Lambda(T) = 6.7\times10^7\ T_8^{+1/2}/n$ years where T_8 is the temperature of the gas in units of 10^8°K. For an open universe with H = 100h km s^{-1} Mpc^{-1}, $t_H = 10^{10}h$ years and cooling will occur if the central gas density exceeds $6.7\times10^{-3} T_8^{-1/2}\ h^{-1}$ cm^{-3}. This should be compared with central gas densities of $4\times10^{-3}\ h^{-1/2}$ cm^{-3} (Coma) typical for the X-ray luminous rich clusters.

If the cluster core gas density has ever exceeded the critical cooling density a quasi static cooling flow situation will be set up. In Cowie and Binney (1977), where this was first noted, this was called radiatively regulated accretion flow. It was independently suggested by Fabian and **Nulsen** (1977)and the thermal stability and rotation of the flow was treated by Mathews and Bregman (1978) and by Cowie, Fabian, and **Nulsen** (1980).

Whether a cluster posseses such a cooling flow or not cannot depend on the properties of galaxies in the central regions of the cluster. For gas at 10^8°K the galaxy potential wells are not sufficiently deep to perturb the gas density and influence the cooling. Consider a galaxy with a massive spherical halo described by a constant rotation velocity V. An isothermal hydrostatic gas atmosphere then obeys the equation

$$\frac{d}{dr}(\rho c^2) = \frac{-GM(r)\rho}{r^2} = \frac{-V^2\rho}{r} \qquad (2)$$

where c is the isothermal sound speed (1200 $T_8^{1/2}$ km/s). This gives in terms of a reference density and radius (ρ_R, r_R)

$$\rho = \rho_R \left\{ \frac{r_R}{r} \right\}^{V2/c^2} . \tag{3}$$

Typical circular velocities even for massive galaxies do not exceed 500 km s^{-1} so that the power law index is less than .17. Even changing the radius by a factor 300, the densities vary by less than 3. Thus the difference between clusters such as Coma, which show no evidence of cooling, and Perseus, which is the only well established case of a cooled flow, is probably best understood in terms of the gas density dependance, rather than the presence of a central stationary and possibly massive galaxy.

Cooled flows are interesting for two main reasons. Reducing the temperature of the gas brings us into regions (both X-ray and optical) where spectral lines of many elements may be observed and cluster gas metallicities determined. Also, once the gas flow does cool (to $T \lesssim 10^7$°K) it may focus onto existing stationary galaxies and the resulting mass accretion flow may significantly influence the evolution of such galaxies.

As regards the first point the cooling flows are unusual in that the spectra are easy to predict theoretically and departures from collisional ionisation are also easily included. The reason for this is that in the central regions of the cluster where the temperature drops below the asymptotic temperature we require, for self-consistency, that gravitational energy input be small and that radiative losses be balanced by the inward enthalpy flux of the accreting material. From integrations through the results of Cowie and Binney we estimate that the gravitational correction is less than 40% at the radius where the temperature begins to decrease ($\sim$1 cluster core radius or 6' for Perseus) and drops rapidly thereafter. We may therefore write the energy balance equation as

$$\frac{5}{2} k \frac{\mathring{M}}{\bar{m}} \frac{dT}{dr} = -4\pi r^2 n^2 \Lambda(T) \tag{4}$$

where k is the Boltzmann constant, $\mathring{M}$ the mass flow rate and m the average mass per particle. The luminosity in a line with emissivity $\varepsilon(T)$ is now

$$L = \int 4\pi r^2 n^2 \varepsilon(T) dr = \frac{5}{2} k \frac{\mathring{M}}{m} \int_0^{T_{max}} \frac{\varepsilon(T)}{\Lambda(T)} dT \tag{5}$$

where we have assumed that ε peaks below the asymptotic temperature. For asymptotic temperatures of 10^8°K this is almost always the case as most lines of interest occur at lower temperatures. Departures from ionisation equilibrium may be approximated by using isobaric time dependent models calculated with an initial temperature equal to the asymptotic temperature.

Equation (5) shows that the mass flow rate is obtainable from observations of a single spectral line as a simple multiplicative constant and in addition that the ratio of spectral lines should be invariant from cooling cluster to cooling cluster.

In Table II we have estimated the luminosity of a number of observable lines using equation (5) and published emissivities given by Raymond and Smith (1977). We also show the available information on line strengths in NG1275. All lines are consistent with a mass flow rate of about 100 $M_\odot$/year.

Table II. Line Luminosities in Radiatively Regulated Flow

		Predicted luminosity (100 $M_\odot$/yr Cosmic Abd.) 10^{42} erg/s	NGC 1275 10^{42} ergs/s	M87* 10^{42} ergs/s
10.65	Fe XXIV*	1.1		≤.37+
11.01	Fe XXIV/XXIII*	1.6		.27+
12.26	Fe XVII*	3.5		
12.83	Fe XX*	3.9		
14.37	Fe XVIII*	2.8		
15.01	Fe XVII*	6.2	4.4+	≤.22+
15.26	Fe XVII*	2.9		
16.77	Fe XVII*	2.2		
17.05	Fe XVII*	5.7		
18.97	O VIII	2.2		1.2+
5.04	S XV*	.34	detected†	
6.18	Si XIV	.78	detected†	
6.65	Si XIII*	.32	detected†	
8.06	Fe XXIV*	.42		
8.42	Mg XII*	.33		

\+ Canizares et al. (1980).

* Lines marked with a star peak at or above 10^7 °K and should not be compared with the observations of M87 where the initial temperature is only 3.10^7 °K.

† Muzhotsky et al. (1980).

We may also infer mass flow rates from detailed observations of optical filament structure by estimating masses in the cooling regions and by considering the cooling times in the asymptotic hot gas. All three methods give a consistent mass flow rate of about 200 $M_\odot$/yr similar to that estimated above for the Perseus/1275 system (Cowie et al. 1980).

4. THE EVOLUTION OF CLUSTER X-RAY SOURCES OVER COSMOLOGICAL TIMESCALES

The observed metallicity of the cluster gas implies that the cluster atmospheres are mainly formed from galaxy ejecta. The small mass of gas compared with the virial mass of the cluster also implies that little primordial material is left from the period of cluster formation. This implies either that galaxy formation is extremely efficient (greater than 90%) or there is some way of removing primordial gas from the cluster. In the first case the early X-ray luminosity of the cluster may be relatively low, while in the second case the luminosity may be high. It is hard to model such initial transient effects. However, we may argue that they die out quickly and that a slowly evolving hydrostatic cluster atmosphere forms rapidly thereafter.

The time dependence of the galaxy mass lossage into the cluster is not clear. On energetic grounds De Young (1978) has shown that there is enough supernova energy to get the galaxy cleared of gas at all times and Bookbinder et al. (1980) have argued that hot winds stream from early galaxies. Sarazin (1979), Norman and Silk (1979) have suggested instead that this energy is radiated away and the gas stored in thermally stable haloes around the galaxies until finally released into the cluster by halo-halo collisions. For homogeneous clusters with high central galaxy densities the cluster luminosity is determined by the central core regions where the interaction processes are fast on cosmological timescales. Both possibilities therefore imply that the cluster atmosphere comes into existence rapidly after cluster formation typically in less than or order 10^8 years.

Once the atmosphere forms it subsequently behaves quasistatically in the evolving cluster potential (c.f. section 2). Cowie and Perrenod (1978) have made time dependent numerical simulations of the evolving X-ray luminosity for a static cluster atmosphere and Perrenod (1978) has generalised this to evolving cluster potentials.

For the static case the atmosphere is hydrostatic and essentially invariant except for the density normalisation which may rise slowly as mass is released into the cluster from the galaxies. This results in a very slow change in cluster evolution. The luminosity in this case varies by less than a factor of 1.4 from $z = 1$ to 0. Perrenod's simulations were based on a cluster potential from White's n-body simulations of the evolution of the cluster galaxies. Here the additional effects of the steepening of the potential well result in a faster evolution. Typically clusters may be 10× more luminous at $z = 0$ than $z = 1$. The two cases reasonably cover the possible range of cluster evolution.

In figure (3) we show the evolution of the bolometric cluster luminosity from $z = 1$ to $z = 0$, and compare it with recent Einstein observations made by Henry et al. (1979). Henry et al. converted measured count rates to luminosity by assuming that the temperature of the cluster gas (7 Kev) does not vary from $z = 0$ to 1, (the

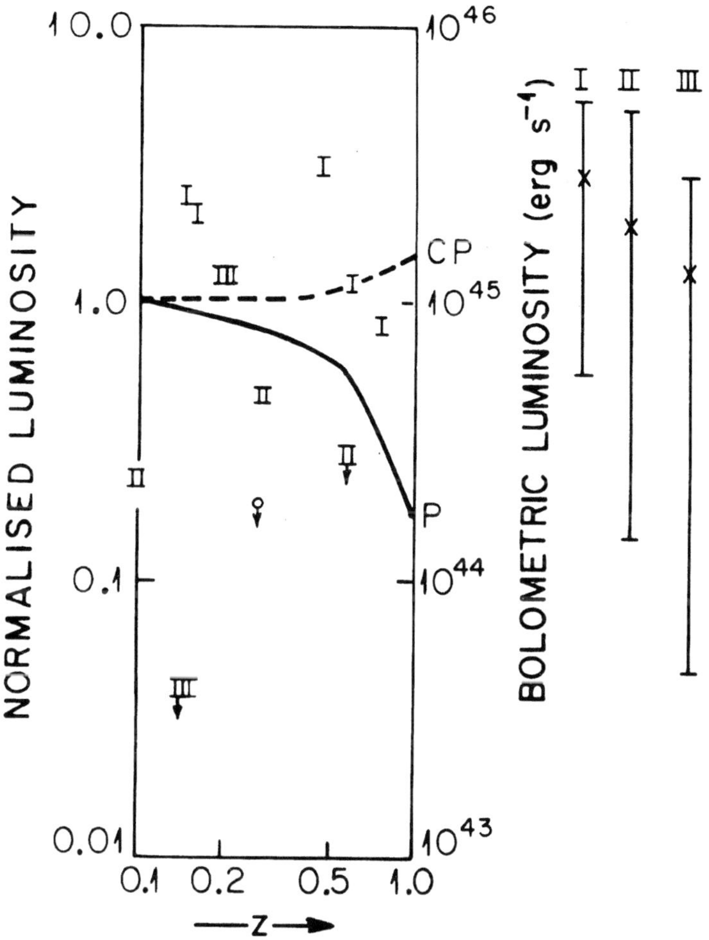

Figure 3 The evolution of the bolometric luminosity as a function of z is shown for models of Perrenod (P) and Cowie and Perrenod (CP). Also shown are the measured luminosities of 11 clusters obtained with Einstein by Henry et al. (1979). These are marked with their Bautz-Morgan classification which correlates roughly with luminosity. Shown at right are the range and mean of luminosities for clusters of the various BM classes at z = 0 from detected clusters in the Ariel V survey derived from the table of McHardy (1978).

variation is less than a factor of 1.5 even for evolving cluster potentials). If the T is high compared with the deredshifted energies at which the observations are made, the luminosity is then related to the flux at measured energy E (Weinberg 1972), as

$$L_B = 4\pi d_L{}^2 \; . \; kT(1+z) \; . \; F(E)$$

where d_L is the luminosity distance in the usual notation. (We have set $q_o = 0$.) The results are not sensitive to the assumed source temperature since $L_B \sim T$ only and T does not differ by more than a factor 2 from this value. Henry et al. quote (0.5 - 4.5 Kev) luminosities, the bolometric luminosity is 2.4 times this value.

It is clear when plotted in this fashion that the data is not yet adequate to distinguish the evolution of the cluster potential and a much larger sample at z = 1 is necessary. What is clear however is that the general treatment of the gas in terms of slowly varying atmospheres is correct and that the evolution of the luminosity is small, probably no more than the factor 10 decrease predicted by the Perrenod models.

5. CLUMPED CLUSTERS

Modelling galaxy atmospheres in clusters is a difficult problem **requiring knowledge of the rate of mass loss and energy injection from the stars, as well as the potential of the galaxy and the ambient intracluster gas properties.**

For central static galaxies the problem may be simplest. Most work has been done on modelling the extended emission around M87 in the Virgo cluster. The original work on this subject postulated that there existed a massive halo in M87 capable of gravitationally confining the observed hot atmosphere (Bahcall and Sarazin 1977, Mathews and Bregman 1977). Most recently Fabricant et al. (1980) have used this model **to suggest that M87 has a halo of $\sim 2 \times 10^{13}\ M_\odot$ stretching to 230 kpc.**

Such a halo is not required by the observations (Binney and Cowie 1980). If there exists a hot ($T \sim 10^8$°K) intracluster gas in Virgo as has been suggested by Davison (1978) and Lawrence (1978), the pressure of this gas may confine the cooler galaxy atmosphere.

To demonstrate that models can be generated which satisfy the observations but yet do not require M87 to be massive, let us postulate that the gravitational potential of M87 is negligible and the galaxy atmosphere isobaric and in pressure equilibrium with its surroundings [$nT = 5.10^4$°K cm^{-3}]. Let us also assume that the early primordial galaxy was not an efficient radiator so that the energy per released particle declined as the galaxy evolved. Then in the absence of mixing the atmosphere will have an onion skin structure and possess a radial-temperature gradient increasing outward to match the intracluster gas temperature. (Higher temperature gas would have buoyed from the cluster.) We may describe the initial form of the atmosphere by specifying the mass of gas in temperature range dT as $M_o(T)dT$. The onion skin atmosphere is a stratified memory of the galaxy evolution.

Such an atmosphere is interesting since the outer hot regions are thermally stable over Hubble timescales ($T = 10^8$°K, $n = 5.10^{-4}$ cm^{-3}, $t_{cool} = 1.3 \times 10^{11}$ yrs) while the central regions cool and condense onto the galaxy. The resulting quasi-static flow is very analagous to the radiatively regulated flow models of section 3. The outer regions form a mass and energy reservoir for the central inflow. It may easily be shown from consideration of the cooling times that at time t the mass of gas in temperature range dT is related to the initial mass distribution by,

$$M(T) = M_o \{T(1 + (2+\alpha)t/t_c)^{1/2+\alpha}\}/(1 + (2+\alpha)t/t_c)^{1+\alpha/(2+\alpha)}$$

where α is defined by $\Lambda(T) \sim T^{-\alpha}$. If $M(T) = KT^{\beta}$, then in the inner regions ($t_c << t$) $M(T) \sim T^{1+\alpha}$ while the gas distribution in the outer regions ($t_c >> t$) remains unperturbed and $M(T) \sim T^{\beta}$. The cooling time equals the Hubble time for gas at temperature $T_R \approx 3.10^7$°K for $nT = 5.10^4$°K cm^{-3} and half cosmic abundance which divides the two regimes. The inward mass flow rate is given by $M(T_R)T_R/t_{Hubble}$.

We may combine the mass distributions with the isobaric assumption to obtain the temperature profile,

$$\frac{T}{T_R} = \{\frac{r}{R}\}^{3/3+\alpha} \qquad r < R \quad (T = T_R \text{ at } r = R)$$

$$\frac{T}{T_R} = \{\frac{r}{R}\}^{3/2+\beta} \qquad R_m > r > R \quad (T = T_{cluster} = 10^8 \text{°K at } r = R_m)$$

We have specified the profile in terms of two parameters which measure the mass of gas released by the galaxy at late times (R) and overall (R_m).

Figure 4 shows the surface brightness predicted by this profile when a uniform cluster background is subtracted off compared with the Einstein results of Fabricant *et al.* The expression for calculating the surface brightness is,

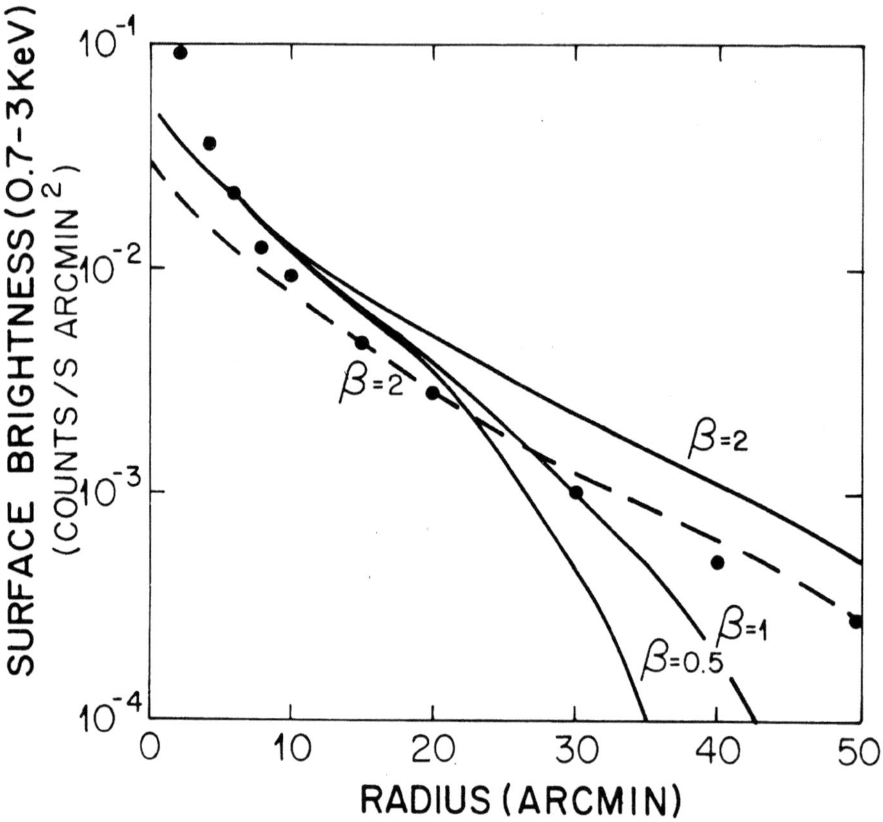

Figure 4 The surface brightness predicted by a cooling isobaric atmosphere around M87 is shown for (nT) = 5×10^4°K cm^{-3} [solid line] and values of (β, R_m) of (2,66'), (1,45'), and (0.5, 36') and (nT) = 4×10^4°K cm^{-3} and (β, R_m) = (2,66'). The latter model provides an excellent fit to the data of Fabricant et al. (1980) outside ∿ 5'. The interior discrepancy should not be taken too seriously since in this region line emission becomes dominant and we have not attempted to vary abundances (1/2 cosmic).

$$S(R_p) = \frac{1}{4\pi} \cdot 2(nT)^2 \int_o^{R_m} \left\{\frac{\varepsilon(T)}{T^2} - \frac{\varepsilon(T_d)}{T_{C1}^2}\right\} \frac{2rdr}{\sqrt{r^2-R_p^2}}$$

evaluated at projected radius R_p. The emissivity was calculated from Raymond and Smith (1977) and convolved through the Einstein effective area to give a count rate in the 0.7 - 3 Kev band. The mass of gas within radius R = 100 kpc ($n = 1.7 \times 10^{-3}$ cm^{-3}, $T = 3.10^7$°K) is 2.10^{11} $M_\odot$, while the total mass released is an order of magnitude larger. These observationally determined losses agree well with theoretical predictions (Bookbinder et al. 1980). The present inflow rate of $M(R)/t_{Hubble} \approx 20$ $M_\odot$/yr and agrees well with mass flow rates inferred from low temperature spectral lines [Table I]. This simple model is in remarkably good agreement with the observations.

For general galaxies in the cluster the problem is greatly complicated by their motion through the cluster. Motion along the orbit may produce sufficient changes to modify the properties of individual galaxies and in addition substantial variation from galaxy to galaxy may be expected. Fabian, Schwarz and Forman (1980) have considered the galaxies M86 and M84 in Virgo. From consideration of the morphology and luminosity (2.10^{41} erg/s) of M86 they conclude that the emission is from a pressure confined gaseous atmosphere released by the galaxy while outside the core. This gas is now being ablated and destroyed as the galaxy re-enters the core. M84 on the other hand appears to lie within the core and possess no sizeable atmosphere. This strongly illustrates how much individuality may be expected in galaxies in the clumped clusters.

ACKNOWLEDGEMENT

This work was supported by NASA Grant NGL31-001-007.

REFERENCES

Bahcall, J. and Sarazin, C: 1977, Ap. J., 213, p. L99.
Bahcall, N.: 1980 in I.A.U. Highlights of Astronomy, ed. P. A. Wayman (Dordrecht: Reidel), p. 699.
Bergeron, J. and Gunn, J.: 1977, Ap.J., 217, p. 892.
Binney, J. and Cowie, L.: 1980, in preparation.
Bookbinder, J., Cowie, L.L., Krolik, J., Ostriker,J.P. and Rees, M.: 1980, Ap.J., 237, p. 647.
Canizares, C., Berg, C., Clark, G., Jernigan, J.G., Kriss, G., Markert, T.H., Schattenburg, M. and Winkler, P.F.: 1980 in I.A.U. Highlights of Astronomy 5, ed. P.A. Wayman (Dordrecht: Reidel), p. 657.
Cavaliere, A. and Fusco Femiano, R.: 1976, Astron. and Astrophys., 49, p. 137.
Cowie, L.L., and Binney, J.: 1977, Ap.J., 215, p. 723.
Cowie, L.L., Fabian, A. and Nulsen, P.E.J.: 1980, M.N.R.A.S., 191, p. 399.
Cowie, L.L. and Perrenod, S.C.: 1978, Ap.J., 219, p. 354.
Cowie, L.L. and Songaila, A.: 1977, Nature, 266, p. 501.
Davison, P.J.N.: 1978, M.N.R.A.S., 183, p. 39P.
De Young, D.S.: 1978, Ap.J., 223, p. 47.
Fabian, A. and Nulsen, P.E.J.: 1977, M.N.R.A.S. 180, p. 479.
Fabian, A., Schwarz, J. and Forman, W.: 1980, preprint.
Fabricant, D., Lecar, M. and Gorenstein, P.: 1980, submitted to Ap.J.
Gull, S.F. and Northover, K.J.E.: 1975, M.N.R.A.S., 173, p. 585.
Henry, J.P., Branduardi, G., Briel, U., Fabricant, D., Feigelson, E., Murray, S.S., Soltan, A. and Tananbaum, H.: 1979, Ap.J., 234, p. L15.
Jones, C., Mandel, E., Schwarz, J., Forman, W., Murray, S.S. and Harnden, F. R., 1979 Ap. J. 224, L21.
King, I.R.: 1972, Ap. J., 174, p. L123.
Lawrence, A.: 1978, M.N.R.A.S., 185, p. 423.
Lea, S.: 1975, Ap.J.Letters, 16, p. L141.
Muzhotsky, R., Holt, S., Boldt, E. and Serlemitsos, P.: 1980, preprint.
Norman, C. and Silk, J.: 1979, Ap.J., 223, p. L1.
Perrenod, S.: 1978, Ap.J., 224, p. 285.
Raymond, J.C. and Smith, B.W.: 1977, Ap.J.Suppl., 35, p. 419.
Sarazin, C.: 1979, Astrophys. Letters, 20, p. 93.
Weinberg, S.: 1972, Gravitation and Cosmology (New York: Wiley).

EXTRAGALACTIC RADIO SOURCES AND X-RAY ASTRONOMY

M. S. Longair *
Mullard Radio Astronomy Observatory
Cavendish Laboratory, Cambridge

1. INTRODUCTION

I have been asked to survey a number of aspects of extragalactic radio sources in which the radio and X-ray observations come together to provide complementary information about physical conditions in active systems. Some of these aspects are already well known, for example the X-ray emission of clusters of galaxies and of Type I Seyfert galaxies, but others are new and have come directly out of high resolution observations made with the Einstein X-ray observatory - for example, the X-ray jets seen in Centaurus A, Virgo A and 3C 273. What I want to do is to put on my radio astronomical hat to begin with in order to isolate those aspects of the physics of extragalactic radio sources which appear to be of particular relevance to X-ray astronomy and specifically to the capabilities of the Einstein Observatory. I begin with a brief survey of the properties of the radio source population.

2. THE RADIO SOURCE POPULATION

One of the most remarkable things about studying extragalactic radio sources is that if you look at all the bright sources in the sky, they span a very wide range of intrinsic luminosities. Figure 1(a) shows the histogram of the luminosity distribution of all the brightest extragalactic radio sources in the northern sky in directions away from the Galactic Plane with flux densities $S_{408} \geq 10$ Jy. This sample was selected at a low radio frequency of 408 MHz. On this luminosity scale, our own Galaxy has $L_{408} \approx 10^{22}$ W Hz^{-1} so that the typical radio sources which are most easily observable have luminosities between 10^3 and 10^8 times that of the Galaxy. The quasars in the sample have been indicated by shaded boxes in Figure 1(a). Figure 1(b) shows a smoothed version of this histogram including a more precise determination of the low luminosity tail from observations of larger samples of nearby galaxies. The figure also indicates where specific well-known

* Present address: Royal Observatory, Blackford Hill, Edinburgh, Scotland.

R. Giacconi (ed.), X-Ray Astronomy with the Einstein Satellite, 241–259.

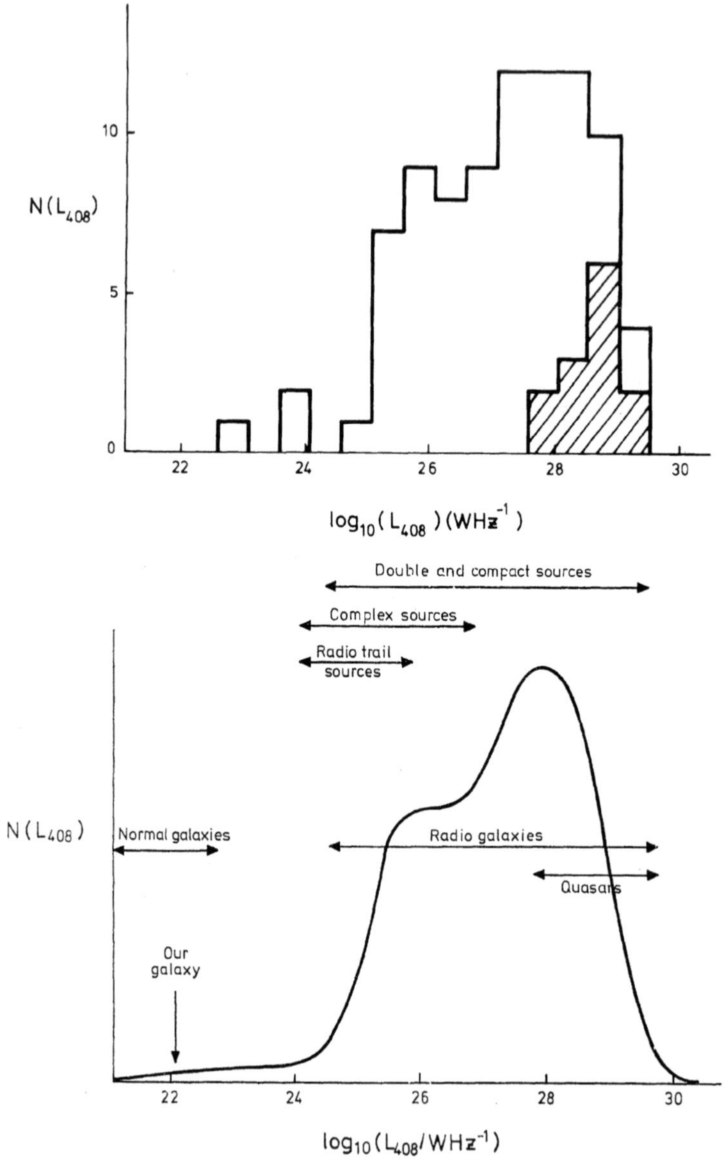

Figure 1(a) The luminosity distribution at 408 MHz for sources having $S_{408} \geq 10$ Jy. Quasars are shown shaded.

(b) The smoothed luminosity distribution at 408 MHz including additional information on low luminosity sources.

types of source are found on this histogram. It should be noted that Seyfert galaxies are in general weak radio emitters and virtually none of them have radio luminosities greater than 10^{24} W Hz^{-1}. None of them have the typical extended radio structures found in typical radio galaxies with luminosities in the range $10^{25} < L_{408} < 10^{29}$ W Hz^{-1}. Virtually all the sources in the low frequency sample have steep radio spectra i.e. $S \propto \nu^{-\alpha}$ with $\alpha \sim 0.8 \pm 0.2$. Similar samples of radio sources selected at high frequencies $\nu \geq 1$ GHz contain much larger fractions of sources with flat radio spectra, $\alpha \sim 0$. A larger proportion of these are quasars and VLBI observations indicate that their angular sizes are $\sim$ milli-arcsec, consistent with the hypothesis that the flattening of the spectrum is due to the superposition of a number of source components in which synchrotron self-absorption is important at high frequencies $\nu > 1$ GHz.

The morphologies of the radio structures of extended extragalactic sources differ widely but I want to make the case that there are a number of persistent themes running through physical discussions of their origin and evolution. These are:

(i) The importance of jets or beams of high energy particles or radiation;

(ii) Particle acceleration, not only in active nuclei but also in more diffuse regions of interstellar and intergalactic space;

(iii) The interaction of clouds of relativistic plasma with the ambient medium.

Let us look at the various types of radio source.

3. TYPES OF EXTRAGALACTIC RADIO SOURCE

In order to put some order into the wide variety of types of radio source, it is convenient to introduce the Classes I and II described by Fanaroff and Riley (1973). Their original intention was to develop some operational definition for distinguishing double sources from the more complex sources. They adopted the definition that if the maxima of the radio brightness distribution are separated by more than half the total dimension of the source measured to the faintest outer contour, the source belonged to Class II. If the separation were less than half the total size, it belonged to Class I. The simple mnemonic to remember the classes is that class II are basically double sources. Figure 2(a) indicates this distinction schematically. Fanaroff and Riley showed that there is a rather sharp distinction in radio luminosity between these classes. For all the bright radio galaxies and quasars available to them in 1973, all class I sources had $P_{408} \leq 10^{25}$ W Hz^{-1} whereas all those with $P_{408} \geq 10^{25}$ W Hz^{-1} belonged to class II. Inspection of figure 1(a) shows that this means that double sources are by far the most common objects in high flux density samples selected at low frequencies.

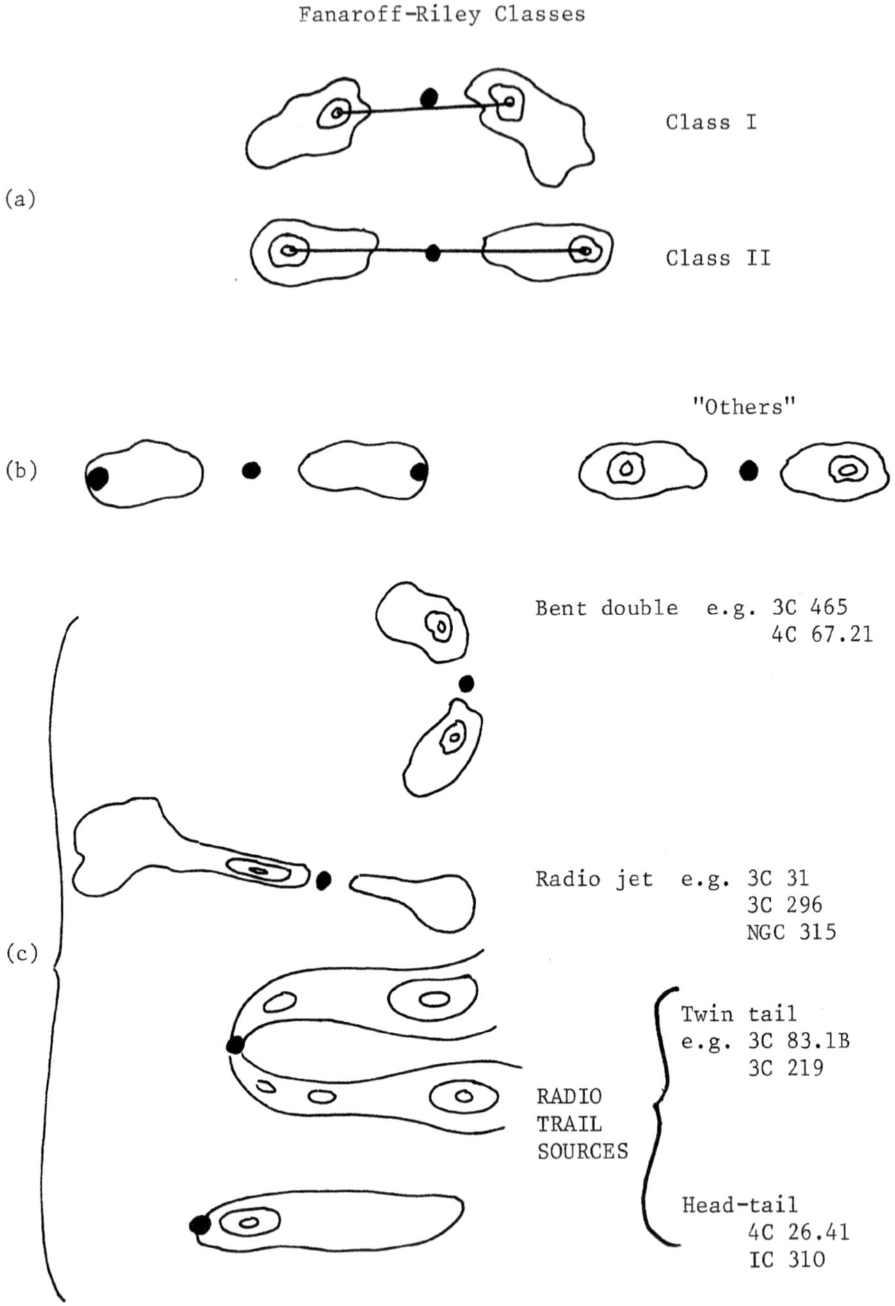

Figure 2. The morphologies of extragalactic radio sources.

Within Class II, we can make a further distinction between those sources in which the radio maxima are intense "hot-spots" at the leading edges of the source components and those in which they are not (Figure 2(b)). A large fraction of the Class II doubles are of the former type and I like to call them "classical double sources". These sources give rise to the most difficult physical problems because it is necessary to explain how such high energy densities of ultrarelativistic electrons and magnetic field were produced at distances which can be $\sim$ 100 to 500 kpc from the ultimate source of energy which is presumed to be in the nucleus of the galaxy or quasar. Many very beautiful maps of these sources have now been produced by the 5-km telescope at Cambridge, the Westerbork Synthesis Radio Telescope and most recently by the VLA. Let us give a brief summary of some of their more important properties.

Generally the radio emission originates from two components symmetrically disposed with respect to the galaxy or quasar with roughly the same intensities. Sometimes the double structure lies within the optical envelope of the associated galaxy but generally the component separations are on a much larger scale, D $\sim$ 100-500 kpc. The most extensive double sources, the giant radio sources, have physical sizes up to about 6 Mpc. The radio structure of Cygnus A is shown in Figure 3 and is typical of classical double sources. Often, the source component contains a compact "hot-spot" towards its leading edge. This type of structure is often called "head-tail" structure, the "head" being the compact component whilst the "tail" is the region of lower surface brightness which extends back towards the nucleus of the galaxy. The heads typically have physical sizes $\gtrsim$ 10 kpc and thus they are not nearly as compact as the milliarcsec sources which are found exclusively in galactic nuclei. The head-tail structure is found in the most powerful of double sources; the heads are relatively weaker or absent in intrinsically weak sources. An important parameter in the theory of double sources is the axial ratio, the ratio of the distance of the hot-spot from the galactic nucleus to the physical size of the hot-spot. In sources such as Cygnus A and 3C 390.3, the axial ratio can have values as large as 30-50.

Double structure is found in sources of all intrinsic luminosities from weak nearby sources such as 3C 76.1 (z = 0.03) to extremely powerful quasars such as 3C9 (z = 2.016). A feature of many double sources is the presence of a component with a flat radio spectrum associated with the galactic nucleus or quasar. The evidence of VLBI suggests that these are similar in character to the milliarcsec sources and in some cases (for example, Cygnus A and 3C111) this very fine structure is elongated in the direction of the extended double components. It is probable that all sources possess such compact central radio components and that those sources in which they are not observed have not yet been studied with sufficient sensitivity. These components are important clues to the energy supply and energy balance of double sources. In a few cases, radio jets are observed extending from the galactic nucleus

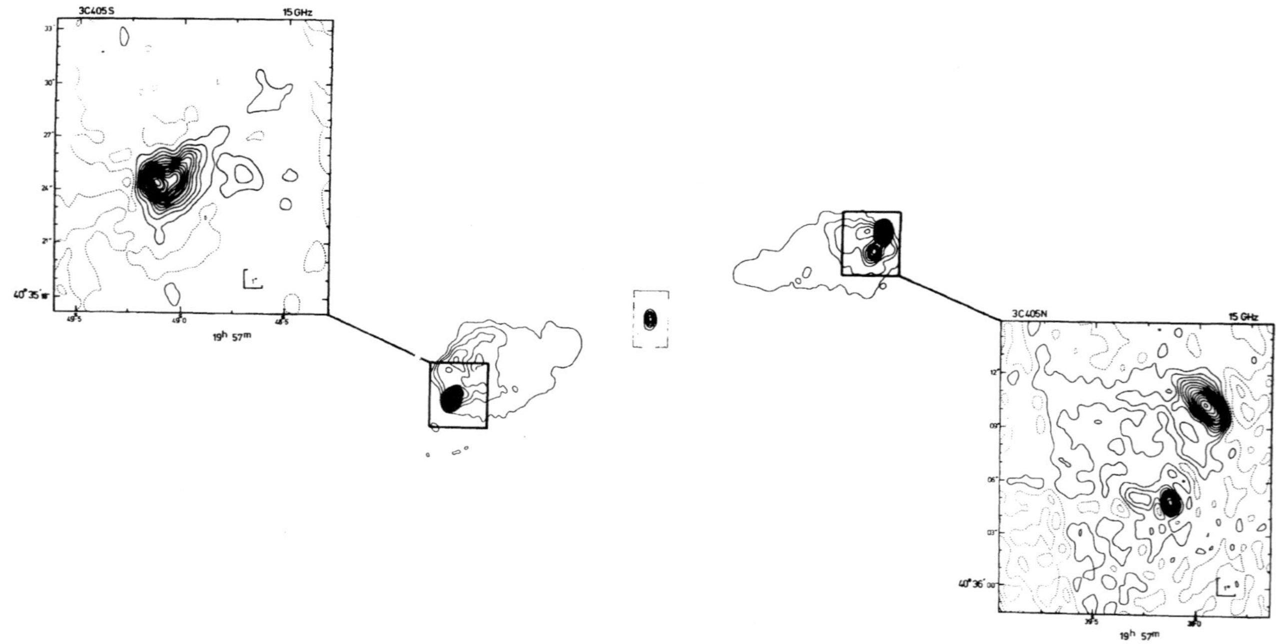

Figure 3. Cygnus A at 5 and 15 GHz. The overall structure was determined at 5 GHz and the high resolution maps of the compact features made at 15 GHz (Hargrave and Ryle 1976).

towards one of the radio components. A good example is that of 3C219 which is otherwise a typical double radio source. A particularly spectacular case is that of NGC6251 in which there is evidence of jet structure on all scales from milliarcsecs to several minutes of arc (Waggett et al. 1977, Readhead et al. 1978).

The objects belonging to Class I come in a much wider range of radio morphologies. Some of these and the names found in the literature are shown schematically in Figure 2(c). The designations of some sources illustrating these types are included on the diagram. Notice that the radio surface brightnesses of these sources are very much less than those of the powerful double sources. Many of these sources are associated with clusters and indeed statistical analyses of well defined samples of sources show that the Class I sources are typically found in richer associations of galaxies than those of Type II (see Section 6.2).

The radio trail sources form a reasonably well-defined class in which the bulk of the emission originates in an extended trail with the general appearance of a wake extending to very large distances behind the associated galaxy. There is often evidence for a compact source associated with the nucleus of the associated galaxy. Radio trail sources appear to be exclusively found in groups and clusters of galaxies, although there is the important selection effect that most of them have been searched for in clusters.

Finally, there are the compact sources. I define compact sources as those with angular size $\theta \leq 1$ arcsec which is roughly the limit at which they can be resolved by Earth-rotation synthesis radio telescopes. Thus, information about their structures must be obtained from long baseline interferometry and from studies of their spectra and variability. In angular sizes, they range from less than 10^{-3} arcsec and there is often evidence for structure on a variety of scales. Many of the sources with flat high frequency spectra have angular sizes $\theta \approx 10^{-3}$ arcsec which is consistent with their radio variability and with the observation on synchrotron self-absorption in the components at frequencies $\nu > 1$ GHz. In many of the brightest of these sources, VLBI observations have shown evidence of changes in structure and, perhaps most interesting of all, have provided evidence that the components are observed to separate at velocities exceeding the velocity of light by factors of 2-10. Besides the important clues which such observations provide on the physics of compact nuclei, they are also important evidence on the mechanism by which energy can be transmitted to the outer components of double sources.

4. MODELS FOR EXTRAGALACTIC RADIO SOURCES

The consensus of opinion is that all these different types of radio source are different aspects of the same basic phenomenon. It is convenient to begin with extended double radio sources since the problems associated with them are most well-defined. The key to models of double sources is the study of the "hot-spots" observed at the leading edges of extended source components. The radio emission mechanism is generally accepted to be synchrotron radiation and hence the problem is to understand how very large fluxes of relativistic plasma and magnetic fields can survive so far from their ultimate source of energy which is the galactic or quasar nucleus. This problem has been studied quantitatively by investigating how long such regions of high energy density could survive in the typical conditions in the intergalactic medium in the vicinity of strong radio sources.

Unless prevented in some way, a cloud of relativistic plasma expands at the relativistic hydrodynamic velocity $c/\sqrt{3}$ and thus it would be expected that such clouds should expand and disperse over timescales of the order of the light travel-time across them. Thus, if we observe hot-spots, they must be continuously replenished with relativistic electrons and magnetic fields unless there exist efficient confinement mechanisms which prevent their expansion. Such confinement mechanisms include thermal pressure confinement, inertial and ram pressure confinement and gravitational binding. For sources such as Cygnus A, the energy densities in the heads are so large, $\varepsilon \geq 10^{-8}$ erg cm^{-3}, that it is unlikely that any of these mechanisms can prevent the dispersal of the hot-spots (Longair, Ryle and Scheuer 1973). The inference is that the hot-spots must be continuously replenished, presumably from the nucleus of the associated optical object.

There are a number of independent pieces of evidence which are consistent with this picture. First, in a number of cases, it has been established that the spectral indices of the heads are smaller than those of the tail regions. According to the synchrotron interpretation of the observations, this suggests that the particles in the heads are younger than the particles in the tails because the effect of synchrotron losses is to steepen the energy spectrum of the electrons and consequently the emitted radio spectrum. This result is consistent with the model in which the particles have been freshly injected into the head regions and the tails represent the debris of previous generations of hot-spots which have dispersed and formed a wake behind the advancing head. In the case of Cygnus A, if the equi-partition value of the magnetic field is adopted, the lifetimes of the particles in the heads are less than the light travel-time from the nucleus and are only a few times larger than the light travel time across the hot-spot itself. These observations are consistent with the continuous acceleration and outflow of relativistic electrons from the heads of the sources.

Second, compact central radio components are observed in a large fraction of all double sources implying that there is continued activity in the nucleus which produces relativistic particles and possibly magnetic fields throughout much of the lifetime of the radio source. Some of them are known to be variable sources and in some cases VLBI observations have shown these sources to be elongated along the axis of the double source.

Third, the observation of superluminal velocities in compact sources gives a general indication that something is being ejected relativistically from the nucleus and this is precisely the type of phenomenon required to transfer energy efficiently from the nucleus to the outer components. In the case of 3C273, the axis of the superluminal expansion is in roughly the same direction as the extended component 3C273A which possesses the typical "head-tail" structure of a double radio source component.

Fourth, a number of jets are observed emanating from the compact central component in the direction of extended source components.

The resulting model is depicted schematically in Figure 4. It is known as the "beam" or "continuous-flow" model. The nucleus is the source of a double, highly collimated relativistic "beam" which burns its way out through the interstellar and intergalactic gas in opposite directions. At the point where the beam encounters the ambient gas, the energy of the beam is converted into relativistic particle and magnetic field energy, producing intense hot spots of radio emission which are identified with the heads of the radio source components. Pressure balance at the interface between the beam and the intergalactic gas defines the velocity of advance of the hot-spot V, $p \approx \varepsilon \approx \rho_e V^2$ where p is the pressure of the beam and ε is the energy density of relativistic particles in the head which are supplied by the beam. $\rho_e V^2$ is the ram pressure associated with pushing back the ambient medium which has density ρ_e.

The thermal particle densities are too small to prevent expansion of the heads and hence they expand laterally leaving behind the advancing head a trail or wake which is identified with the "tails" of the radio components. When the energy supply from the nucleus is exhausted, the beam will cease to supply energy to the components and all that will survive will be the diffuse tail regions which can come into thermal pressure balance with the ambient intergalactic gas. These sources may be identified with "relaxed" or diffuse double sources.

It should be emphasised that the model is based on a patchwork of observations of many different types of source. Definite statements about the lifetimes of relativistic particles and the cold particle densities in different regions of sources can only be made in a few cases. Systematic observations of much larger numbers of sources with high resolution at a number of frequencies will establish how widespread the properties described above are (see e.g. Burch 1978).

The basic model is sufficiently flexible to account for most types of extended extragalactic radio sources. For example, bent-doubles may result from motion of the ambient intergalactic gas relative to the radio emitting regions. An extreme case of this may occur in the radio trail sources in which the energy densities in relativistic particles and magnetic fields are low and the overall radio structure may be determined by the motion of the galaxy relative to the intergalactic gas in the cluster. Other "anomalous" features may be explained if

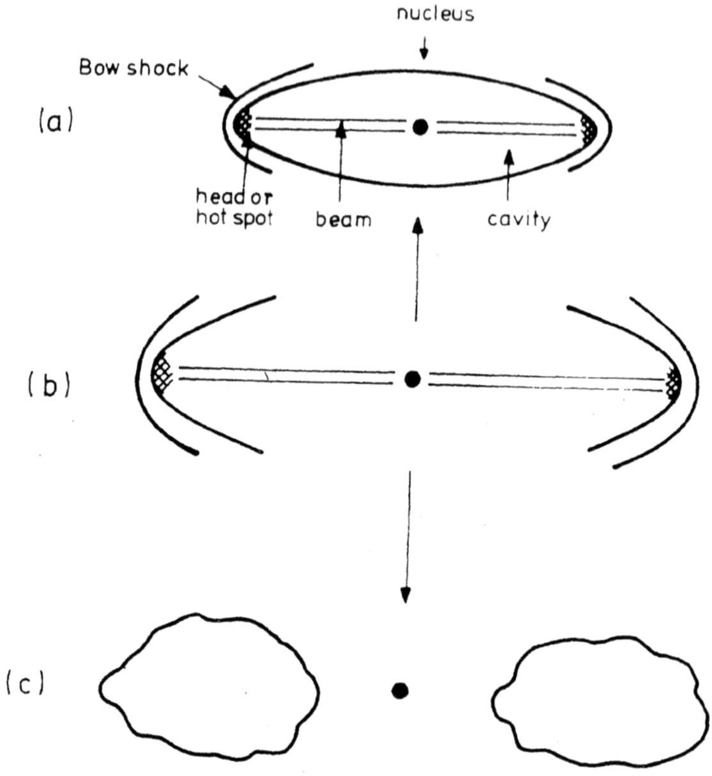

Figure 4. A schematic diagram illustrating the evolution of the "beam" or "continuous flow" model of extragalactic radio sources.

the beam is intermittent so that "multiple" heads can be produced or if the beam were to precess. There may be instabilities at the interface between the intergalactic gas and the beam which could account for more of the detailed structure.

If we adopt this model, we can make considerable progress in defining the physical conditions in the double sources. For example, Julia Riley and I (1979) recently showed how limits to the velocity of advance of the heads of the classical double sources could be obtained from the "distant quotient" $Q = D_1/D_2$ where D_1 and D_2 are the distances of the hot-spots from the central source of energy. If the components moved relativistically $v \sim c$, we should observe a range of values of D_1/D_2 because of the time-delay of radio signals across the source when observed in projection. The observed distribution of Q for classical double sources and that predicted for different values of ν are shown in Figure 5. It can be seen that on average the velocities of advance of the heads must be less than 0.2c.

This analysis also showed that there is a high probability ($\gtrsim$ 50%) of observing a "hot-spot" in each component and therefore the supply of energy to the outer components must be "on" throughout essentially the whole lifetime of the source.

5. PHYSICAL PROBLEMS

There are, however, many physical problems which must be solved before the model can be considered more than a rationalisation of a disconnected set of observations.

(i) The origin of the large amounts of energy involved is far from being understood. An advantage of the continuous flow model is that it minimises the energy production rate from the central object because the total energy requirements of the extended regions, which pose the greatest energetic problems, are supplied throughout the lifetime of the source which is typically $\sim 10^7$ - 10^8 years. Black holes models involving accretion discs are a highly plausible source but the origin of the relativistic electrons in nuclei is uncertain.

(ii) The origin of the double structure in galactic nuclei is not certain. Attractive hydrodynamic models have been described by Blandford and Rees (1974) but electromagnetic models associated with the "flaring" of relativistic discs also seem to be acceptable. (Figure 6)

(iii) We also have to be careful about making good use of the energy liberated by the nucleus. If the material ejected from the nucleus exerts a significant pressure, then on expanding to the dimensions of the "hot-spots", a large fraction of the energy would be lost adiabatically. The way round this problem is to send the

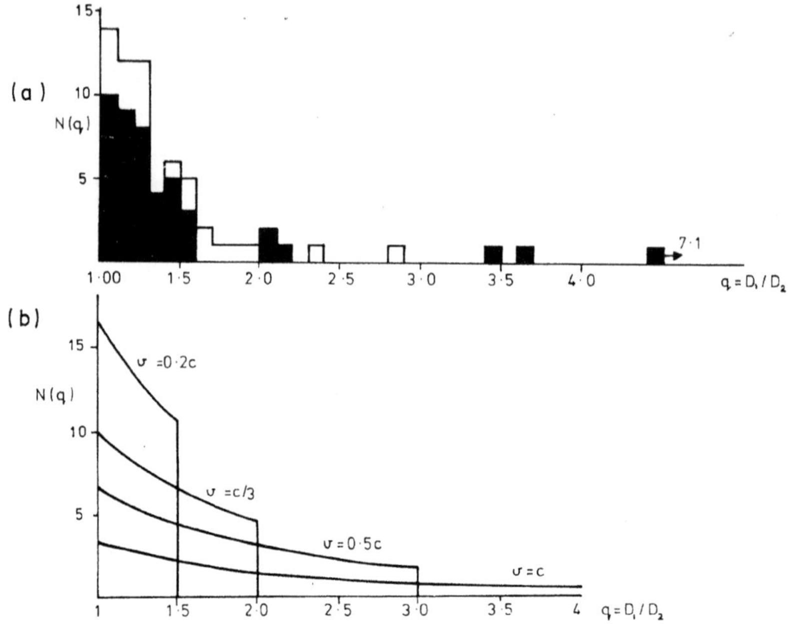

Figure 5(a) The observed distribution of the separation quotient q, for classical double sources. (b) The expected distribution of 2 for different values of ν, normalised to the total number of sources in (a).

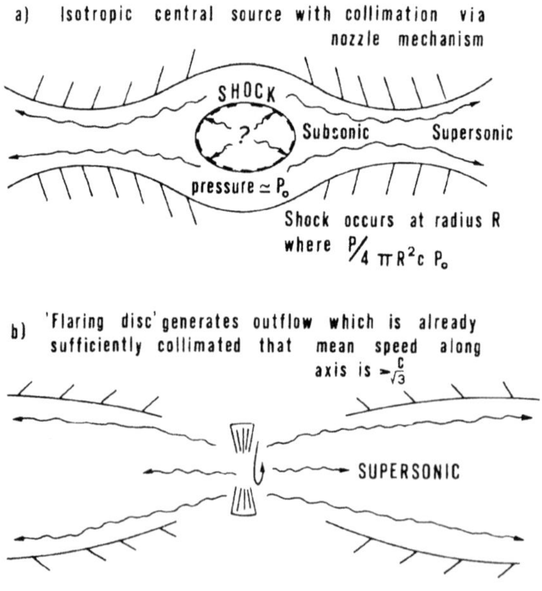

Figure 6. Two models for the beams in double radio sources (Rees 1976)

energy out in the form of a jet or beam in which case most of the energy is in the form of kinetic energy.

(iv) This energy must be converted into relativistic particle and magnetic field energy. Until recently, it was not clear how this could be achieved but the work of Gull (1974) on the origin of the magnetic field in young supernova remnants and of Bell (1978) and Blandford and Ostriker (1978) on the acceleration of charged particles in strong shock fronts suggest ways in which this may be achieved. Gull makes a convincing case that the magnetic fields in young supernovae such as Cassiopeia A can be attributed to Rayleigh-Taylor instabilities in the ejected remnant when deceleration becomes significant. Any seed field in the shell is sheared and tangled by the instability until there is roughly as much energy in the field as in random motions. In the case of acceleration of particles, the above authors appeal to first-order Fermi acceleration of particles which continually cross and re-cross the shock front. Exactly how these ideas are adapted to double radio sources is not yet clear but there are several realistic possibilities. An important step is the development of a proper hydrodynamical model for the passage of the beam through the intergalactic medium including as much real physics as possible.

(v) Beam instabilities threaten to destroy the beams from the nucleus, even if they are ultrarelativistic. The problem is the relativistic version of the Kelvin-Helmholtz instability (Turland and Scheuer 1976). The study of these phenomena is of considerable importance because it may provide clues to the material within the beam as well as the properties of the ambient medium. Thus, some beams seem to be visible, as illustrated in Figure 2(c) whereas the beams assumed to be present in classical double sources are invisible. This may all be related to the stability of the beams in the sense that if the beam is visible it must have become unstable already.

(vi) We are not clear what the material of the beams is. It could be cold matter i.e. matter with $mc^2 > kT$, relativistic material or strong electromagnetic waves, all permeated by a magnetic field. It does not matter too much for much of the dynamical discussion which of these possibilities is correct. Nonetheless, it would be very important if the nature of the beam could be defined more precisely.

(vii) Finally, we have not given an adequate discussion of where the cold matter in the beam comes from. There must be some cold matter present because the radio waves are depolarised at frequencies $\sim$ 1-2 GHz by internal Faraday depolarisation. Is the cold matter ejected from the nucleus? Is it in fact the beam? Is it entrained from the surrounding medium? These topics need more theoretical study.

The above points have been made mostly in relation to the Class II sources but they have their counterpart in the physics of the Class I sources as well. These sources are intrinsically weak radio emitters and consequently the energy densities are much smaller. Correspondingly, the problems of confinement are much less as explained in Section 4. In fact the typical pressures found in the intergalactic gas in rich clusters of galaxies are more than adequate to confine the diffuse radio components by thermal pressure. However, a number of topics recur.

(i) In the longest radio trail sources, the particles have lifetimes against inverse Compton losses due to scattering of photons of the microwave background which are shorter than the dynamical age of the source. Thus, we must find an acceleration mechanism for the particles in these very diffuse regions.

(ii) Why are so many jets visible in the weak sources? It may be that the instabilities develop much more rapidly if the power in the beam is small i.e. the beam cannot penetrate so far through the interstellar or intergalactic gas. This idea was tested by Birkinshaw, Laing, Scheuer and Simon (1978) who plotted the location of the first maximum in the radio brightness distribution from the galactic nucleus as a function of radio luminosity for Class I sources (Figure 7). It can be seen that there is a correlation in the sense expected if stronger beams can penetrate further before becoming unstable.

We can list a further range of problems concerning jets in extragalactic radio sources.

(i) In some cases the magnetic field runs parallel to the jet (M87, 3C219, 4C32.69) and in others perpendicular to it (M84, 3C449). In some cases, such as NGC315, 3C31 and NGC6251, the direction of the magnetic field changes direction along the jet. Presumably these phenomena could be explained by various combinations of shearing and expansion motions.

(ii) Why are so many jets asymmetric? In sources such as 3C449, 3C31, NGC6251, M87, 3C219, 4C32.69 and 3C66B, the jets are all one-sided. Is this due to relativistic beaming as may be the case for the superluminal sources? Is the ejection mechanism itself asymmetric? Or are the physical conditions on either side of the galaxy different? I am not aware of convincing answers to these questions.

(iii) Is limb-brightening observed in any of these jets? So far the results look negative but it is a difficult experiment since few of the jets are clearly resolved.

(iv) We now know of some very powerful jets. 4C32.69, observed by Potash and Wardle (1980) using the VLA, is about 100 times more luminous than the jets described above and is associated with a quasar with classical double radio structure. The

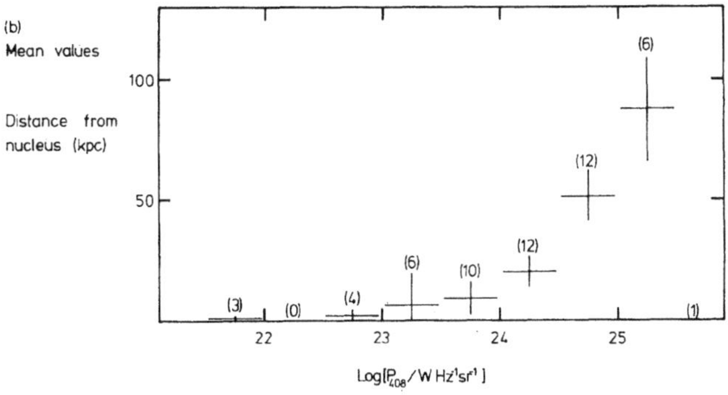

Figure 7 Mean values of the distance from the nucleus to the first radio maximum as a function of luminosity for Class I radio sources (Birkinshaw, Laing, Scheuer and Simon 1978).

Figure 8

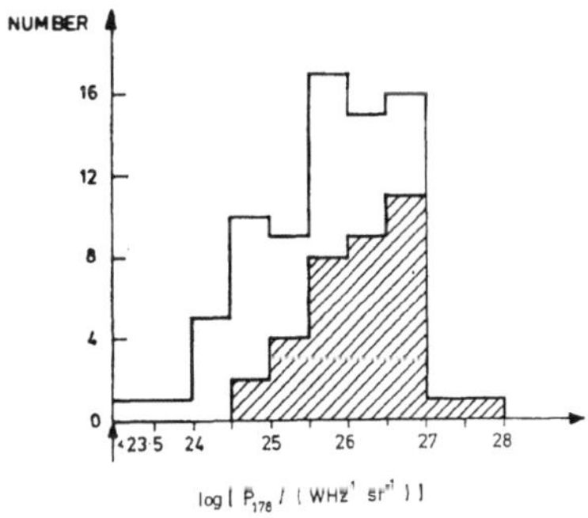

(a) The distribution of luminosity for 3CR radio galaxies - open boxes have class B spectra; shaded boxes have class A spectra.

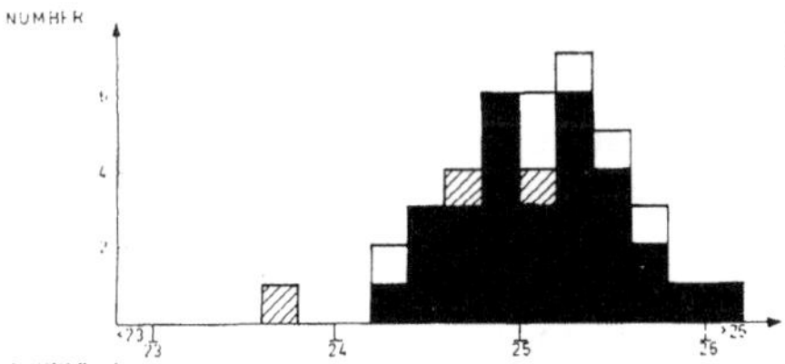

(b) Luminosity distribution for Class A radio galaxies - blank boxes are unresolved sources, filled boxes are double sources and shaded boxes are F-R class I sources.

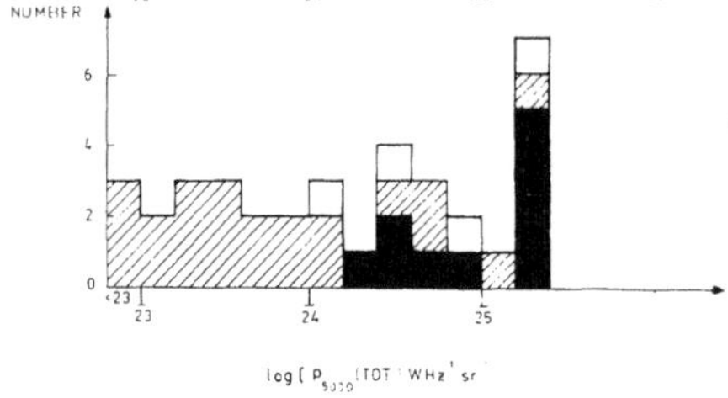

(c) Luminosity distribution for class B radio galaxies - notation as in (b).

energy densities in the beam are very large, $\varepsilon_{min} \geq 3 \times 10^{-10}$ erg cm^{-3}; why is such a powerful jet observed in this classical double and not in others? Plainly, there is a need for a systematic study of such sources with the very high sensitivity available at the VLA.

6. INTERACTION WITH X-RAY OBSERVATIONS

First of all, let us look at the jets. We expect some sources to have X-ray jets by virtue of extrapolating the radio-optical spectrum into the X-ray waveband. This seems certain to be the explanation of the X-ray emission from the jet in M87 where the X-ray knots are coincident with the brightest of the radio and optical knots. The importance of this observation is that it makes the case for the continuous injection and re-acceleration of particles even stronger than was possible from the radio and optical observations. This is because the half-lives of the electrons which radiate at X-ray wavelengths are much shorter, being inversely proportional to the electron's energy. It also seems likely that the X-ray emission from the jet of 3C273 has the same origin. Van Breugel and Miley (1980) have discovered similar faint optical jets in 3C31 and 3C66B which are coincident with the radio jets. The nature of the optical emission from these regions is not yet clear.

A significant question is whether the X-ray and optical emission could be associated with the dissipation of the energy of the beams which power the outer components of the radio source. In most cases, this is unlikely because the inferred particle densities in the vicinity of radio source components are too low to be strong sources of thermal bremsstrahlung. For example, to be visible optically $V \sim 22$, the particle densities in the hot-spots in Cygnus A would have to be ~ 1 cm^{-3} which far exceeds the limit imposed by Faraday depolarisation. This argument applies equally strongly at X-ray wavelengths. Thermal bremsstrahlung would probably only be important if very hot dense bodies were ejected in beams from the galactic nucleus.

A second important aspect of the X-ray observations is the measurement of the magnetic field strength from components which are simultaneously sources of synchrotron radio emission and inverse Compton X-ray emission due to scattering of the microwave background radiation. This test is already being performed on the inner lobes of Centaurus A.

Third, there are various aspects of the environment of the radio source, both on the large scale and within its nucleus which offer important observational opportunities for X-ray astronomy.

6.1 The Optical Spectra of Radio Galaxies

Hine and I (1979) have studied the optical spectroscopic properties of a complete sample of bright radio galaxies in order to discover if there are significant relations between the optical and radio properties.

The radio galaxies were selected from the 3CR catalogue - the radio data was of uniformly high resolution and sensitivity (taken with the Cambridge 5-km telescope) and the optical information from a literature search. Radio galaxies may be divided into two broad classes, those which possess strong emission line spectra (class A) and those which do not (class B). We established the following relations:

(i) The class A radio galaxies are stronger radio emitters than class B. Specifically, for low radio luminosities, only $\sim$ 10-15% of radio galaxies possess strong emission line spectra whereas at the highest radio luminosities, this percentage reaches $\sim$ 70% (see Figure 8(a)).

(ii) Galaxies of class A have stronger central radio components in the sense that there must be a correlation between the intensities of the non-thermal emission at radio and optical wavelengths.

(iii) Class A radio galaxies are almost exclusively classical double sources whereas those of class B can be associated with a wide range of radio morphologies (see Figure 8(b and c)).

(iv) There is a correlation between the breadth of the emission lines and the relative intensity of the non-thermal radio luminosity of the nuclear component.

Spectroscopically, Class A radio galaxies are similar to Seyfert galaxies except that the radio galaxies show weaker lines of FeII, have stronger forbidden lines of O^{++} and have steeper Balmer decrements. The Class A radio galaxies can be divided into broad-line radio galaxies (BLRG) and narrow-line radio galaxies (NLRG) corresponding to the distinction between Type I and Type II Seyfert galaxies. However, the radio galaxies are almost exclusively giant elliptical galaxies whereas the Seyfert galaxies are mostly spirals.

Because Seyfert galaxies are known to be strong X-ray emitters it is of great interest to extend these X-ray studies to complete samples of radio galaxies, of both class A and class B. Such a study would involve the correlation between X-ray luminosity and other properties such as radio luminosity, radio morphology, optical spectral type (class A, class B, BLRG, NLRG), optical luminosity etc.

6.2 The clustering of galaxies about extragalactic radio sources

There have always been problems in defining quantitatively a group or cluster of galaxies and the question of whether radio galaxies belong preferentially to clusters of galaxies. Seldner and I (1979) have studied this problem using cross-correlation functions to describe the clustering of galaxies about radio galaxies with redshifts $z \leq 0.1$. The position of each radio source was cross-correlated with the Lick counts of galaxies and spatial enhancement factors of the density of galaxies about each source derived. A linear scale of clustering was set up from galaxies selected at random in the Universe (enhancement factor = 1) to rich clusters of galaxies (enhancement factor $\gtrsim$ 13). The details of the statistical results are shown in Table 1. On average, 3CR radio galaxies have enhancement factors $\gtrsim$ 4. However, the clustering varies with the physical properties of the radio galaxy. Broadly speaking, those radio galaxies with class A spectra and double radio structure lie in much smaller enhancements of galaxy density than those with class B spectra and complex radio morphology. In fact, if classical double sources like Cygnus A are considered, they appear to lie in regions of galaxy density which are no more enhanced than galaxies selected at random. Thus, there are certainly not rich clusters of galaxies about most classical double sources. Since there must be hot intergalactic gas present at least to the limits of the radio structures, these massive galaxies must be able to hold on to their own extended atmospheres. These should be X-ray emitters similar in X-ray morphology to Cygnus A and their X-ray angular sizes should be about 3-5 arcmin.

In addition, observations of the class B complex radio galaxies are important in order to detect the intergalactic gas in those systems which belong to richer groups and weak clusters.

7. CONCLUDING REMARKS

This survey has indicated just some of the ways in which radio and X-ray observations of active galaxies can produce a synthesis which is greater than the sum of their individual parts. My own view is that it is of most importance to undertake systematic studies of properly defined samples of radio galaxies and quasars so that one is not always studying the one or two well-known exotic objects such as M87 and 3C273 which always exhibit peculiarities. The goals should be to quantify the frequency of occurrence of X-ray emission from various components of active galaxies and the relative importance of different physical processes. The most important programmes involve:

(i) the environment of the radio source, certainly in the case of cluster sources, but especially in those cases where the cluster of galaxies is much weaker;

(ii) the components themselves via inverse Compton scattering. The most extended sources are candidates for scattering of the microwave background radiation but compact central sources where the energy densities of infrared and optical photons are high are also strong candidates;

(iii) the detailed study of X-ray jets and their relation to their radio and optical counterparts;

(iv) Empirical studies of the relation between X-ray activity and radio and optical properties for radio galaxies.

TABLE I

Enhancement Factors $B_{gg}*/B_{gg}$ for various sub-samples

		n	$B_{gg}*/B_{gg}$	
Optical Spectral Class	A	12	2.3 ± 0.7	Strong emission lines
	B	24	4.5 ± 0.9	No strong emission lines
Fanaroff-Riley Class	I	12	4.6 ± 1.3	
	II	15	2.2 ± 0.8	
Fanaroff-Riley	"Classical Doubles"	8	0.7 ± 0.8	
	Others	7	4.0 ± 1.2	

REFERENCES

Bell, A.R., 1978. Mon. Not. R. astr. Soc., 182, 147.
Birkinshaw, M., Laing, R.A., Scheuer, P.A.G. and Simon, A., 1978. Mon. Not. R. astr. Soc., 185, 39P.
Blandford, R.D. and Ostriker, J.P., 1978. Astrophys. J., 221, L29.
Blandford, R.D. and Rees, M.J., 1974. Mon. Not. R. astr. Soc., 169,395.
Burch, S.F., 1978. Mon. Not. R. astr. Soc., 186, 343.
Fanaroff, B.L. and Riley, J.M., 1974. Mon. Not. R. astr. Soc., 167,31P.
Gull, S.F., 1974. Mon. Not. Astr. Soc., 161, 47.
Hargrave, P.J. and Ryle, M., 1976. Mon. Not. R. astr. Soc., 175, 481.
Hine, R.G. and Longair, M.S., 1979. Mon. Not. R. astr. Soc., 188, 111.
Longair, M.S. and Riley, J.M., 1979. Mon. Not. R. astr. Soc., 188,625.
Longair, M.S., Ryle, M. and Scheuer, P.A.G., 1973. Mon. Not. R. astr. Soc., 164, 243.
Longair, M.S. and Seldner, M., 1979. Mon. Not. R. astr. Soc., 189,433.
Potash, R.I. and Wardle, J.F.C., 1980. (to be published)
Readnead, A.C.S., Cohen, M.H. and Blandford, R.D., 1978. Nature, 272, 131.
Rees, M.J., 1976. "The Physics of Non-Thermal Radio Sources", (ed. G. Setti), 107, D. Reidel Publishing Co.
Turland, B.D. and Scheuer, P.A.G., 1976. Mon. Not. R. astr. Soc., 176, 421.
Van Breugel, W.J.M. and Miley, G.K., 1980. (to be published).
Waggett, P.C., Warner, P.J. and Baldwin, J.E., 1977. Mon. Not. R. astr. Soc., 181, 465.
Wall, J.V., Pearson, T.J. and Longair, M.S., 1980. Mon. Not. R. astr. Soc., (in press).

X-RAY IMAGING OBSERVATIONS OF ACTIVE GALACTIC NUCLEI AND QUASARS

J. Patrick Henry
Harvard-Smithsonian Center for Astrophysics
Cambridge, Massachusetts 02138 U.S.A.

ABSTRACT

I summarize the results obtained from X-ray imaging observations of active galactic nuclei and quasars with the Einstein Observatory. The X-ray luminosities of these objects form a continuous sequence with the least active objects, Seyfert 2 galaxies, being the least luminous and the most active objects, quasars, being the most luminous.

X-rays are detected from the jets in 3C273 and M87. The entire electromagnetic spectrum of both of these jets can be interpreted as syncrotron emission from radiating particles with the higher energy particles suffering radiation losses.

For quasars there is a correlation between radio emission and X-ray emission. For a given optical luminosity the average X-ray emission of a radio loud quasar is $\sim$3 times higher than that of a radio quiet quasar. Allowing for the differences in X-ray luminosity between radio quiet and loud quasars, and low and high redshift quasars, we estimate that all the X-ray background can be produced by quasars brighter than $m_B \approx 21.5$. This means that the optical log N - log m_B relation for quasars cannot be extrapolated with the same slope much beyond $m_B \approx 20$. Most quasars are not variable on time scales of hours.

I. INTRODUCTION

It is generally agreed that what is happening in the nuclei of objects such as Seyfert galaxies, BL Lac's and quasars is qualitatively the same. Radio and optical observers have accumulated a large body of data from which it is possible to infer details about the structure and radiation mechanism of these objects. However, our understanding becomes less and less as we attempt to describe the conditions closer and closer to the central energy source.

R. Giacconi (ed.), X-Ray Astronomy with the Einstein Satellite, 261–278.

X-ray studies of these objects promise to help determine the conditions very near the central source. This is because the X-ray luminosities are comparable to those in other wavebands and X-ray variability, when observed is extremely rapid. Both of these imply that the X-rays are being produced very close to the central power source.

In this paper I will review the X-ray properties of active galactic nuclei and quasars that we have learned from imaging observations made with the instruments on-board the Einstein Observatory. In practical terms this means that I will limit my discussion to Seyfert galaxies and quasars. I will also give a preliminary discussion of the observation of X-ray jets coincident with the optical and radio jets in 3C273 and M87.

To get oriented, Figure 1 shows the range of observed luminosities for several classes of objects. Normal galactic nuclei like the Milky Way and M31 have luminosities in the range 10^{35} to 10^{40} erg s^{-1} (Van Speybroeck 1980). Seyfert 2 galaxies, which do not exhibit broad permitted emission lines, have been observed at present to have luminosities of between 5 x 10^{40} to 5 x 10^{43} erg s^{-1} (Kriss and Canizares, 1980). The luminosities of Seyfert 1 galaxies, which do have broad permitted emission lines, overlap Sy 2's at the low end and extend to overlap the QSO's at the high end; see Figure 1.

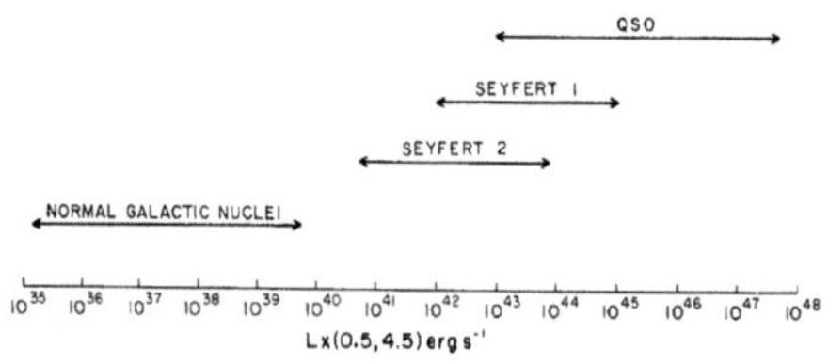

Figure 1. The X-ray luminosities of several classes of extragalactic objects.

QSO's are the most luminous X-ray sources known (Tananbaum et al. 1979, Ku and Helfand 1980) with luminosities ranging from 10^{43} to 6 x 10^{47} ergs s^{-1}. There is a clear trend for the X-ray luminosities to merge together in the manner that would be expected from the similar behavior of the nuclear optical magnitudes. This only emphasizes the basic similarity of these objects.

II. SEYFERTS

Seyfert 1 galaxies were known to be X-ray sources prior to the launch of the Einstein Observatory. An excellent review of their properties is contained in a paper by Andrew Wilson (1979). A review of their spectra is given by Holt (1980) in these proceedings.

The X-ray brightest Sy 1 and one of the closest is NGC 4151. In Figure 2 we show the field around this object as obtained with the HRI on the Einstein Observatory (cf Giacconi et al. 1979 for a description of the observatory). In addition to NGC 4151 there is a QSO WE 1207+39W4 with redshift 1.8 (Arp, private communication) and NGC 4156.

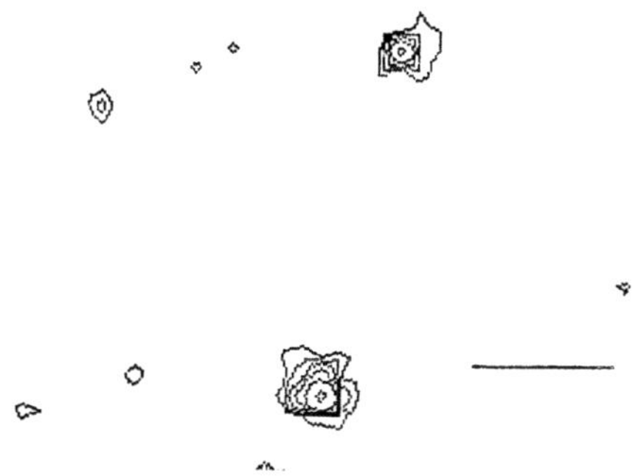

Figure 2. The field around NGC 4151 obtained with the HRI. North is up and east is to the left. The bright source is NGC 4151. NGC 4156 is to the NE, and the other source is the QSO WE 1207+39W4. The bar is 2 arc minutes long. The exposure time for this image was 7377s. The contour levels are at 6, 12, 24, 48, 96, and 310 counts per 256 arc-second^2.

In Figure 3, I show the radial distribution of X-rays around NGC 4151 obtained with the HRI. The 1.54 keV point response function, obtained from ground calibrations is superimposed. The intrinsic size of the X-ray emitting region must be less than about 1".5 FWHM. This is smaller than the forbidden line region, which has been resolved optically (Walker 1968), and shows that the X-rays are not coming from that region. Of course, this was suspected to be so since X-ray variability was observed on one day timescales (Mushotzky et al. 1978) and was claimed by Tananbaum et al. 1978 on 1000 s time-scales. These fast variations imply an emission region much smaller than the forbidden line region.

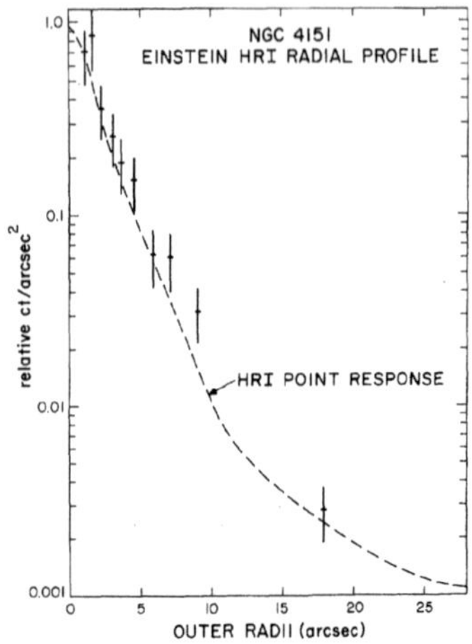

Figure 3. The NGC 4151 radial distribution.

With the increased sensitivity available with the Einstein Observatory we are searching for variability of Seyfert galaxies on timescales shorter than one day. In Figures 4, 5 and 6 I show our observations with the IPC of 3 Sy 1 galaxies. There is no convincing evidence for variability for NGC 4151 or 3C120 although we could have detected a 25% change in 1000s for 3C120. We do see evidence for variability in 10,000 s from NGC 6814. This implies that the size of the emission region is of order 10^{-4} pc.

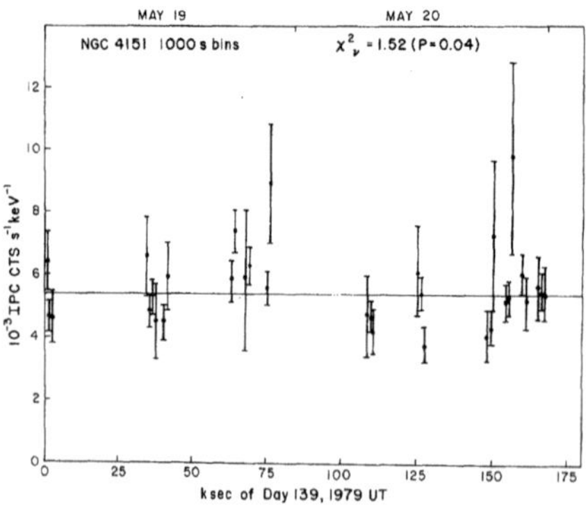

Figure 4. The NGC 4151 X-ray light curve.

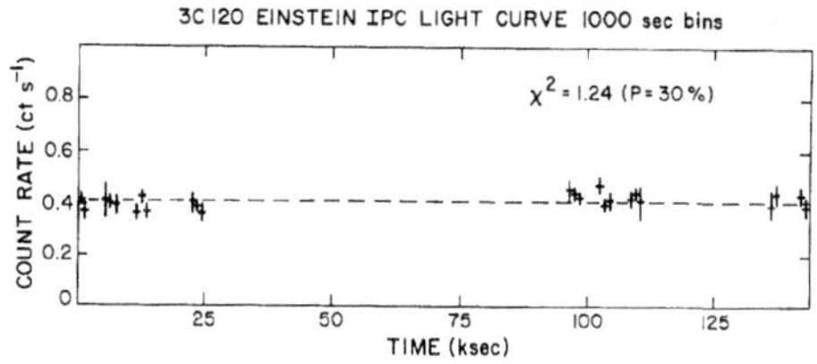

Figure 5. The 3C120 X-ray light curve.
Time 0 = UT 1979 March 8 07:18.

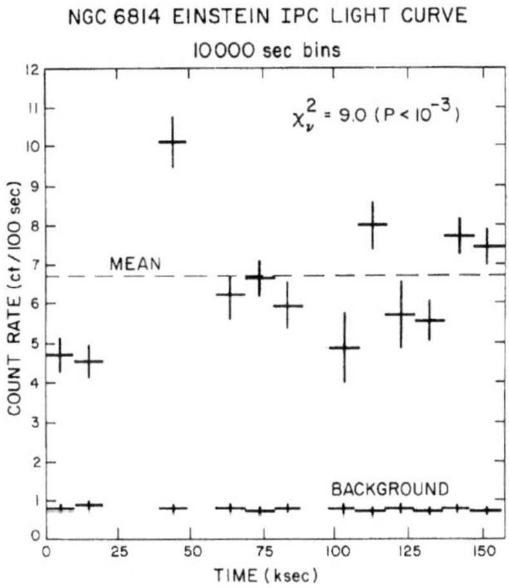

Figure 6. The NGC 6814 X-ray light curve.
Time 0 = UT 1979 April 12 00:53.

Such fast variability also constrains the emission mechanism. Consider a region radiating by thermal bremsstrahlung. Then the electron scattering optical depth is $\tau_{es} \approx 50 \Delta L_{42} \Delta t_3^{-1} T_9^{1/2} C^{-1}$ where ΔL_{42} is the change in luminosity (in units of 10^{42} ergs s^{-1}) which occurred within a time Δt_3 (in units of 10^3 s). T_9 is the gas temperature in units of 10^{9}°K and C is the clumping factor ($C = \langle n_e^2\rangle / \langle n_e\rangle^2$. For NGC 6814 this implies that the electron scattering optical depth is of order 15 and hence optically thin bremsstrahlung is not the emission mechanism.

If this short time scale variability occurs in all Seyferts then we don't expect to observe any X-ray emission lines due to Compton smearing. As reported by Holt at this conference, X-ray emission lines have not been observed from any Seyfert galaxies.

We are now beginning to study the X-ray properties of Sy 2 galaxies. I have listed in Table 1 the X-ray luminosities of the Sy 2 galaxies observed by Kriss and co-workers at MIT and Ku and co-workers at Columbia. There are now (Jan. 80) 6 detections and 8 upper limits. Prior to the launch of Einstein there were several narrow emission line galaxies known to be X-ray sources which were later reclassified as Sy 2 (cf McClintock et al. 1979). Almost all of the new ones are from the list of Weedman (1977) and there is no doubt of their classification. Most are detected below 0.5 keV so their spectra are not very cutoff like NGC 4151. Since these sources are fainter than Sy 1 galaxies, there is no detailed information available as yet.

Table 1

X-Ray Luminosities of Seyfert 2 Galaxies

Name	$L_x(0.5, 4.5)$	Name	$L_x(0.5, 4.5)$
MKN 1	$<1.8 \times 10^{41}$	MKN 348	4.2×10^{41}
MKN 34	$<3.1 \times 10^{42}$	MKN 463	$<2.6 \times 10^{42}$
MKN 78	7.6×10^{41}	MKN 507	8.1×10^{42}
MKN 176	$<6.3 \times 10^{41}$	III Zw 55	$<6.3 \times 10^{41}$
MKN 270	$<8.4 \times 10^{40}$	NGC 1068	7.2×10^{41}
MKN 273	$<2.8 \times 10^{42}$	NGC 3227	1.7×10^{41}
MKN 298	2.6×10^{42}	NGC 6764	$<8.6 \times 10^{40}$

The lower luminosities of the Sy 2 with respect to Sy 1's can be explained with the simple picture described by Osterbrock (1978). In this picture the gas distribution about the central power source consists of two components, a 10^{-1} pc cylindrical component and a 10^{2} pc spherical component. The narrow emission lines come from the spherical component while the broad lines come from the cylindrical component. Since the X-ray emission is likely to be powered by accretion onto a compact object much smaller than the cylindrical component, only those galaxies that have this component (i.e. Sy 1) are likely to have the gas available to produce large X-ray luminosities.

III. THE X-RAY JETS IN 3C273 and M87

In this section I will describe the preliminary analysis of our X-ray observations of the two best studied jets at other wavelengths. The work on the 3C273 jet is being done by Henry, Feigelson, Willingale and Briel (1980); that on M87 by Schreier and Feigelson (1980).

In Figure 7 I show the HRI image of M87. The two obvious sources are coincident with the nucleus and knots A + B to a few arcseconds. We cannot resolve knot A from B but observations at other wavelengths indicate that A is the dominate component. The luminosity of the nucleus and knot in the 0.25 to 4.5 keV band is about 10^{41} erg s^{-1} if the radiation is isotropic.

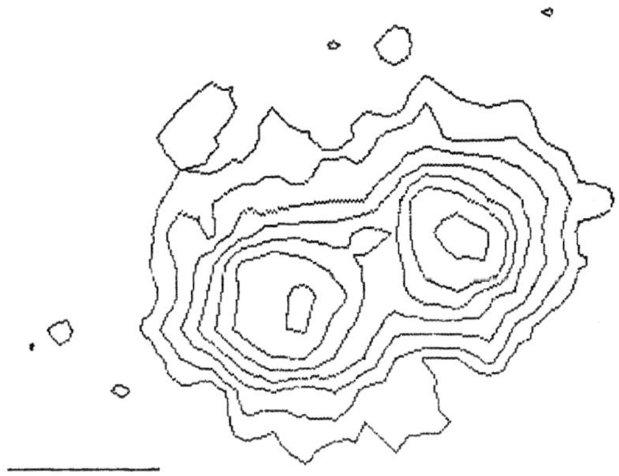

Figure 7. The field around the M87 nucleus obtained with the HRI. North is up and east is to the left. The M87 nucleus is the source at the center of the picture. The bar is 10 arc seconds long. The exposure time for this image was 69092 s. The contour levels are at 20, 28, 39, 51, 65, 82, and 115 counts per 4 arcsecond^2.

The HRI image of 3C273 is shown in Figure 8. This figure shows the over exposed source from the nucleus. The optical jet is at a position angle of 223° extending from about 10" to 25" from the nucleus, but any emission is overwhelmed by the radiation from the nuclear source. To search for the jet we have subtracted away a properly normalized ground calibration image of a point source obtained at an energy of 1.49 keV. In Figure 9 we have plotted the residual counts in the annulus extending from 10" to 20" from the nucleus as a function of azimuth. There is a 5σ peak at 223°, the position angle of the jet. Since no other bin is more than 2.3σ from 0, we regard this as a detection of the jet. The luminosity of this feature in the 0.25 to 4.5 keV band is about 3 x 10^{43} erg s^{-1} if the radiation is isotropic. The flux ratio of the jet to central source is 0.3%, which is about what is observed in the optical.

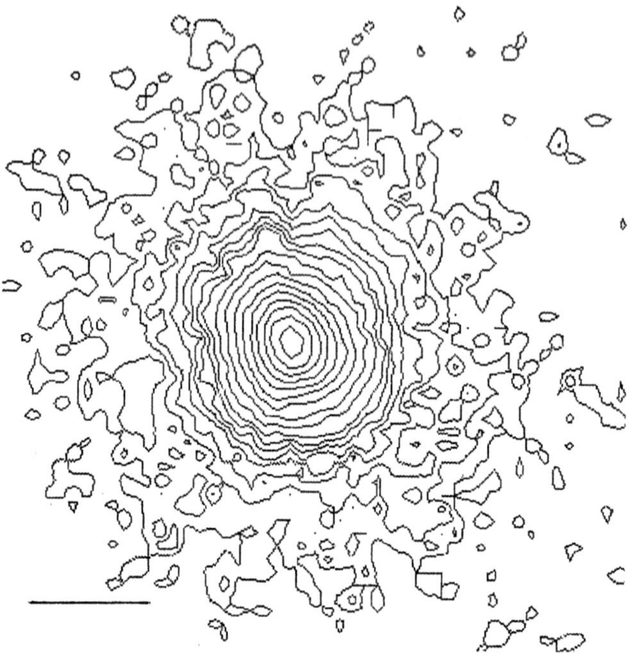

Figure 8. The field around 3C273 obtained with the HRI. North is up and east is to the left. In this display any emission from the jet is lost in the wings of the point response function. The bar is 10 arcseconds long. The exposure time for this image was 77987 s. The contour levels are at 5, 10, 16, 23, 33, 45, 58, 73, 104, 147, 208, 293, 415, 587, 829, 1173, and 1659 counts per arcsecond^2.

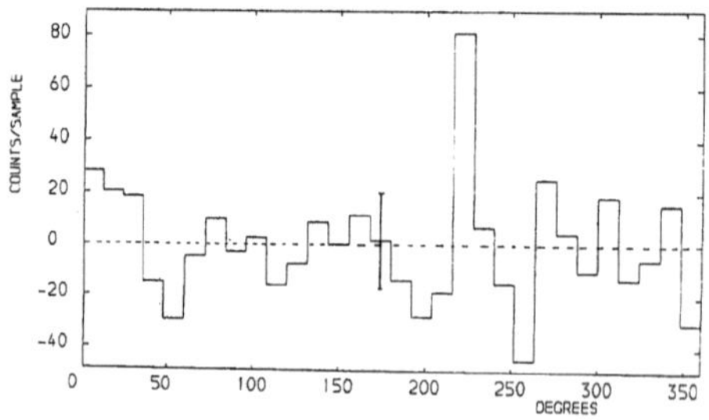

Figure 9. The azimuthal distribution of counts from 3C273 minus the counts from a calibration point source. The radial interval that these counts are accumulated over was 10 to 20 arcsec.

In Figure 10 I plot the observed spectra of the two jet sources from radio to X-ray. The optical emission from the M87 jet is known to be syncrotron radiation from the same particles producing the radio, because the optical polarization position angle as a function of position along the jet is offset by a constant amount from the radio, presumably due to Faraday rotation (Schmidt, Peterson, and Beaver, 1978). The optical X-ray spectral index is 0.8 power steeper than the radio which is indicative of syncrotron losses by the higher energy electrons relative to those at lower energy.

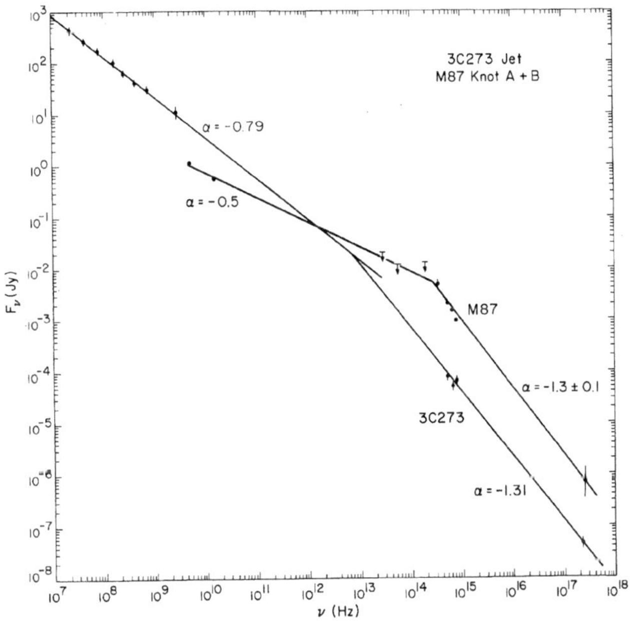

Figure 10. The electromagnetic spectrum of the jets in 3C273 and M87.

The spectrum of 3C273 is remarkably similar. There is a half power break in the spectral indexes and the radio is polarized (Conway and Stannard, 1972). It is not known whether the optical radiation from the jet in 3C273 is polarized, but the upper limit on the integrated polarization is consistent with that of M87.

All of this strongly suggests that the entire electromagnetic spectrum of these two jets can be explained by a syncrotron process with the higher energy electrons suffering radiation losses. In 3C273, with a half power break, the time scale for these losses would be much longer than the electron acceleration time scale so that the electron injection can be considered continuous. For M87, with a break which is greater than a half but less than that which would occur for an instantaneous injection (1.2), the injection time scale would be comparable to the loss time scale. Such processes could occur in the mechanism suggested by Rees (1978) in which a modulation of the jet velocity creates relativistic shocks as the faster material overtakes the slower. Particles are accelerated in these shocks.

On the most important questions about jets is the velocity of the material in them. Many models assume relativistic motion (Rees 1978, Blandford and Königle 1979). However, some observations of double jets imply much slower speeds (Perley, Willis and Scott, 1979). We can use our observations to place some limits on the jet velocity.

Suppose that the jet is moving from the nucleus at a constant velocity βc along an angle θ to the line of sight. In the rest frame of the jet, the break frequency ν'_b where the spectral index steepens is related to the time t'_b since the injection of relativistic particles began or ceased (assuming either a continuous or instantaneous injection respectively) and the magnetic field H' by

$$\nu'_b (H')^3 (t'_b)^2 = 1.12 \times 10^{24} \text{ (cgs units).} \qquad (1)$$

Primed quantities are measured in the rest frame of the jet. The observed ν_b and t_b are related to the primed quantities by the Doppler shift and time dilation: $\nu'_b = \gamma(1-\beta \cos \theta)\, \nu_b$ and $t'_b = t_b/\gamma$ where γ is the relativistic factor. We estimate the magnetic field from equipartition of particle and magnetic energies assuming the magnetic field direction is parallel to β.

$$H'\text{min} = [6\pi K \frac{I'}{S'} \frac{(1+k)}{|\sin\theta|^{3/2}}]^{2/7} \qquad (2)$$

Perley, Willis, and Scott 1979). Here I' is the rest frame surface brightness integrated from 10MHz to 10GHz, S' is the path length through the emitting volume assumed equal to S the observed width, k is the ratio of energy in protons to electrons assumed equal to 1 and K is $c_2 c_{12}(2\alpha + 1)/c_1 c_3 c_8 (2\alpha + 1)$. The c-factors are defined and tabulated by Pacholczyk (1970) and α is the spectral energy index. Now from Blandford and Königle (1979) $I'(\nu) = I(\nu)\sin\theta[\gamma(1-\beta \cos \theta)]^{(2+\alpha)}$ where $I'(\nu)$ is measured by a comoving observer whose line of sight is perpendicular to the jet velocity. For a jet there are two relevant sizes, the distance along it or across it, which means that there are two characteristic times $t_e = \ell/c\beta \sin \theta$ where ℓ is the observed jet length or $t_d = S/c$. Using either of these two times for estimates for t_b we obtain one equation in the two unknowns γ and Θ. This permits us to constrain these parameters.

For either estimate of t_b, two values of θ are permitted for a given γ. If the characteristic time is obtained from the distance along the jet than $\gamma > 5$ for M87 and $\gamma > 17$ for 3C273. If the characteristic time is obtained from the distance across the jet than $\gamma > 1$ for M87 and $\gamma > 1.2$ ($\beta > 0.5$) for 3C273. Our data thus require relativistic beaming models for 3C273; the M87 jet could also be relativistic. It is interesting that $\theta \approx \frac{1}{\gamma}$, which is what one would expect from relativistic motion, is permitted. Relativistic motion would also explain why there are no counter jets in these sources.

IV. QUASARS

It is known that quasars are luminous X-ray sources (Tananbaum et al. 1979, Ku and Helfand 1980, Zamorani et al. 1980). We have now (Jan. 1980) observed a large enough sample (of previously known plus 20 X-ray selected) that we can start to study how the X-ray luminosity depends on other properties such as strong radio emission or redshift. We still have not observed complete samples, but we're almost there; for example we have observed about three-fourths of 3CR statistical sample. Since we have not observed a complete sample yet, we cannot derive luminosity functions. Instead we parameterize our results by the apparent energy spectral index between 2500 Å and 2 keV, α_{ox}.

The motivation for doing this is to be able to use the constraint that the integrated surface brightness of any class of objects cannot exceed the observed X-ray background. We can then use an effective α_{ox}, defined by $\langle L_x(2keV)/L_{opt}(2500Å)\rangle \equiv 10^{-2.605\alpha_{ox}^{eff}}$, combined with quasar number counts done in the visible, which hopefully are complete, to calculate the background produced by quasars to any apparent magnitude. We hope that this method will introduce the fewest uncertainties.

Our first analysis, (Tananbaum _et al_. 1979) derived a value of α_{ox}^{eff} of 1.25 ± 0.05 by taking the sample average of the quasars that we had observed. This implys that the integrated contribution of quasars brighter than $m_B \approx 20.3$ equals the X-ray background. We have assumed the Braccesi et al. (1980) number counts for the number of quasars per square degree with magnitude less than m_B, $N(<m_B)$:

$\log N(<m_B) = a\left(\frac{m_B - m_o}{2.5}\right)$ where $a = 2.16$ and $m_o = 18.33$. However, there are known systematic errors in the determination of α_{ox}^{eff} because while most quasars are radio quiet, the first sample used to find α_{ox}^{eff} contained mostly radio loud objects.

Zamorani _et al_. (1980) have determined new values for α_{ox}^{eff}, computed by the maximum likelihood method of Avni _et al_. (1980) as shown in Table 2. Our definition of a radio loud object is one where the 5GHz - 2500Å energy index is greater than 0.35. This seems to divide the data clearly into two parts. Note from Table 2 that the radio loud all Z value agrees with our previous determination and that the effective index for radio loud quasars is different from radio quiet ones at the 2.5σ level. This is the most significant difference and I will discuss it later in the paper. It is also interesting to note that for Z<1 the index is independent of whether the quasar is selected by radio, optical, or X-ray emission. In fact Grindlay et al. (1980) show that a sample of six X-ray selected quasars are indistinquishable in any way from other low redshift quasars.

Table 2

α_{ox}^{eff} For Quasars

	All Z	Z<1	Z>1
Radio loud	1.27 ± 0.03	$1.25^{+0.05}_{-0.04}$	$1.30^{+0.08}_{-0.06}$
Radio Quiet	$1.46^{+0.05}_{-0.07}$	$1.35^{+0.05}_{-0.08}$	$1.62^{+0.08}_{-0.16}$
X-Ray Selected	---	1.28 ± 0.03	---

Since most quasars are radio quiet and have Z>1, the true value of α_{ox}^{eff} is around 1.5. This implies that quasars brighter than $m_B = 21.5$ can make up the entire X-ray background which means that the number counts turn over beyond about $m_B \approx 20$ to avoid exceeding the background. This has been examined by Kron and Koo (private communication) who are doing counts of blue objects to about 23rd magnitude. Based on a preliminary examination of their data, they find that the counts do flatten at these faint magnitudes. Murray (1980), in another paper in this session, shows that $\alpha_{ox}^{eff} = 1.5$ also yields the observed number of identified (i.e. $m_B \lesssim 20$) QSO's in a Einstein deep survey field.

From the previous discussion it seems likely that integrated emission from quasars make up a large fraction of X-ray background in the 1-3 keV band. An observational consequence of this assertion is that the average of many quasar spectra should be the same as the background spectrum which is a power law over our band with an energy index 0.4. Most quasars are so faint that it is impossible to fit detailed spectra. It is possible to determine a hardness ratio. In Figure 11 I show the histogram of the hardness ratios of 27 X-ray bright quasars. The radio quiet ones are indicated by the shading. The average hardness ratio of 1.13 corresponds to an index of 0.4 which is the background index. The width of the distribution is due mainly to counting statistics. Some caution is called for here because most of the above sample is radio loud while most quasars are radio quiet. However, the average hardness ratio of the radio quiet objects alone is close to the overall average.

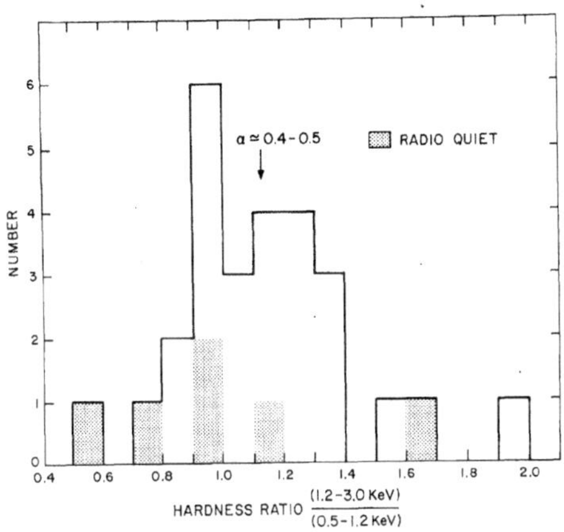

Figure 11. The distribution of the hardness ratio for 27 quasars with more than 100 net counts.

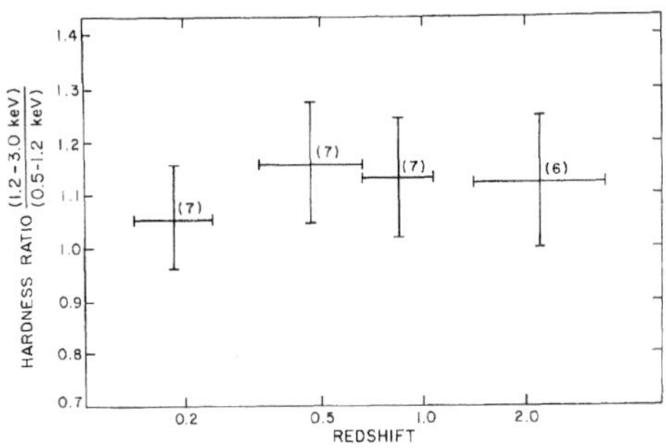

Figure 12. The weighted average of the hardness ratio in four different redshift bins. The number of objects in each bin is indicated.

It is also possible to get some idea of a composite quasar spectrum by examining the hardness ratio as a function of redshift. This is shown in Figure 12. Since a redshifted power law spectrum is also a power law of the same slope, the composite quasar spectrum could be described by a power law of energy index 0.4 at least to 10 keV because the hardness ratio appears constant at least to redshift as large as 2.5. This is also consistent with the background spectrum. These data can also be used to test for absorbtion by heavy elements in the intergalactic medium as a function of redshift. Any epoch where the medium contains absorbing material would produce a deviation in the hardness ratio from the average value assuming a power law quasar spectrum.

I now want to examine the correlation of X-ray emission with radio emission mentioned previously. In Figure 13 I show a plot of the radio optical energy index α_{ro} vs the optical X-ray energy index α_{ox}. A clear correlation can be seen which is significant at greater than 3σ. Of course the radio quiet quasars are only known to be in the lower right portion of this diagram but these objects also appear to follow the trend.

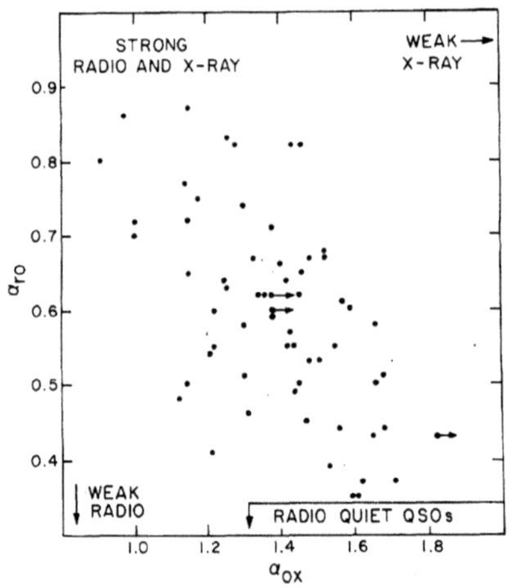

Figure 13. α_{ro} vs α_{ox} for 62 radio loud quasars. More than 90% of the radio quiet quasars have α_{ox} within the region bracketed at the bottom of the figure.

It is possible to use these data to eliminate some simple models of X-ray emission such as pure synchrotron, which would have $\alpha_{ro} = \alpha_{ox}$, or synchrotron with a break, which would have $\alpha_{ro} + C = \alpha_{ox}$ where C is some constant. The only objects that are consistent with these models lie in the upper left corner of Figure 13. These tend to be the optically violent variables and in fact pure synchrotron models have been proposed for them (Blandford and Königle 1979).

We interpret these results as indicating that there are at least two components to the X-ray emission from quasars. One component is present in every quasar. A second component is related to the radio emission process and is only present when there is such emission. Using the same variability arguments that we did previously for Seyferts, we will show below that at least some of the X-rays do not come from an optically thin thermal bremstrahlung source. So what seems to be left for the X-ray emission mechanism is some sort of Compton process.

The existence of a correlation between α_{ro} and α_{ox} requires some caution in deriving the α_{ox}^{eff} for radio loud quasars. In order to obtain an unbiased estimate of α_{ox}^{eff} for these quasars the distribution of the sample α_{ro} must be the same as that of a complete sample. In other words our sample of the α_{ox} distribution will not be unbiased unless our sample of α_{ro} is unbiased. Our sample is in fact deficient in objects with $\alpha_{ro} < 0.55$ as compared with the complete samples of Fanti et al. (1977). This bias, along with the correlation shown in Figure 13, means that the values in Table 3 for radio loud quasars are lower limits. This bias does not invalidate our basic conclusion that radio loud quasars are more X-ray luminous than radio quiet quasars because a Kolmogorov-Smirnoff test gives a probability of $\sim 5 \times 10^{-4}$ that the two samples are drawn from the same parent population. The α_{ox}^{eff} used in the background calculation is also not appreciably affected because most quasars are not radio loud.

Finally, I want to discuss the X-ray variability of quasars on time scales less than one day. Zamorani and Maccacaro (1980) have examined ~ 30 quasars observed with the IPC. They found no variability by factors of 50% on time scales of 1000 s. This means that a typical quasar is variable less than 5% of the time with these amplitudes and time scales.

We do see variability in two quasars on short time scales. Tananbaum et al. (1979) report that the flux from OX 169 changed by a factor of ~ 3 in $\Delta t \sim 6 \times 10^3$s. This is a ΔL of $\sim 2 \times 10^{44}$ erg s^{-1}. This large amplitude variation suggests that there is only one source of energy, rather than several small sources.

I show in Figure 14 the observations of 3C273 which covered three days. We observed this source simultaneously with the HRI and the MPC. The HRI observations, at a mean energy of about 0.7 keV, exhibit an 8% change in about 5×10^4s. The change is significant at the 6.6σ level. The same change is observed with the MPC, with a mean energy of about 4.5 keV. The amplitude is 4% and the change is significant at the 4.4σ level. Thus $\Delta L \approx 10^{45}$erg s^{-1}.

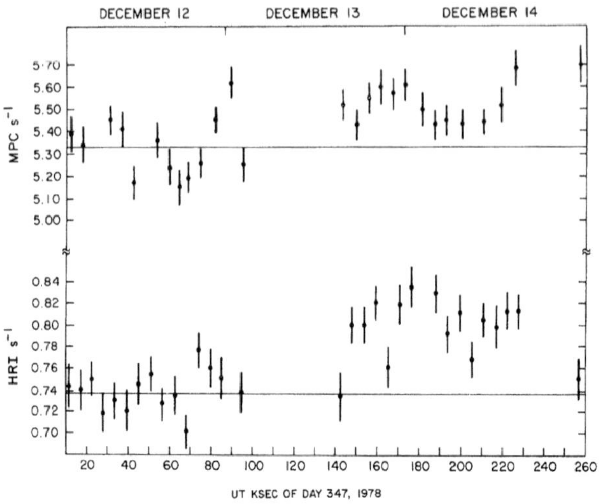

Figure 14. The 3C273 X-ray light curve. The upper panel is the MPC data and the lower panel is the HRI data. The horizontal lines in each panel is the average rate during the first 100 ksec of the observation.

In a manner similar to what was done for the Seyfert galaxy NGC 6814 I find that the electron scattering optical depths assuming thermal bremsstrahlung radiation, are about 1000 $T_9^{-1/2}$ C^{-1} for 3C273 and about 1600 $T_9^{-1/2}$ C^{-1} for OX 169. These very large optical depths imply that optically thin thermal bremsstrahlung is not the emission mechanism.

I can also use an argument originally due to Cavallo and Rees (1978) to determine the minimum efficiency, η, with which matter is converted to energy $\eta \gtrsim 1 \times 10^{-4}\ \Delta L_{42}\ \Delta t_3^{-1}\ (\tau es + 1)^2/\tau es$, where the symbols are as defined before. Since τes is unknown, (the estimate above was for a thermal bremsstrahlung process which may not be the actual process) a strict limit may be placed on η by setting $\tau es = 1$ which minimizes the right hand side. In this case $\eta \gtrsim 1\%$ for 3C273 and OX 169. While this limit assumes no relativistic motions, it does indicate that gravitational accretion onto a compact object rather than supernova explosions, for example, may be the energy source for these quasars.

I wish to thank the members of the Einstein consortium institutions, particularly C. Canizares, M. Elvis, G. Kriss, W.H.-M. Ku, E. Schreier, and G. Zamorani, for sharing their data with me prior to publication. This work was supported by NASA Contract NAS8-30751.

References

Avni, Y., Soltan, A., Tananbaum, H., and Zamorani, G. 1980, Ap.J., in press.

Blanford, R.D. and Königle, A. 1979, Ap.J., 232, 34.

Braccesi, A., Zitelli, V., Bonoli, F., and Formiggini, L. 1980, Astr. Ap., in press.

Cavallo, G. and Rees, M.J. 1978, M.N.R.A.S., 183, 359.

Conway, R.G. and Stannard, D. 1972, Nature Phy. Sci., 239, 22.

Fabian, A.C. 1979, Proc. R. Soc. Loud. A., 366, 449.

Fanti, C., Fanti, R., Lari, C., Padrielli, L., van der Laan, H., and deRuiter, M. 1977, Astr. Ap., 61, 487.

Giacconi, R., et al. 1979, Ap. J., 230, 540.

Grindlay, J.E., Steiner, J.E., Forman, W.R., Canizares, C.R., and McClintock, J.E. 1980, Ap.J., in press.

Henry, P., Feigelson, E., Willingale, R., and Briel, U. 1980, in preparation.

Holt, S. 1980, these proceedings.

Kriss, G.A., and Canizares, C.R. 1980, Ap.J. in press.

Ku, W.H.-M. and Helfand, D.J. 1980, Nature, in press.

McClintock, J.E., van Paradijs, J., Remillard, R.A., Canizares, C.R., Koski, A.T., and Veron, P. 1979, Ap.J., 233, 809.

Murray, S. 1980, these proceedings.

Mushotsky, R.F., Holt, S.S., and Serlemitsos, P.J. 1978, Ap.J. (Letters), 225, L115.

Osterbrock, D.E. 1978, Proc. Natl. Acad. Sci. USA, 75, 540.

Pacholczyk, A.G. 1970, Radio Astrophysics (Freeman, San Francisco).

Perley, R.A., Willis, A.G., and Scott, J.S. 1979, Nature, 281, 437.

Rees, M.J. 1978, M.N.R.A.S., 184, 61p.

Schmidt, G.D., Peterson, B.M., and Beaver, E.A. 1978, Ap.J. (Letters), 220, L31.

Schreier, E. and Feigelson, E. 1980, in preparation.

Tananbaum, H., Peters, G., Forman, W., Giacconi, R., Jones, C. and Avni, Y. 1978, Ap.J., 223, 74.

Tananbaum, H., Avni, Y., Branduardi, G., Elvis, M., Fabbiano, G., Feigelson, E., Giacconi, R., Henry, J.P., Pye, J.P., Soltan, A., and Zamorani, G. 1979, Ap.J. (Letters), 234, L9.

Van Speybroeck, L. 1980, these proceedings.

Walker, M.F. 1968, Astr. J., 73, 854.

Weedman, D.W. 1977, A. Rev. Astron. Astrophys., 15, 69.

Wilson, A. S. 1979, Proc. R. Soc. Loud. A., 366, 461.

Zamorani, G., Henry, J.P., Maccacaro, T., Tananbaum, H., Soltan, A., Avni, Y., Liebert, J., Stocke, J., Strittmatter, P.A., Weymann, R.J., Smith, M.G., and Condon, J.J. 1980, Ap.J., in press.

Zamorani, G. and Maccacaro, T. 1980, private communication.

OPTICAL COUNTS AND SPACE DISTRIBUTION OF QUASARS

Maarten Schmidt
Palomar Observatory,
California Institute of Technology

A substantial fraction of sources detected at radio wavelengths or in X-rays are quasars. Work at optical wavelengths is always needed to confirm the identification as a quasar and to determine the redshift. Some quasars show no radio emission and, possibly, no X-radiation. Hence, purely optical surveys for quasars yield the parent population from which radio and X-ray quasars derive.

For a derivation of the space distribution of quasars based on optical surveys, see Schmidt (1978). Much additional information will be available soon, upon the completion of the Palomar Bright Quasar Survey which covers 10,000 square degrees to an average magnitude B = 16.2 (for early, tentative results, see Green and Schmidt, 1978). Some preliminary results are given below.

The surface brightness of quasars brighter than B = 18 is about 70 times that of quasars brighter than B = 16. The ratio corresponding to a uniform distribution in an empty Friedman universe would be 5.6. If the co-moving space density is an exponential of cosmic time t,

$\rho = e^{-K(t-t_o)/t_o}$, where t_o is the present epoch, then the observed ratio of counts requires K = 18.

A more detailed investigation of our survey shows that K depends on optical absolute luminosity, in the sense that it is larger for higher luminosity. This is confirmed by a preliminary comparison with Braccesi quasars and with objective prism survey results. Lower evolution for quasars of low luminosity is required in order to avoid quasar surface densities at B around 22^m that exceed star counts.

R. Giacconi (ed.), X-Ray Astronomy with the Einstein Satellite, 279–280.

Quasars that are radio sources show evolution characterized by K = 13 for those with steep radio spectra, and K = 5 for flat radio spectra. In this case there appears to be no dependence on optical luminosity.

We hope to study the space distribution of X-ray quasars on the basis of Einstein observations of quasars of the Palomar Bright Quasar Survey. These observations will also allow an improved determination of the distribution fraction of L_x/L_{opt}, which in turn will lead to a more precise prediction of the X-ray background contributed by quasars.

References

Green, R.F. and Schmidt, M.: 1978, Astrophys. J. (Letters) 202, L1.

Schmidt, M.: 1978, Physica Scripta 17, 329.

THE NUMBER-INTENSITY DISTRIBUTION FOR EXTRAGALACTIC X-RAY SOURCES

S. S. Murray
Harvard-Smithsonian Center for Astrophysics

1. INTRODUCTION

Since the discovery of the isotropic X-ray background in 1962 by Giacconi et al. (1962), the obvious question regarding its origin has been whether this is due entirely to the superposition of unresolved sources or if there is a truly diffuse component as well. For example, hot gas as suggested by Silk (1973), or Field and Perrenod (1977) or Compton scattering of the μ-wave photons by relativistic electrons as suggested by Felton and Morrison (1966).

Observations of the spectrum of the background, most recently by Marshall et al. (1980) using the HEAO-1-A-2 experiment seems to support the hot gas interpretation with a temperature of ~45 keV. However, with the capabilities of the Einstein Observatory (Giacconi et al. 1979a), it is now possible to directly observe individual sources at low enough intensity to determine the contribution of discrete sources to the background. Various arguments have already been made which show that the summed spectra of individual sources can produce the observed background spectrum. For example, Cavaliere et al. (1980) have recently shown that non-thermal sources with an evolving luminosity function can synthesize the apparent observed spectrum. Our approach to this problem has been to conduct high-sensitivity surveys in limited regions of the sky and to observe actual source counts. We have already reported early results based on two such surveys in a paper by Giacconi et al. (1979b). We review those previous observations and present added data which permit a more refined analysis of the number-intensity distribution of extragalactic source. The results indicate that the discrete source contribution to the background is large, and consistant with making up all of the observed flux. These latest results are preliminary and due to the combined efforts of many co-workers at CfA and will be

R. Giacconi (ed.), X-Ray Astronomy with the Einstein Satellite, 281–295.

presented in more detail in a paper by Murray et al. (1981).

2. PREVIOUS STUDIES

We begin this review with one of the first extragalactic source count distributions in X-ray astronomy, which is based on the UHURU catalog (Giacconi et al. 1972) and is shown in figure 1 (Matilsky et al. 1973). Source intensities span less then a decade and only 33 sources were observed. One 1 UFU corresponds to 1.12 x 10^{-11} erg cm^{-2} s^{-1} in the 1-3 keV band (used for "Einstein" data) assuming a background-like spectrum with photon index 1.4. Since these are the most intense and presumably nearest extragalactic sources it is assumed that they follow the power law distribution in N(>S) with index -1.5 expected for a Euclidean geometry.

In addition to direct source counts, point-to-point fluctuations in the X-ray background can be analyzed to yield information about the source distribution particularly at a flux level just below source detection. Holt et al. 1974, Fabian 1975 and Schwartz, Murray and Gursky 1976 found that the number of sources obtained from the Matilsky analysis was overestimated by about a factor of two. This was shown by Murray (1977) to be due to the effects of source confusion and noise which are particularly important at the lowest fluxes observed. A similar analysis was carried out for the source count data obtained with the Ariel V all sky survey. Warwick and Pye (1978) used the Ariel V catalog to measure the normalization of the number-intensity distribution and find

$$N(>S) \simeq 15 \pm 3\ S^{-1.5}\ \mathrm{ster}^{-1} \text{ with S in UHURU ct/s.}$$

The corresponding result for analysis of the 4U catalog gives

$$N(>S) \sim 13 \pm 3\ S^{-1.5}\ \mathrm{ster}^{-1}$$

clearly in agreement with the Ariel result. For comparison with the Einstein data which is in the 1-3 keV band rather then the 2-10 keV band of the earlier surveys we assume an average source spectrum which is background like so that at $S > 1.12 \times 10^{-11}$ erg cm^{-2} s^{-1} (1-3 keV) there are 11.8 ± 1.8 sources $ster^{-1}$. Moreover, based on both the source count data and fluctuation analysis, the index of the power law distribution is 2.5 (differential) and extends to at least fluxes as low as 0.2 UHURU cts/s.

TABLE 1

HRI DEEP SURVEYS

REGION	TYPICAL EXPOSURE TIME (seconds)	SOLID ANGLE (Square Degrees)
Draco	30,000	0.5
Eridanus	50,000	0.3
Pavo	100,000	0.3
U.Minor	300,000	0.1

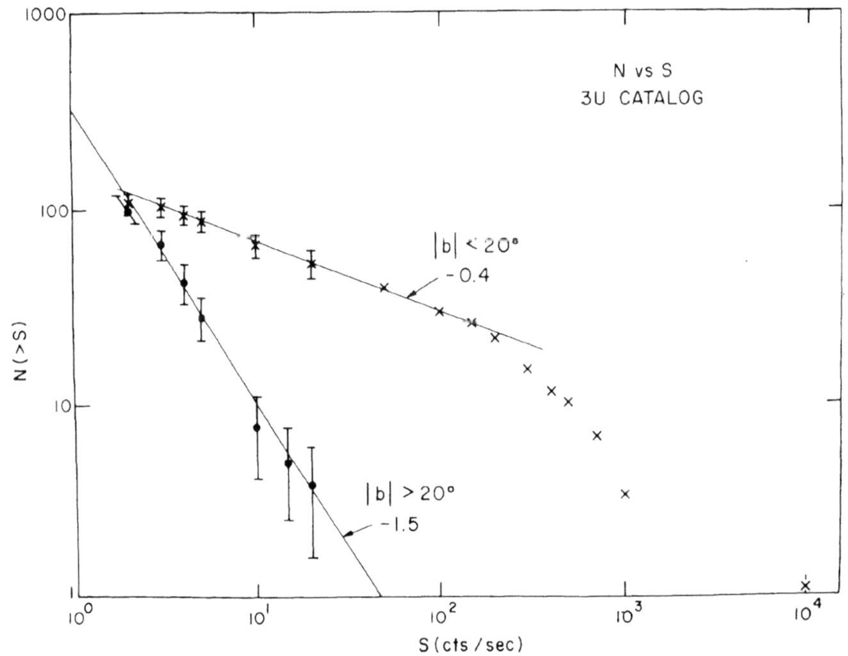

Figure 1: Number-Intensity distribution for sources in the UHURU catalog from Matilsky et al. 1972.

Identifications of the sources in both the UHURU and Ariel catalogs indicates that approximately ~1/3 of the source are associated with rich clusters of galaxies, ~1/3 with active nucleus galaxies particularly Seyferts and ~1/3 remain unidentified.

3. EINSTEIN RESULTS

With the Einstein Observatory we have extended the source detection sensitivity by a factor of about 1000 and obtained source locations with an accuracy of at least 1 arc min. Sources at large redshifts, that is $z \geq 1$ have been detected. Source free areas are selected for the surveys consistant with operational constraints of the observatory. A representative viewing pattern (Figure 2) shows the Eridanus survey area. Both the moderate resolution IPC and high resolution HRI detectors are used. The HRI instrument fields overlap, and we have taken overlapping IPC exposures as well to insure that intrinsic detector irregularities do not give rise to spurious sources.

The deep survey regions thus far observed and analyzed are listed in Table 1, the range of exposure times reflect the details of the observing program plus post-facto recovery of spacecraft attitude. The solid angles reflect the viewing sequences carried out and the degree of overlap. In the case of the Ursa Minor survey, only one HRI field was covered at the exposure of 300,000 seconds. Additional UMn fields have the more typical 50,000 to 100,000 second exposures but these have not yet been analyzed and are not included in the following analysis. A 300,000 second HRI exposure is at the practical sensitivity limit for the Einstein Observatory for point source detection ($S_{Min} \simeq 10^{-14}$ erg cm^{-2} s^{-1}).

Each survey field is processed using standard computer programs (c.f. Giacconi et al 1979b) which reconstruct the X-ray image and scan it for potential sources. The detection algorithm uses a sliding window/centroid analysis in two dimensions to establish existence and source strength. Locations are based on a weighted centroid derived from a maximum likelihood solution for the combined detector/telescope point response function.

The existence criteria for a source is based on a threshold number of counts in a detection bin such that the expected number of spurious sources is less then 1 per field of view. For the IPC this generally corresponds to a

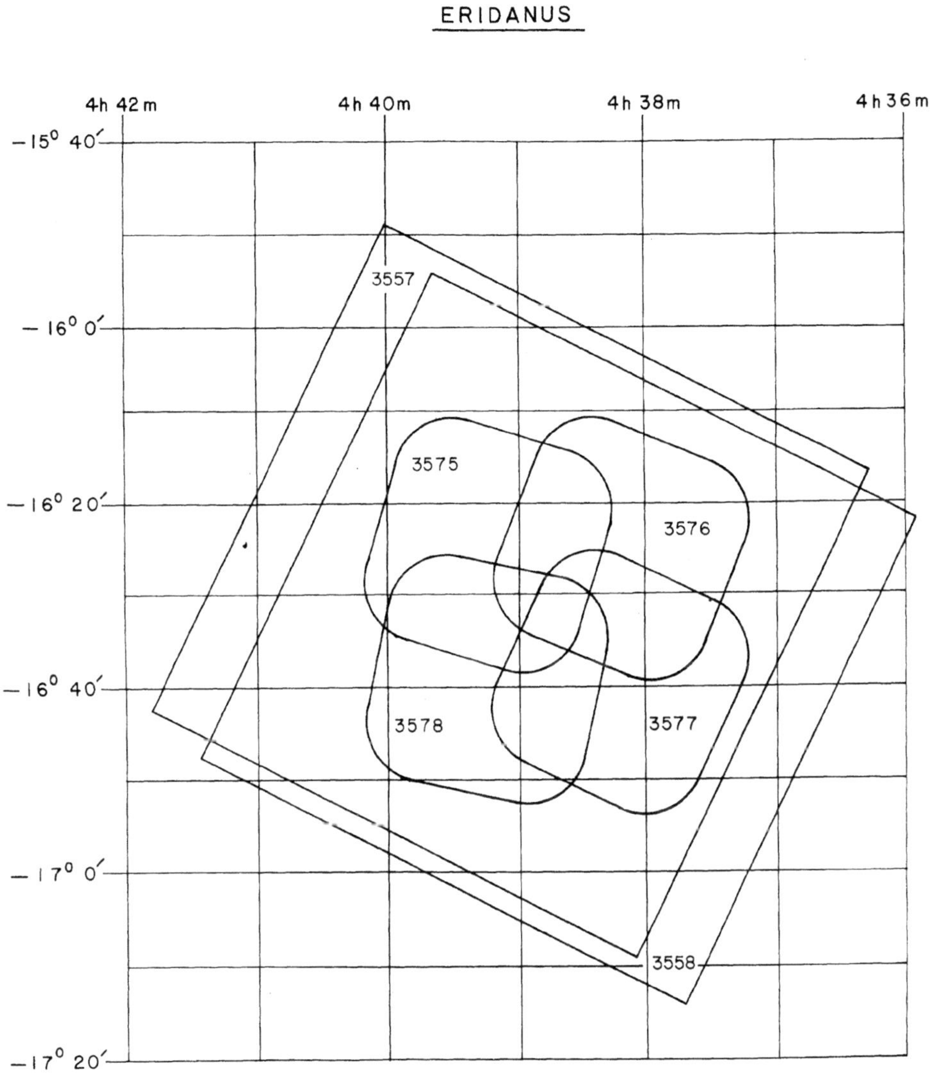

Figure 2: Einstein fields of view for the deep survey in Eridanus.

Gaussian equivalent of about 3.5σ existence and for the HRI about 4σ.

For the purposes of source counts and the number-intensity relationship, it has been shown for example by Murdock et al. (1973) that one should restrict analysis to sources which have intensities measured at the 5σ level or better. Since this is generally more stringent then our existence requirement, the number of sources available is reduced. For example, in the Draco and Eridanus fields (Giacconi et al. 1979b) we detected 43 sources, but only 14 are above the 5σ level in terms of intensity measurement.

In their first analysis of source counts, Giacconi et al. started with these 14 sources. Their next step was to distinguish stars from extragalactic objects. Figure 3 shows some of the finding charts for sources in Eridanus and illustrates the technique used. HRI observations lead to positional accuracy of about 10 arc sec. and the error circles for such sources typically have less then 2 possible optical counterparts down to a limiting magnitude of 21 to 22. In the case of IPC observations, location errors are somewhat larger ~60 sec and there are typically several optical objects for each source. To distinquish galactic from extragalactic objects all obvious stars are eliminated such as 1E043838-1534.8 where the X-ray position coincides with a G0 star. In other cases the B magnitude and B-V color for each optical candidate are measured from either the Palomar Observatory Sky Survey prints or more recently obtained deep plates. These are compared with the results from Vaiana et al. (1980) for stars of various spectral classes to see if the ratio of X-ray to optical luminosity is consistent with a stellar identification. In addition, where possible, candidate objects are examined for proper motion using the POSS and more recent plates. Finally the X-ray sources are checked for flaring during the observation period. These data are used to test for relatively nearby white dwarf or M type stars which can have large L_x/L_{opt} ratios. In many cases we have benefited from actual optical spectra which have been provided by observers including Liller, Sargent, Murdin, Griffiths, and others. In this manner, we have obtained identifications for several sources. For example, in the Eri field, in addition to the G0 star, there is a QSO at $z = 0.5$ with $M_\beta = 19.8$. A third identified source is also a QSO at $z = 1.96$ and M_β -17.8. However, this source was not detected at 5σ and is not included in the source count analysis.

From the set of 14 sources above the 5 detection level, 4 have been associated with stars leaving 10 which

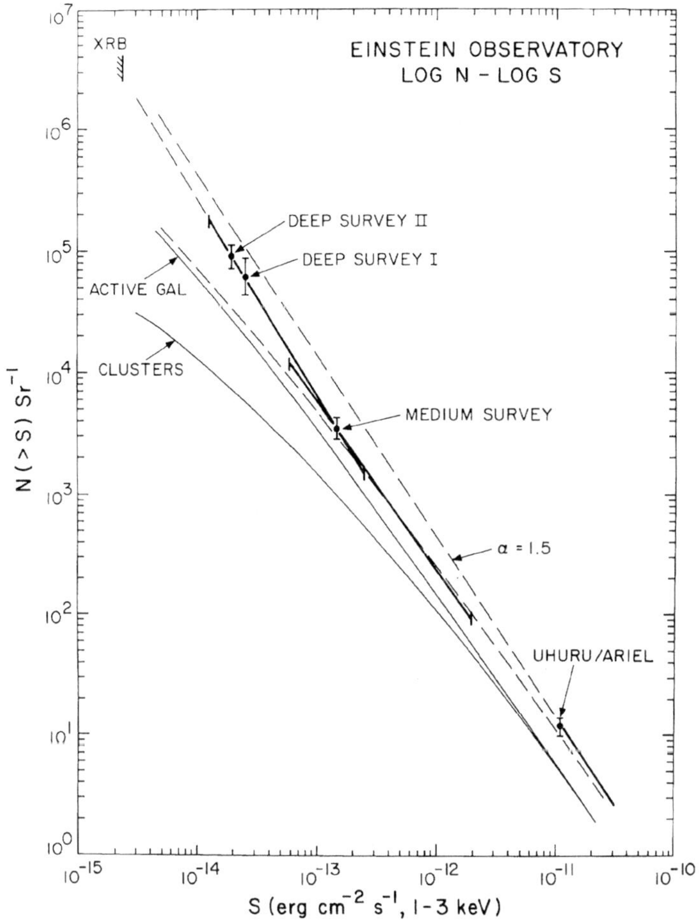

Figure 3: Log N(>S) vs log S for extragalactic X-ray sources. The UHURU/ARIEL all sky survey result is shown as well as the results from Einstein. Deep Survey I is the integral point for Giacconi et al. 1979b. Deep Survey II is the current work, the line drawn has index α = 1.64 and spans the flux range of the deep surveys. The Medium Survey is from preliminary analysis by Maccacaro (1980), the line drawn has index α = 1.4. Also shown are the expected source counts for clusters and active galaxies (assuming no evolution and q_o = 0) and their sum. If the deep survey distribution with index α = 1.64 is continued to lower fluxes the X-ray background is reached at the flux indicated by XRB.

are likely to be extragalactic. The contamination of this set of sources by galactic objects is presumably small. Gorenstein and Schwartz (1980) have discussed the bimodal distribution of deep survey X-ray sources with regard to their B-magnitude in support of this conclusion. The resulting extension to the number-intensity distribution using these sources is shown in Figure 4. Giacconi et al. have discussed the implications regarding the number of sources and their contribution to the background. Their results are based upon an assumed slope of -1.5 for an integral power law distribution. Taking the flux limit of 2.6×10^{-14} erg cm^{-2} s^{-1} (1-3 keV) approximately 25% of the extragalactic background is accounted for by discrete sources.

We have now extended the deep survey program as indicated in Table 1 with observations in Pavo and Ursa Minor. At least two additional surveys are planned - in Cetus and in Canes Venatici (the Braccesi fields) during the remaining lifetime of the obseratory. The Pavo survey results are discussed in detail by Griffiths et al. (1981) in which the optical identifications of individual sources are described. In this survey, maximal use of the HRI to locate precisely the sources detected with the IPC, has resulted in many more possible identifications then for the early surveys. Of 28 detected sources there are HRI positions for 22 with error circles of radius 5 arc sec to ~15 arc sec. Four definite QSO's have been discovered from the optical follow-ups, of these two meet the 5σ significance criteria and are used in source-count analysis. None of the remaining Pavo sources are used.

The Ursa Minor field includes an extremely deep HRI exposure of ~300,000 seconds. This is essentially at the sensitivity limit for the Einstein Observatory. Such an exposure requires about one week of continuous real-time observations due to the viewing efficiency of the satellite. Constraints such as power, thermal, and control system gas consumption preclude significantly longer observations. In addition, the detector background dominates the source counts so that sensitivity increases only as the square root of exposure time. Finally systematic variations in gain, quantum efficiency, and distortions are not sufficiently well known at the levels required by such a long exposure time, and become statistically significant even at the 5σ detection level. We detect a total of 28 sources in the Ursa Minor survey fields. As in the case of the Pavo survey, most are detected with the HRI and have small positional errors. The details of this survey will be presented at a later time when more complete optical studies are carried out.

Four sources meet the $\gtrsim 5\sigma$ statistical significance criteria for use in source count analysis. This survey region reaches a flux limit of ~1.3 x 10^{-14} erg $cm^{-2} s^{-1}$ which is a factor of two fainter than the previous Einstein deep survey reported by Giacconi et al (1979b).

4. NUMBER-INTENSITY DISTRIBUTION

The data now available from Einstein deep survey fields yields 15 sources which are probably extragalactic and meet the $\geq 5\sigma$ significance requirement in their intensity determination. These are listed in Table 2 along with relevant data on the flux and possible optical/radio counterparts. Seven sources have been identified of which 4 are quasars with redshifts fom 0.5 to 2.6 as indicated in the table. For the 8 sources with no identifications, 4 have relatively large error circles since they were detected only by the IPC. The remaining 4 sources which are unidentified have no object brighter then 19th magnitude and thus far no optical spectra have been obtained.

These sources span a flux range of about 20 and we have fit the observed intensity distribution to a power law of the form:

$$\frac{dN}{dS} = As^{-(\alpha+1)}$$

Using the maximum likelihood method to obtain best values for the index α and normalization A. The small sample size and the effects of noise on the flux measurements require corrections to be applied to the derived parameters. This is described by Murdock et al. (1973). We have performed a series of Monte-Carlo experiments to confirm the applicability of Murdock's corrections to this sample of data and find no discrepancy. Our resulting best fit is shown in figure 3. The deep survey data span the flux range 1.3 x 10^{-14} erg $cm^{-2}s^{-1}$ to 2.5 x 10^{-13} erg $cm^{-2}s^{-1}$ in the 1-3 keV band. The plot is an integral distribution of N(>S) versus S. The results of the deep survey are indicated by the line of slope 1.64 extending from 1.3 x 10^{-14} to 2.5 x 10^{-13} erg $cm^{-2}s^{-1}$. The point at 2 x 10^{-14} erg $cm^{-2}s^{-1}$ indicates the normalization ~(9 x 10^{4} sr^{-1}) and its error. The parameters are correlated, so that all of the allowed indicies to the power law result in lines which pass through the point shown. For comparison the previous integral point from Giacconi et al. (1979b) is plotted at a flux of 2.6 x 10^{-14} erg $cm^{-2}s^{-1}$.

TABLE 2

5σ DEEP SURVEY SOURCES

Constellation	Position		Detector	Comments
Draco	$17^h07^m22^s$	71°04'06"	IPC	no ID, M_B>19
	17 09 35	71 03 54	IPC	no ID, M_B>17.6
	17 10 51	70 55 18	HRI	no ID, M_B>21.5
	17 11 22	70 53 54	HRI	no ID, M_B>19.1
	17 12 10	71 14 16	HRI	radio,galaxy, $M_B \simeq 20.5$
Eridanus	04 37 46	-16 20 18	IPC	radio,galaxy, $M_B \simeq 21$
	04 37 58	-16 23 30	IPC	no ID, M_B>18.4
	04 38 09	-16 21 54	HRI	no ID, M_B>21.5
	04 38 22	-16 42 12	IPC	no ID, M_B>19.2
	04 38 26	-16 35 32	HRI	QSO,Z=0.5, M_B=19.8
Pavo	21 21 21	-68 06 34	HRI	QSO, Z=2.4, M_B=20.0
	21 11 35	-67 47 51	HRI	QSO, Z=2.6, M_B=20.5
U.Minor	14 11 54	73 12 17	HRI	galaxy, M_B=19.8
	14 11 11	73 12 02	HRI	no ID, M_B>20.8
	14 12 31	73 03 45	HRI	QSO, Z=1.9, M_B=18.8

15 Total Sources,
7 with possible identification of which
4 are QSO's.

We have also shown preliminary results of a medium survey being carried out with the Einstein Observatory (Maccacaro et al. 1980, 1981). In this program serendipitous sources which are detected in IPC fields are systematically collected and sky coverage is accumulated taking into account the exposure times of the observations. In figure 3 the medium survey source distribution is shown extending from ~6 x 10^{-14} to ~2 x 10^{-12} erg $cm^{-2}s^{-1}$. About 20 square degrees of sky are covered at the higher flux limit and ~15 sources were found meeting the statistical requirements and likely to be extragalactic. This segment of the distribution has integral index 1.4 with normalization indicated by the point at 1.5 x 10^{-13} erg $cm^{-2}s^{-1}$. We note the good agreement between the two Einstein surveys in their region of overlap.

The data point at ~1.12 x 10^{-11} erg $cm^{-2}s^{-1}$ in figure 3 is the result obtained by the earlier all sky surveys of UHURU and Ariel V. The solid line extending towards higher flux is the index 1.5 power law observed and it is continued to lower fluxes as a dashed line.

5. CONTRIBUTION OF DISCRETE SOURCES TO THE XRB

The observed source distribution from 1.3 x 10^{-14} erg $cm^{-2}s^{-1}$ to 2 x 10^{-12} erg $cm^{-2}s^{-1}$ can be integrated to give the measured flux in the 1-3 keV band due to discrete sources. We obtain a total emission of 7.0 x 10^{-9} erg $cm^{-2}s^{-1}sr^{-1}$ in the 1-3 keV band down to the deep survey flux limit. This is to be compared with the total isotropic background intensity estimated by Schwartz (1979) to be 1.9 x 10^{-8} erg $cm^{-2}s^{-1}sr^{-1}$ in the same band. Discrete sources therefore account for >35% of the extragalactic background. If the observed deep survey distribution is extended to a flux of 2.4 x 10^{-15} erg $cm^{-2}s^{-1}$ then 100% of the background would be reached, thus there must be some turnover in the source counts beyond the survey limit and before this limiting flux, which is less then a factor of six lower.

6. DISCUSSION

As shown in figure 3, the number of sources detected at the Einstein sensitivity limit is substantially below the extension of source counts from the UHURU limit if a 1.5 power low index (Euclidean) is used. Since we are observing sources with high redshifts (from the QSO's identified) this turnover in log N - log S is not surprising. We have calculated the expected source count distribution for a Friedmann cosmology with $q_o = 0$ assuming that the sources do not evolve and are described by the local luminosity functions for clusters of galaxies (c.f. Schwartz 1978) and for Seyfert galaxies (c.f. Tananbaum 1978, Elvis 1978). The contribution of clusters at the faint flux limit is very small due to their high intrinsic luminosities. Active galaxies (represented by the Seyferts) extend to lower luminosities, so that at low fluxes the typical sources are at low redshift and the cosmological turnover is less pronounced. At the Einstein deep survey flux limit the observed number of sources exceeds the expected number of unevolved sources by about a factor of ~3. This is well outside of errors in either the observed source counts or the extrapolation from the UHURU limit. Thus, there is strong evidence from the total source counts that a population of evolving sources contributes significantly to the discrete sources observed at low fluxes.

In the intermediate survey flux interval, the combined contribution of sources (clusters and active galaxies) matches the observed distribution both in total numbers and in the intensity versus flux distribution. However, whereas at the UHURU flux limit there are as many clusters detected as active galaxies, the mixture becomes predominantly active galaxies at this lower flux level. Maccacaro et al. (1981) have carried out a more detailed analysis of the intermediate survey in terms of the observed source distribution and comparison with the composition of sources. They conclude that in the flux interval of the medium survey, the source population cannot be made consistent with the lower sensitivity surveys without taking cosmological expansion into account.

In addition to the increased number of sources detected at low flux, the observed power law index for the source distribution requires strong source evolution. The differential source distribution can be expressed as

$$\frac{dn}{dS} = \left.\frac{dn}{dS}\right)_E (1+z)^{-(5+3\gamma/2)} \quad f(z)$$

where $\left.\frac{dn}{dS}\right)_E$ is the Euclidean form which is a power law of index 5/2; the expression involving (1+z) is the cosmological term for a Friedmann Universe with $q_o = 0$ and energy spectral index γ; and f(z) is an evolutional term often written as either $(1+z)^k$ or $e^{A\tau}$ with τ the Hubble time (z/1+z).

Taking the logarithmic derivative of dn/dS and comparing with the observed power law index we have

$$(\alpha+1) = 2.5 + (3.1-k)\frac{d\ \log(1+z)}{d\ \log S}$$

where we have taken $f(z) = (1+z)^k$ and $\gamma = 0.4$

For a $q_o = 0$ model

$$\frac{d\ \log(1+z)}{d\ \log S} = \frac{-z(1+z/2)}{1.7z^2+1.4z+2}$$

We observe $z_{typical} \gtrsim 0.25$, this gives $\frac{d\ \log(1+z)}{d\ \log S} \lesssim -0.1$, and

$$k > 3.1 - \frac{(1.5-\alpha)}{0.1}$$

For no evolution (k=0) the observed index α must be as flat as $\simeq 1.2$. We observe $1.6 \pm .3$ which is inconsistant with the non-evolving source hypothesis.

7. CONCLUSIONS

A self consistant picture of the source count data can be made if many of the deep survey sources are found to be QSO's which are already known to be evolving strongly at fluxes which are comparable to the survey limits. (Taking the correspondence between L_x and L_{opt} from Tananbaum et al. 1979 and Zamorani et al. 1980). In this case the QSO contribution becomes significant at X-ray fluxes below 10^{-12} erg cm^{-2} s^{-1} and dominates at fluxes below 10^{-13} erg cm^{-2} s^{-1}. The source count data shown in figure 3 can be divided into 3 regions. At the highest flux level we are sampling a Euclidean universe of nearby sources where there are virtually no evolving sources present. The intermediate flux region (medium survey) samples sources where cosmological expansion is significant but for which evolution is not yet important. The lowest flux level survey reaches far enough out in redshift so that evolution is important and in particular the QSO population contributes significantly. Within a factor of 6 less in flux, the X-ray background intensity places a limit on total source counts and requires a turnover in the evolution of X-ray emitting sources. Such a turnover in QSO counts may already be observed in the recently reported works of Kron (1980), Bahcall and Soneira (1980), and Bonelli et al. (1980).

Sources already detected in the Einstein survey account for over 35% of the 1-3 keV background flux. This observation strongly suggests that most of the background is due to discrete sources since an extrapolation based on QSO optical data can easily reach the total flux required. Marshall et al. (1980) have suggested that, on the basis of the types of sources already known, a diffuse component to the background of ~20% is consistent with their HEAO-1 spectral data. The source count data we have obtained thus far do not rule out such a component. Observations of greater sensitivity and surveys of larger areas will be required to determine the extent of the discrete source contribution to the background. We are extending the medium survey with Einstein to cover about 200 square degrees of sky. The deep surveys will be completed for two regions (Cetus and U.Minor) but for reasons given above they will be no more sensitive than those already done. We must look to future X-ray observatories such as AXAF to extend sensitivity to the desired limits.

ACKNOWLEDGMENTS

This work was supported under NASA contract NAS 8-30751. I would like to thank my co-workers at CfA whose efforts have been of great importance in obtaining and analyzing the survey data from the Einstein Observatory. I particularly appreciate the efforts, help, and discussions I have had with Drs. R. Giacconi, R. Griffiths, T. Maccacaro, D. Schwartz, and G. Zamorani. J. Bechtold, E. Farmer, and M. Fener have contributed to the seemingly endless task of data processing necessary for this research. I thank Mrs. M. Sullivan for her patient prodding to have this manuscript completed and K. Gilleece for typing it.

REFERENCES

Bahcall, J.N. and Soneira, R.M. 1980 preprint.

Bonoli, F., Braccesi, A., Marano, B., Merighi, R., and Zitelli, V. 1980 Submitted to Astr. and Ap.

Cavaliere, A., Danse, L., DeZotti, G., and Franceschini, A. 1979, Astr. and Ap., 79, 169.

Fabian, A.C. 1975, Mon.Nat.R.Astr.Soc., 172, 149.

Felton, J.E. and Morrison, P. 1966, Ap.J., 146, 686.

Field, G. and Perrenod, S.C. 1977, Ap.J., 215, 717.

Giacconi, R., Gursky, H., Paolini, F.R., and Rossi, B.B. 1962, Phys.Rev. Letters, 9, 439.

Giacconi, R., Murray, S., Gursky, H., Kellogg, E., Schreier, E., and Tananbaum, H. 1972, Ap.J., 179, 281.

Giacconi, R., Branduardi, G., Briel, U., Epstein, A., Fabricant, D., Feigelson, E., Forman, W., Gorenstein,

P., Grindlay, J., Gursky, H., Harnden, F.R., Jr., Henry, J.P., Jones, C., Kellogg, E., Koch, D., Murray, S.S., Schreier, E., Seward, F., Tananbaum, H., Topka, K., Van Speybroeck, L., Holt, S.S., Becker, R.H., Boldt, E.A., Serlemitsos, P.J., Clark, G., Canizaris, C., Markert, T., Novick, R., Helfand, D., and Long, K. 1979a, Ap.J., 230, 540.
Giacconi, R., Bechtold, J., Branduardi, G., Forman, W., Henry, J.P., Jones, C., Kellogg, E., van der Laan, H., Liller, W., Marshall, H., Murray, S.S., Pye, J., Schreier, E., Sargent, W.L.W., Seward, F., and Tananbaum, H. 1979b, Ap.J., 234, L1.
Gorenstein, P. and Schwartz, D.A. 1980, in "Highlights of Astronomy", Patrick A. Wayman (ed.) Vol. 5, p. 763.
Griffiths, R., et al. 1981 in preparation.
Holt, s.S., Boldt, E.A., Serlemitsos, P.J., Murray, S.S., Giacconi, R., Kellogg, E.M. and Matilsky, T.A. 1974, Ap.J., 188, L97.
Kron, R.G. 1980 in "Two Dimensional Photometry" (ESO Workshop) P.O. Lindblad and H. van der Laan (eds.), Geneva.
Maccacaro, T. 1980 private communication.
Maccacaro, T. et al. 1981 in preparation.
Marshall, F.E., Boldt, E.A., Holt, S.S., Miller, R.B., Mushotzky, R.F., Rose, L.A., Rothschild, R.E., and Serlemitsos, P.J. 1980, Ap.J., 235, 4.
Matilsky, T., Gursky, H., Kellogg, E., Tananbaum, H., Murray, S., and Giacconi, R. 1973, Ap.J., 181, 753.
Murdock, H.S., Crawford, d.F., and Jauncey, D.L. 1973, Ap.J., 183, 1.
Murray, S.S. 1977, CfA preprint.
Murray, S.S. et al. 1981 in preparation.
Schwartz, D.A., Murray, S.S., and Gursky, H. 1976, Ap.J., 204, 315.
Schwartz, D.A. 1978 Proc. IAU/COSPAR Symposium on X-Ray Astronomy, Innsbruck, Austria.
Silk, J. 1973, Ann.Rev.Astr.Ap., 11, 269.
Tananbaum, H., Avni, Y., Branduardi, G., Elvis, M., Fabbiano, G., Feigelson, E., Giacconi, R., Henry, J.P., Pye, J.P., Soltan, A., and Zamorani, G. 1979, Ap.J., 234, L9.
Vaiana, G.S., Cassinelli, J.P., Fabbiano, G., Giacconi, r., Golub, L., Gorenstein, P., Haisch, B.M., Harnden, F.R., Jr., Johnson, H.M., Linsky, J.L., Maxson, C.W. Mewe, R., Rosner, R., Seward, F., Topka, K., and Zwaan, C. 1980 Submitted to Ap. J.
Warwick, R.S. and Pye, J.P. 1978, Mon.Nat.R.Astr.Soc., 183, 169.
Zamorani, G., Henry, J.P., Maccacaro, T., Tananbaum, H., Soltan, A., Avni, Y., Leibert, J., Stocke, J., Strittmatter, P.A., Weymann, R.J., Smith, M.G., and Condon, J.J. 1980, submitted to Ap.J.

X-RAY OBSERVATIONS AND THE COSMIC DISTANCE SCALE

Paul E. Boynton
University of Washington

The physics of hot gas in the cores of distant clusters of galaxies can be studied in considerable detail through a specific combination of X-ray, microwave and optical observations. In principle, the X-ray luminosity of the gas can be calculated from these data, providing a measure of the cluster distance. We examine these ideas and their application to Abell 2218, the only cluster for which even marginally suitable data yet exist.

I. INTRODUCTION

Understanding the "global" dynamics of the Universe is a primary goal of modern cosmology. Within the comfortable confines of orthodoxy this quest boils down to a determination of, q_0, the "deceleration parameter" which describes the slowing of the expansion rate of the Universe, which in turn is specified by the Hubble constant, H_0. It is in a sense remarkable that the assumptions of homogeneity and isotropy when imposed on the Einstein equations yield such a simple description of the totality of existence. I wonder if Lemaître or Friedmann ever lay awake in the night considering the colossal conceit implicit in the presentation of a Universe in just two parameters. After all, any worthwhile subsystem (e.g. an iguana) is well known to require at least three parameters to provide even a modest description. Yet I find it still more remarkable that such a simple formulation of the World continues for a half-century to fascinate the scientific community. How many colleagues would sit through a lecture entitled "A New Way to Measure the Mass and Spring Constant of a Simple Harmonic Oscillator"? Surely the comparison is strained, but I suspect my own judgement on the matter is less than objective.

The values of H_0 and q_0 may be inferred from measurement of distance and redshift of remote objects. There are many ways to establish a cosmic distance scale; all have strengths and weaknesses. The method I will discuss is appropriate to this meeting since it depends crucially on X-ray imaging, which is now routinely possible with the Einstein Observatory.

R. Giacconi (ed.), X-Ray Astronomy with the Einstein Satellite, 297–309.

The first problem is to decide both formally and operationally on a suitable definition for distance measure. A familiar choice is "luminosit distance", suggested by the inverse square law:

$$D_L = [L/(4\pi F)]^{1/2} \tag{1}$$

In a cosmological context, the luminosity distance may be expressed as

$$D_L = c\,[zq_o + (1 - q_o)(1 - (2q_o z + 1)^{1/2})]/H_o q_o^2 \tag{2}$$

for the Lemaître-Friedmann models (Weinberg 1972). A more transparent fon results when written as a power series in the redshift parameter, z:

$$D_L = c\,[z + (1-q_o)z^2/2 + \ldots]/H_o \tag{3}$$

For sources at small redshift, the customary form of the Hubble relation $z = H_o D_L$ is recovered, whereas for values of z approaching unity, the correction due to deceleration becomes important. Clearly with a means for independently measuring the redshift and luminosity distance, both H_o and q_o could be determined. After many years of astronomical experience our optical tools for measuring redshift are well developed, but what means do we have to infer D_L? The answer I have chosen to stress involves observations of clusters of galaxies.

Since the detection of clusters of galaxies as X-ray sources, the existence of a hot, intracluster gas has been suggested as the source of emission through thermal bremsstrahlung (Gursky et al. 1971). Subsequent observations of X-ray emission lines in many clusters support this view (Serlemitsos et al. 1977). Moreover, Sunyaev and Zel'dovich (1972) recognized immediately that the inverse Compton cooling of such a hot intracluster medium ($\sim 10^8$°K) through interaction with the Primordial Background Radiation (referred to in the current epoch as the "3°K radiation") can result in a measurable perturbation to the composite spectrum of photons scattered by and transmitted through the optically thin cluster core gas. In simplest terms, this photon-conserving "heating" of the cosmological black body radiation is roughly equivalent to a shift of the thermal spectrum to shorter wavelengths, producing a relative reduction in radiation intensity (sometimes referred to as the "microwave decrement") at wavelengths longward of the black body peak at $\lambda = 1$mm, and a consequent increase in intensity at submillimeter wavelengths.

Over the past decade a substantial amount of telescope time has been devoted to a continuing attempt to detect and measure the Sunyaev-Zel'dovich effect at centimeter to millimeter wavelengths (Parijskij 1973, Lake and Partridge 1977, Birkinshaw, Gull and Northover 1978, Rudnick 1978, Perrenod and Lada 1979). The combined results of this survey of 23 Abell clusters have been reviewed by Lake (1979), who notes that only two clusters, A576 and A2218 have been reported to have significant indications of a microwave signature; but for the latter cluster agreement on the sign of the effect is not unanimous. However, our X-ray observations of A2218 suggest a microwave decrement as large as any yet reported.

Recently, several researchers (Gunn 1978, Cavaliere, Danese and De Zotti 1979, Silk and White 1978, Boynton and Murray 1978) have independently proposed a scheme for combining microwave decrement measurements with X-ray observations which enables the cluster distance to be calculated directly from these data, bypassing the standard "chain" of distance indicators and thereby providing a new means for establishing the cosmic distance scale. This method is particularly interesting because it is feasible for clusters at truely cosmological distances, $z > 0.1$, yet is in principle immune to the effects of possible source evolution which have historicaly thwarted attempts to extend the Hubble diagram toward sufficiently high redshift to determine the deceleration parameter, q_o.

Briefly, this new technique may be summarized as follows. In the Rayleigh-Jeans limit, the fractional microwave decrement, $\Delta T/T = \Delta I/I$, is proportional to the electron temperature and density, and to the characteristic dimension (line-of-sight depth) of hot gas: $\Delta I/I \propto n_e T_e r_c$, where r_c is the cluster core radius. In these same variables, the thermal bremsstrahlung luminosity of the gas in an X-ray energy band ΔE, centered on energy $E < kT$ may be expressed as $L_x \propto n_e^2 T_e^{-1/2} r_c^3$. From these expressions one finds $L_x \propto (\Delta I/I)^2 T_e^{-5/2} r_c$. The luminosity distance is therefore expressible in terms of five observables: X-ray flux, the microwave decrement, electron temperature, the cluster redshift, and a characteristic angular size of the scattering region,

$$D_L = C_L F_x (\Delta I/I)^2 T_e^{-5/2} \theta_c / (1 + z)^2 \tag{4}$$

where we have used the relation, $D_L = (1+z)^2 D_A$ between luminosity distance and angular size distance in order to write $\theta_c = r_c / D_A$ (cf Weinberg 1972). The coefficient, C_L, in equation 4 depends primarily on fundamental constants, but weakly on element abundances in the gas. In this way one sees in principle how to calculate the distance to a particular cluster if these observables are accessible to practical observation. The extension of this technique to higher redshift clusters is limited only by their detectability; and immunity from cluster evolution is provided by making no assumptions regarding class characteristics, but rather the distance estimate is based on source-by-source observations.

But, as we shall see, a more detailed examination of this prescription reveals that measurement of the X-ray surface brightness profile is a key element as well, making this study depend crucially on the ability to record X-ray images, and which is currently possible only with the Einstein observatory. Having dwelt on advantages, we also admit to difficulties which lie primarily with practical implementation. These problems are best illustrated by the following review of recent work on A2218 by Bob Schommer, Steve Murray and myself (1980).

II. CLUSTER OBSERVATIONS

Abell (1958) classified A2218 as a richness class 4, distance class 6 cluster, subsequently determined to have a redshift of 0.164 (Spinrad

1979), and catagorized by Leir and van den Bergh (1977) as Bautz-Morgan class II although it contains a large cD galaxy at $\alpha = 16^h\ 35^m\ 37^s$, δ = 66 18' 44" (epoch 1950, Bruzual and Spinrad 1979). The fundamental data for our study consist of X-ray images obtained by the Imaging Proportional Counter (IPC) and High Resolution Imager (HRI). Details of the detector systems and capabilities of the Einstein Observatory have been published b Giacconi et al. (1979). An IPC exposure of 1535 seconds has yielded a net count of 347±25 photons in the 0.5 to 3.0 keV energy band, within 10 arcminutes from the source centroid. Assuming an HI column density of $0.3X10^{21}$ cm, we calculate a flux (0.5 to 3.0 keV) of $5.4X10^{-10}$ erg $cm^{-2}s^{-1}$ for thermal bremsstrahlung spectrum with kT between 5 and 50 keV (consistent with the IPC spectral data).

An observed core diameter of roughly 2 or 3 arcminutes indicated the necessity of an HRI image to provide information on the distribution of X-ray surface brightness. In our $3.6X10^4$ second HRI exposure of the cluster, a secondary source is evident at the edge of the NW quadrant, which apparently coincides with a star located at $\alpha = 16^h\ 34^m\ 0^s.5$, $\delta = 66°$ 24' 22". A two-dimensional cross-correlation of our HRI map (12 arcsecond resolution) with the cluster galaxy positions given by Dressler (1976) reveals no significant X-ray emission from galaxy sites other than from th giant cD galaxy near the cluster center, which contributes less than 6% of the total flux.[1] For the remainder of our analysis, those image elements associated with the cD galaxy (four 12 arcsecond bins) have been excised. A subsequent calculation of the source centroid yields $\alpha = 16^h\ 35^m\ 42^s.4$, $\delta = 66°$ 18' 02" (1950) with an uncertaintly of about 10 arcseconds. This location is distinct from that of the central cD galaxy, the two being separated by roughly one arcminute. Departure from circular symmetry abou the centroid was examined by partitioning the HRI image into 30 azimuthal bins. From a χ^2 of 7 calculated on deviations from the mean counts/bin (1 degrees of freedom) we infer no statistically significant departure from symmetry. Although this test is not particularly stringent, the radial surface brightness profile shown in figure 1 with 15 arcsecond resolution indicates a smooth, centrally condensed intracluster gas distribution similar to that observed for A1795 and A85 (Jones et al. 1979), and strongly suggests spherical symmetry. For comparison, in figure 2 we show the surface density of galaxies computed from member coordinates given by Dressler (1976).

For our purpose, we tentatively adopt the microwave decrement measurement of Birkinshaw and his colleagues, $\Delta I/I = 4X10^{-4}$ observed with a beam width of 4.5 arcminutes (FWHM) and beam switching angle of 15 arcminutes. For a discussion and comparison of the measurements of all observers refer to the review by Lake (1979).

III. POLYTROPIC GAS MODELS

Of those observables required for the distance calculation we have discussed only the direct measurement of the X-ray flux and the microwave decrement. Moreover, determination of the electron temperature in the

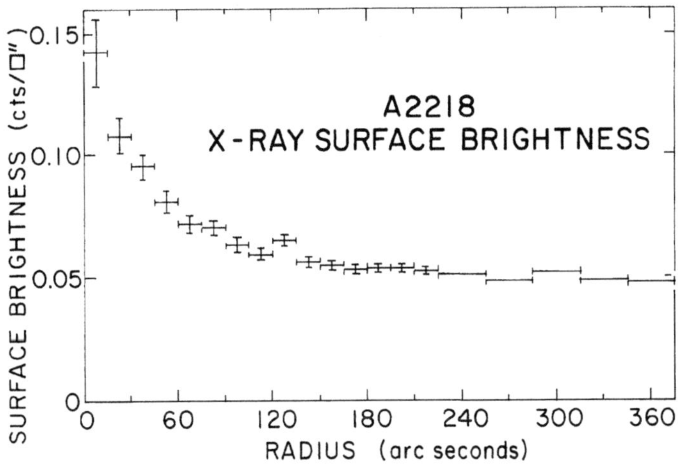

Figure 1. X-ray surface brightness profile for Abell 2218.

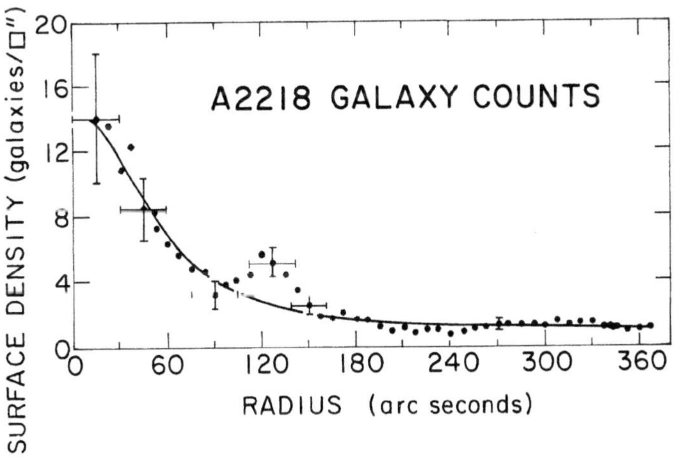

Figure 2. Surface density of galaxies. Dots represent density in 30 arcsecond, overlapping radial bins. Typical error bars (for independent bins) are shown. The curve indicates the best fit King model.

intracluster medium would seem out of reach with current instrumentation since non-imaging detectors cannot distinguish sources at the flux level of A2218 because of source confusion and background domination. On the other hand the imaging detectors aboard Einstein have insufficient response above 3 keV to be able to provide estimates of cluster gas temperatures (typically in the 7-8 keV range) with precision useful for this purpose.

Finally, we also have yet to provide even an adequate definition of the characteristic angular radius of the cluster core.

A partial remedy for these shortcomings is to recognize that more information on the "shape" of the gas distribution than expressed in the simple parameter θ_c is available in the X-ray and optical images of the source and is required for a correct formulation of the distance estimator The integral along the line of sight through the cluster gas expressing th microwave decrement as a Compton scattering process, $\Delta I/I \propto \int T_e n_e \hat{r} \cdot d\underset{\sim}{r}$, mus also include averaging over the plane of the sky according to the approximately gaussian response function of the radio antenna, $G(\theta;\theta_B,\theta_T)$, with half-power half-beam width of θ_B, and beam throw θ_T. Invoking spherical symmetry, we rewrite the integral as a dimensionless form factor which embodies the "shape" of the projected radial dependence of n_e/n_o and T_e/T_o as well as that of the antenna pattern:

$$\Delta I/I = C_{\mu} n_o T_o r_c \int \theta_c^2 \left[\int (n_e/n_o)(T_e/T_o)(\hat{r}/r_c) \cdot d\underset{\sim}{r} \right] \theta/\theta_c \, G(\theta/\theta_c ; \theta_B/\theta_c, \theta_T/\theta_c) \, d(\theta/\theta_c)$$

or,

$$\Delta I/I = C_{\mu} n_o T_o r_c [ff]_{\mu} \qquad (5)$$

Similarly, we may write the X-ray flux as:[2]

$$F_x = C_x n_o^2 T_o^{-1/2} r_c^3 \int (n_e/n_o)^2 (T_e/T_o)^{-1/2} (r/r_c)^2 d(r/r_c) \; /D_L^2$$

or,

$$F_x = C_x n_o^2 T_o^{-1/2} r_c^3 [ff]_x /D_L^2 \qquad (6)$$

Consequently, the distance estimator may be written in the form:

$$D_L = C_L' (\Delta I/I)^2 (1 + z)^{-2} F_x^{-1} T_o^{-5/2} \theta_c \; [ff]_x/[ff]_{\mu}^2 \qquad (7)$$

We now see that the calculation of distance requires knowledge of the radial dependence of both n_e/n_o and T_e/T_o in order to evaluate the form factor ratio, as well as knowledge of the other observables on the right hand side of equation 7.

In order to proceed we are forced to additional assumptions beyond spherical symmetry for the gas distribution:

1. The potential well in which the gas is trapped is "traced" by the mass distribution of galaxies and is also spherically symmetric.
2. The gas is not clumped. That is, perturbations in density and temperature on scales much smaller than the core radius are negligible.
3. The galaxy distribution is well approximated as that of an isothermal gas sphere out to several core radii.
4. The cluster core gas is in approximate hydrostatic equilibrium and is well described by a polytrope relation.

Spherical symmetry and hydrostatic equilibrium seem to be indicated by the circular symmetry of the X-ray image and the smooth, centrally condensed form of the surface brightness profile. But the fine-scale structure of the source is also an issue. If the gas is not clumped, then the form factor ratio, which is approximately $\langle n^2\rangle/\langle n\rangle^2$ for the situation we will meet, expresses only the smooth radial dependence of density given by the large-scale profile. On the other hand, gas clumping could profoundly increase the value of $\langle n^2\rangle/\langle n\rangle^2$ and render the distance estimate meaningless if derived under the assumption of no clumping. It is therefore essential to detect clumping if present at a significant level. The obvious sites of enhanced density would be the local potential minima presented by the cluster galaxies. As already mentioned, examination of the cross-correlation between galaxy positions and X-ray surface brightness indicates that the average perturbation in surface brightness at galaxy sites is less than 5% (90% confidence) and is therefore negligible. A more general yet fairly stringent test for surface brightness structure is provided by constructing an array of image bins with typical dimension of 15 arcseconds defined in a circular-symmetric pattern about the source centroid. Bin to bin fluctuations are found to be consistent with counting statistics, yielding a chi-square of 149 on 150 degrees of freedom. Formally this result indicates a 90% confidence upper limit of about 20% on excess variance due to clumping on any scale. For moderate clumping contrast ($\langle n^2\rangle/\langle n\rangle^2 > 4$), this upper limit on excess fluctuation translates to a lower limit on the number of clumps per 15 arcsecond bin, and is roughly equal the number of X-ray counts per bin (10-20). Thus any putative "clump" population exceeds the surface density of cluster galaxies by an order of magnitude.

We also subscribe to the notion that at least the central portion of the galaxy distribution may be described as an isothermal gas sphere (Rood et al. 1972), and that this distribution is well approximated by the relation:

$$(\rho/\rho_o)_{gal} = [1 + (\theta/\theta_c)^2]^{-3/2} \tag{8}$$

out to several core radii (King 1972). In imposing these simplifications we follow the work of Cavaliere and Fusco-Femiano (1978) who have already formulated a self-consistent description of an equilibrum gas-galaxy system with freedom for accomodating various energy injection and transport mechanisms by choosing a polytropic relation between gas density and pressure: $P \propto \rho^{\gamma}$. They find that

$$(\rho/\rho_o)_{gas} = T/T_o = 1 - 1.5\tau[(\gamma-1)/\gamma]\ \ln[1 - (\theta/\theta_c)^2] \tag{9}$$

where $\tau = \mu m_H \langle v_{\parallel}^2\rangle/kT$ is essentially the ratio of galaxy to gas scale heights. This characterization of the intracluster gas provides a concise description of the radial dependence of electron density (for fully ionized matter) and temperature through three parameters γ, θ_c (or r_c), and τ. By posing a thermal bremmstrahlung model for X-ray emission, the surface brightness profile may be computed

$$B(\theta) = A_X \int (n_e/n_o)^2 (T_e/T_o)^{-1/2} (\hat{r}/r_c) \cdot d\underset{\sim}{r} + B_{X\,bkg} \tag{10}$$

where the detector background is expressed as an equivalent brightness. Then by performing a least squares relaxation of this function onto the data in figure 1, the three parameters of interest (in addition to the X-ray background and an overall amplitude for the surface brightness variation) can, in principle, be determined.

Neither Cavaliere and Fusco-Femiano in their studies of the Coma cluster nor Gorenstein et al. (1977,1978) in their related treatment of Virgo and Perseus clusters, possessed data of adequate quality to provide a meaningful relaxation of all model parameters. In this first attempt at a full solution we demonstrate the considerable diagnostic capability of this technique in revealing the physics of the gas state for suitably chosen clusters.

We find in practice however, that parameters determined by fitting to the X-ray surface brightness alone may be highly correlated. This problem is controlled by providing an additional constraint on θ_c through the measured distribution of galaxies displayed in figure 2. Thus, we extend the fitting procedure to include a simultaneous least squares minimization of the measured (projected) surface density of galaxies relative to the King model:

$$\sigma(\theta) = A_\sigma[(1 - (\theta/\theta_c)^2]^{-1} + B_{\sigma\,bkg} \tag{11}$$

where again the amplitude and background are free parameters. The total sum of squares is then minimized on 28 optical and X-ray data points through the adjustment of 7 parameters to give a χ^2 of 26 on 21 degrees of freedom. The values of the gas parameters determined by the fit are both reasonable and interesting. The polytropic index $\gamma = 1.05\pm0.1$ indicates an isothermal gas at least out to several core radii. This result contradicts the assumption of adiabaticity by Birkinshaw, Gull and Northover (1978) in their treatment of the intracluster medium in A2218. A value of τ less than unity indicates the gas is somewhat hotter and therefore more extended than the galaxy distribution. The particular value $\tau = 0.6\pm0.1$ for A2218 is reminiscent of that assigned to the Coma cluster by Cavaliere and Fusco-Femiano (1978), and leads directly to a central gas temperature given the radial velocity dispersion of the member galaxies: $T_o = \mu m_H \langle v_\parallel^2 \rangle / \tau$. Spinrad (1979) finds $\langle v_\parallel^2 \rangle^{1/2} = 1400\pm200$ km/sec, yielding $T_o = 15\pm5$ keV, which is substantially hotter than the "typical" 6-8 keV measured gas temperature for nearby clusters (Mushotzky et al. 1978). Furthermore, this high temperature is consistent with the 1 to 2 mK antenna temperature decrement claimed for A2218 by Birkinshaw, Gull and Northover (1978), and revised somewhat higher by Schallwich and Wielebinski (1978) in correcting for radio emission from a source in the cluster core region. Finally, the inferred angular core radius of ~60 arcseconds is consistent with previous optical determinations (Dressler 1979, Spinrad 1979).

Many questions arise in assessing the validity of the assumptions which form the foundation of this technique for determining gas parameters. For instance, do the member galaxies truly trace the potential in which the gas lies? The best overall evidence is perhaps the consistency of derived results; however, we note (without interpretive comments) a common perturbation to both X-ray surface brightness and galaxy surface density which occurs at a radius of ~130 arcseconds in figures 1 and 2. These features are not traceable to any significant excess in particular azimuthal bins which were summed to produce these radial plots. Incidently, these two high bins were not incorporated in the least squares fit.

We have resorted to thermal bremsstrahlung model fitting in order to complete the set of parameters which must be evaluated in order to calculate the cluster distance according to equation 7. Within the range of validity of our basic assumptions, this procedure has successfully yielded estimates of these remaining quantities: the central gas temperature, T_o, cluster core angular radius, θ_c, and the radial dependence of both n_e/n_o and T_e/T_o, enabling the form factor ratio to be evaluated; all based largely on a modest image composed of ~1400 detected X-rays.

IV. DISTANCE MEASURE

We rewrite the distance formula using the definition of τ, and partition observables which are directly and independently measured from those inferred from the data through least-squares fitting:

$$D_L = C_L' (\Delta I/I)^2 F_x^{-1} (1 + z)^{-2} \langle v_{\parallel}^2 \rangle^{-5/2} (\tau^{5/2} \theta_c [ff]_x / [ff]_{\mu}^2) \quad (12)$$

The latter set are correlated to various degrees, and for computational convenience are lumped together in one function, $\beta(\gamma,\tau,\theta_c) = \tau^{5/2}\theta_c [ff]_x / [ff]_{\mu}^2$. By considering β as equivalent to a parameter in the fitting procedure, a confidence interval may be computed directly by using the prescription of Lampton et al. (1977). This value, when combined with independently measured quantities, $\Delta I/I = 4X10^{-4}$, $F_x = 5.4X10^{-10}$ ergs $cm^{-2} sec^{-1}$, and a redshift of 0.164, yield a distance of ~1000 Mpc for A2218, which can be interpreted as a value of about 50 km sec Mpc for the Hubble constant.

In a sense this result is quite remarkable. A breakdown in any one or combination of assumptions could have resulted in a value for H_o much more than an order of magnitude removed from the nominally accepted range of 50-100 km sec^{-1} Mpc^{-1}. On the other hand, given the controversy over the measurement of the Sunyaev-Zel'dovich effect, it is premature to attach much significance to this result. At the present time we consider a quantitative appraisal of the capabilities of this distance estimation technique the major issue. Error propagation characteristics of the distance formula follow directly from equation 12:

$$\mathrm{var}D / D^2 = 4\, \mathrm{var}(\Delta I/I) / (\Delta I/I)^2 + 25/4\, \mathrm{var}\langle v_{\parallel}^2 \rangle / \langle v_{\parallel}^2 \rangle^2 + \mathrm{var}F_x / F_x^2 + \mathrm{var}\, \beta / \beta^2 \quad (13)$$

Clearly, the microwave decrement and radial velocity dispersion are major contributors. The expected variance in radial velocity dispersion measurements follows directly from the properties of the χ^2 distribution, $\mathrm{var}\langle v_{\parallel}^2\rangle / \langle v_{\parallel}^2\rangle^2 = 2/N$, where N is the number of member galaxies observed. Even for a nominal 30% precision in the other factors, the radial velocit contribution dominates the error budget unless $N > 50$. For the data at hand $N = 20$ and $\mathrm{var}D / D^2 = 1$. In fact, even for a "best case" situation which $\Delta I/I$, F_x, and β might be measureable to 10% and $N = 100$, one is confronted with $\mathrm{var}D / D^2 = 0.25$. Although this distance measuring metho has some distinct advantages (works at $z > 0.1$, is evolution independent, does not require source "calibration"); clearly there are disadvantages a well: supporting optical observations are tedious (N large), and the meth is intrinsically imprecise (precision, not accuracy is the issue). Application to roughly 30 suitable clusters would be necessary to provide formal one-sigma error bound of 20% on H_0.

Difficulty with adverse error propagation is not specific to this particular implementation of the general distance estimation scheme discussed in the introduction. Even if T_0 could be measured directly to a precision of 15%, a fractional distance uncertainty of 0.5 is still indicated. But what about q_0, and the possibility of extending this meth to large values of redshift, limited only by the X-ray visibility of the source?[3] Of course, source evolution may imply low gas density and/or temperature at higher redshift ($z > 1$) which affect visibility beyond impairment due to inverse square law and redshift effects. To provide so sense of the magnitude of the dependence of D_L on q_0, we consider the distance ratio of sources at $z = 0.1$ and $z = 1.0$ for two values of the deceleration parameter,

$$[D_L(z = 1.0)/D_L(z = 0.1)]_{0.05} = 1.2\ [D_L(z = 1.0)/D_L(z = 0.1)]_{0.5}$$

and find only a 20% effect in comparing the Einstein-deSitter case ($q_0 = 0.5$) with a highly "open" Universe ($q_0 = 0.05$). If we extend this comparison to $z = 2$, the effect is roughly 50% and might be observable if clusters are X-ray visible (or even exist) at $z = 2$. In this case we probably could not rely on optical data for redshifts or velocity dispersion measurements. But at $1+z = 3$, a 15 keV source spectrum looks like a keV spectrum at the detector, suggesting the possibility of direct measurement of source temperature from broad-band spectra and of the redshift from iron line observations. Although this project is beyond the capability of the Einstein observatory, the CCD detector and imaging optic planned for the Advanced X-ray Astronomy Facility (AXAF) represent a major advance in technology. If Abell 2218 were placed at $z = 2$, AXAF could provide an image with a comparable number of photons to that used in this study, with improved spatial resolution (and energy resolution as well in the band from 0.1 to 8 keV) with an exposure time of 2×10^5 seconds. If suitable clusters could be located,with sufficiently high values of $n_e T_e$ to provide a measurable microwave decrement, observation of 10 to 20 would be required to determine q_0 with a formal uncertainty of 0.2.

V. DISCUSSION

The approximate agreement between the distance to A2218 computed here and the customary value based on the Hubble law (H = 50-100 km sec^{-1} Mpc^{-1}) lends support to the general validity of the method and its underlying assumptions. However, the unfortunate error propagation characteristics of this calulation render its application impractical within the limitations of current technology. This judgement is based not only on the large number of clusters which must be surveyed in order to provide interesting precision in the determination of H_0 or q_0, but also the fact that each member of the list must have suitable characteristics: symmetry, approximate hydrostatic equilibrium, polytropic entropy profile, absence of finescale structure, and sufficiently high central temperature to provide a detectable S-Z effect. Since suitable clusters may comprise only a small fraction of the cluster population, the necessary scope of such a survey is compounded.

On the other hand, by simply inverting the distance measuring problem, the methodology described here becomes a rather powerful diagnostic technique for the study of the physical state of the intracluster medium, and consequently of dynamical evolution when extended over a range of cluster redshift. That is, by assuming an independent evaluation of the Hubble constant, the "distance relation" supplement by the X-ray image and $\Delta I/I$ may be solved for the central density and temperature. This same analysis determimes the gas state in terms of γ, τ and θ_c, which in turn define a temperature and density profile, and lead to estimates of the total gas mass, the thermal energy content and cooling time as well. This application is discussed further by Boynton, Schommer and Murray (1980)

The fact remains that the task of locating clusters with measurable Sunyaev-Zel'dovich effect is even more arduous than an X-ray survey seeking clusters with suitable properties for this type of analysis. Given the nature of the observing problem, the integration time for microwave detection may exceed that for X-ray detection by several orders of magnitude (over a thousand hours of radio telescope time already been devoted to the study of A2218). Yet only a frustratingly small fraction of clusters so surveyed have given even marginal evidence for inverse Comptonization. Consequently, as another extension of this analysis technique, we have also formulated a cluster preselection process utilizing only X-ray observations.

We conclude that although cosmological distance measure through combined X-ray and microwave observations of custers of galaxies may be currently impractical, closely related techniques for the study of intracluster gas at cosmological distances might prove worthwhile for cluster classification and evolution studies, as well as facilitating pursuit of the Sunyaev-Zel'dovich effect. In any case, for the applications discussed here, the imaging capability of the Einstein Observatory more than compensates for its limited energy bandwidth.

This research is funded by the National Science Foundation, the National Aeronautics and Space Administration, and the Alfred P. Sloan Foundation; but for support, I thank my friends at The Center For Astrophysics.

FOOTNOTES

[1]At Dresler's magnitude limit, only well within one core radius (~60 arcseconds) is the mean projected galaxy separation less than 12 arcseconds.

[2]For simplicity, we are ignoring the temperature dependence of the Gaunt factor in this discussion,but have included it in our actual analysis.

[3]For fixed gas density and temperature, the amplitude of the Sunyaev-Zel'dovich effect is redshift independent so long as the cluster core is over-resolved (Sunyaev 1977).

REFERENCES

Birkinshaw, M., Gull, S. F. and Northover, K. J. E. 1978, Mon. Not. R. Astr. Soc., 185, 245
Boynton, P. E. and Murray, S. S. 1978, HEAO-B Guest Investigator proposa
Boynton, P. E., Schommer, R. and Murray, S. S. 1980, in preparation
Bruzual, G. and Spinrad, H. 1979, private communication
Cavaliere, A., Danese, L. and De Zotti, G. 1979, Astr. Ap., 75, 322
Cavaliere, A. and Fusco-Femiano, R. 1978, Astron. Astrophys. 70, 577
Dressler, A. 1976, Ph.D. thesis
Dressler, A. 1979, private communication
Giacconi, R., et al. 1979, Ap.J. 230, 540
Gorenstein, P., Fabricant. E., Topka, K., Harnden, F. R. and Tucker, W. 1979, Ap.J. 224, 718
Gorenstein, P., Fabricant, D., Topka, K., Tucker, W. and Harnden, F. R. 1977, Ap.J. 216, L95
Gunn, J. E. 1978, in Observational Cosmology (eds. Maeder and Tamman), Geneva Observatory
Gursky, H., Kellogg, E., Murray, S., Leong, C., Tananbaum, H. and Giacconi, R. 1971, Ap.J. 167, L81
Jones, C., Mandel, E., Schwarz, J., Forman, W., Murray, S. S. and Harnden, F. R. 1979, Ap.J. 234, L21
King, I. R. 1972, Ap.J. 174, L123
Lake, G. 1979, Proceedings IAU Symposium 92, G. O. Abell and P. J. E. Peebles, eds., "Objects of High Redshift", p. 305.
Lake, G. and Partridge, R. B. 1977, Nature, 270, 502
Lampton, M., Margon, B. and Bowyer, S. 1976, Ap.J. 208, 177
Leir, A. A. and Van den Bergh, S. 1977, Ap.J. Suppl. 34, 381
Mushotzky, R. F., Serlemitsos, P. J., Smith, B. W., Boldt, E. A. and Holt, S. S. 1978, Ap. J. 225, 21

Parijskij, Yu. N. 1973, Sov. A.J. 16, 1048
Perrenod, S. C. and Lada, C. J. 1979, Ap. J. 234, L173
Rood, H. J., Page, J. L., Kinter, E. C. and King, I. R. 1972, Ap.J. 175, 627
Rudnick, L. 1978, Ap.J. 223, 27
Schallwich, D. and Wielebinski, R. 1978, Astron. Astrophys. 71, L15
Serlemitsos, P. J., Smith, B. W., Boldt, E. A., Holt, S. S. and Swank, J. H. 1977, Ap.J. 211, L63
Silk, J. I. and White, S. D. M. 1978, Ap.J. 226, L103
Spinrad, H. 1979, private communication
Sunyaev, R. A. 1977, Proceedings I.A.U. Symposium 74, D. L. Jauncey, ed. "Radio Astronomy and Cosmologie", p. 327.
Sunyaev, R. A. and Zel'dovich, Ya. B. 1972, Comments Ap. and Space Physics, 4, 173
Weinberg, S. 1972, "Gravitation and Cosmology" J. Wiley and Sons, Inc., New York

GALAXY FORMATION IN A VIOLENT UNIVERSE

Jeremiah P. Ostriker
Princeton University Observatory

ABSTRACT

Galaxies appear to have formed at a certain epoch, to display certain characteristic mass and size scales, and to have originated from quite small perturbations. A scenario is put forward to explain these facts which depends primarily on one assumption: the explosive energy released on the death of massive stars in forming stellar systems will propagate into the intergalactic medium. It then follows under certain circumstances, that a dense cooled shell will form with mass many times greater than the original "seed" system, whereupon gravitational instability and fragmentation of the shell can lead to the formation of new stellar systems. For $z \lesssim 5$ and for masses $10^8 \lesssim M/M_\odot \lesssim 10^{12}$ a very large amplification of original perturbations is possible with galaxies naturally forming in small groups having velocity dispersions of order 100 km/s (the velocity at cooling of radiative shells). Explosions before $z \approx 5$, which cool primarily due to the inverse Compton interaction with background radiation, will lead to production of very massive stars rather than galaxies.

The most interesting speculative predictions (none of which are unique to this model) are that: a) intergalactic space will be dusty with significant absorption likely for objects seen from $z > 2.5$; b) unvirialized groups should lie on two-dimensional surfaces; and c) black holes with masses 10^2-10^4 $M_\odot$ may be common and detectable.

INTRODUCTION

Why is it that there exist characteristic values for the physical scales of galaxies although no such scales are seen in larger astrophysical systems (cf. Peebles 1980)? The accumulated evidence strongly indicates that galaxies have typical masses of 10^{11}–$10^{12} M_\odot$, radii of order 10^4 pc, and that they usually form in small groups with velocity dispersion $\sim$ 100 km/s at the relatively recent epoch of $z < 5$. Gravity, by itself, contains no preferred scales. Recent work (for example,

R. Giacconi (ed.), X-Ray Astronomy with the Einstein Satellite, 311–323.

Rees and Ostriker 1977, Binney 1977, Silk 1977) has emphasized that clues to the physical scales may be found by consideration of thermal and gas dynamical processes. Thus, for example, if we require that radiative losses of a collapsing gas cloud be significant within a dynamical time, then we find that, regardless of initial conditions, the natural length and mass of a typical galaxy is derived.

It has been generally argued (for a recent summary see Carr and Rees 1979) that the scale sizes for luminous stars, planets and other cosmic bodies can be derived from the basic physical constants. When we study star formation in the galaxy, we examine the ensemble of non-linearly interacting physical processes occurring in a unit volume of the interstellar medium and attempt to calculate the rate of formation of gravitationally bound objects. It is customarily assumed that all memory of initial conditions, the precise spectrum of perturbations in the original collapsing galaxy, has been forgotten.

The initiation of bound galactic systems condensing out of the intergalactic medium has normally been treated in a totally different manner. Here one usually considers an arbitrary and unknown spectrum of initial fluctuations, tracks the growth, decay and interactions of various modes, asking when the typical perturbation of galactic scale will have achieved a sufficient overdensity to collapse and initiate star formation (cf. review by Gott 1977). In an alternate version due to Doroskevich et al. (1979) one follows larger scale perturbations (10^{15} $M_\odot$) which, at an epoch dependent on the initial amplitudes, form shocks and then fragment to galactic scales.

In this talk I explore a point of view towards galaxy formation closer in spirit to the way star formation is viewed. In particular, I hope to demonstrate that there is a natural hydrodynamic instability by which galaxy formation begets more galaxy formation and that the resulting properties of galaxies are relatively independent of the number and size of the "seeds" which may have initiated the process. The basic idea, which might be called "explosive amplification" is as follows:

1. If a bound stellar system (of mass M) forms at an epoch $z < 35$, the massive stars will have a lifetime short compared to the current Hubble time.
2. The explosive energy released on the death of these stars, E, measured in terms of an efficiency, $\varepsilon \equiv E/Mc^2$, can be estimated by a variety of arguments [cf. Bookbinder et al. 1980, $\varepsilon = 10^{-3.1}$ $(M/L)^{-1}$] at 10^{-5} to 10^{-4}.
3. If this energy propagates out into the intergalactic medium as an adiabatic blast wave which eventually cools when the shock velocity is v_c, then a mass will be swept up in the Oort "snowplow" phase of order $M \times \varepsilon(c/v_c)^2$ or an amplification η of order 10^2-10^3.
4. If this happens in a timescale short compared to the current Hubble time, and if the dense cooling shell is unstable to fragmentation and formation of new bound systems, there will be rapid amplification of the amount of bound material per unit volume.

It appears that all of the above conditions will be satisfied at certain epochs; there is an effective hydrodynamic amplification. In the rest of this talk I plan to enumerate some of the important physical processes, note the cosmological "windows" within which given mass scales will amplify significantly, and describe the physical environment and fossil remains left by each epoch.

The most basic of these points is (2), the energy released in galaxy formation. Of the various ways of arriving at the proposed efficiency the simplest is to consider the metallicity of elliptical galaxies and the spheroidal parts of spiral galaxies. For the massive systems which are associated with most of the cosmical mass the metallicity is about solar $Z \approx 0.02$. Then simple arithmetic shows that converting two percent of the mass of a galaxy from helium to carbon (or heavier) will produce approximately 10^{51} ergs per present solar luminosity, much of it in supernova explosions and all of it on a timescale comparable to the dynamical times in the central regions of these systems ($\sim 10^7$ yrs). A similar result is obtained more indirectly by extrapolating back the luminosity function in spheroidal systems to the time when massive stars were present (cf. Ostriker and Thuan 1975). An idea of the significance of this number is that an ordinary galaxy with luminosity L_* produced 10^{60} to 10^{61} ergs in supernovae, an energy output comparable to a quasar. Alternatively, since at an early epoch quasars were as common as massive galaxies, one can derive an efficiency of 10^{-4}– 10^{-5} from quasar input alone.[1]

There are several different epochs to be considered. Before $z \approx 30$ the Compton drag force (cf. Peebles 1971) will prevent significant hydrodynamical rearrangement of matter. Before $z \approx 10$ the cooling time was less than the lifetime of the most massive and short-lived stars.

A. The first period considered is thus $10 \gtrsim z \gtrsim 5$ which might be designated the Compton Cooling Epoch since gas heated in this interval immersed in the background radiation can be effectively cooled to the black body radiation temperature. Because of this efficient cooling and the related high densities in the cooling shells, the Jeans mass is very low and we expect that fragmentation will lead, not to normal galaxies, but to massive star formation. The remains of this epoch are condensed remnants and a certain degree of chemical pollution.

B. In the next interval $z \approx 5$, the Gas Cooling Epoch, amplification of small scale density fluctuations to galaxy dimensions proceeds rapidly. Masses greater than $\sim 10^{12}\ M_\odot$ will not be able to cool and those less than $\sim 10^9\ M_\odot$ are not Jeans unstable. Thus objects of normal galactic scale are made. The expanding blast waves overlap when a fraction of order $1/\eta$ of the gas has been formed into galaxies.

C. In the next epoch $5 \gtrsim z \gtrsim 4$ one expects that the IGM will resemble the present interstellar medium with cool ($T = 10^{4\circ}$) clouds embedded in a much hotter medium. Galaxy formation increases the pressure in the IGM causing the most massive clouds to collapse and the least massive ones to evaporate. Most of the cool gas is processed into galaxies, but a comparable mass ($\Omega_{gas} \sim 0.05$) is left in a hot IGM.

D. Then ($4 > z$) adiabatic expansion lowers the pressure more rapidly than explosions can increase it. The galaxy formation rate decreases rapidly and one is left with a spectrum of pressure bounded low mass clouds with properties close to those inferred by Sargent et al. (1980) from the QSO absorption lines.

The present work is the result of a collaboration with L. L. Cowie at Princeton. We have discussed several specific aspects of this general problem before. The overlapping of intergalactic blast waves (part of phase B) was investigated in Schwarz, Ostriker and Yahil (1975) and the initial parts of individual blast waves in Bookbinder et al. (1980). An early version of phases C and D was presented by J.P.O. at the 1979 Copenhagen conference on the Universe at Large Redshift, and much of the relevant theoretical background appropriate to the analogous galactic interstellar medium has been developed in a series of papers by Cowie, McKee and Ostriker in various combinations.

PHYSICAL PROCESSES

Since we are only interested in the universe for epochs $z < 10$ and believe that the present closure parameter Ω_0 is ~ 0.2, we will for the sake of definiteness, utilize an open Friedman model with density $\rho = 1.8 \times 10^{-29}\ \Omega_0 h^2 (1 + z)^3$ gm/cm^3 and radiation temperature $T_{BB} = 2.7(1 + z)$°K where h is the Hubble constant in units of 100 km/s Mpc and at any epoch, n_0 is the hydrogen particle density with helium equal to $0.1\ n_0$ by number. The explosion will be expected to go through the same sequence of stages as a supernova explosion in the interstellar medium (cf., for example, Chevalier 1974): a) constant velocity of ejecta until swept up mass equals ejecta mass; b) adiabatic expansion of a spherical blast wave (Sedov-Taylor solution):c) sudden formation of dense cooling shell; d) approach to pressure equilibrium with the surrounding medium. Here we are interested in phases b) and c). If we express the energy input from a representative explosion as $10^{61} E_{61}$ ergs, then the shock radius during the adiabatic phase is $R = 3.3\ E_{61}^{1/5} [h^2 \Omega_0 (1 + z)^2]^{-1/5} t_{10}^{2/5}$ Mpc after a time of $t_{10} \times 10^{10}$ years. For gas of this composition and temperature, 10^5°K $< T <$ 10^6°K, the cooling time will be (for eras A and B, respectively)

$$\tau_{cool,10} = \frac{240}{(1 + z)^4}, \quad \frac{0.9 T_5^{\ 2.2}}{h^2 \Omega (1 + z)^3} \tag{1}$$

due to inverse Compton or gas-radiative processes. The conditions at the time of dense shell formation are determined by equating the cooling time to the age:

$$R_{cool} = 30(1+z)^{-2.2} E_{61}^{-0.2} \Omega_0^{-0.2} h^{-0.4} \quad , \qquad (2a)$$

$$3.3\ E_{61}^{+0.29} \phi^{-0.40} \quad \text{Mpc}$$

$$\tau_{cool,10} = 240(1+z)^{-4} \quad , \quad 1.0\ E_{61}^{0.24} \phi^{-0.51} \quad \text{yrs} \qquad (2b)$$

$$v_{cool} = 5(1+z)^{+1.8} \Omega_0^{-0.2} E_{61}^{-0.4} \quad , \qquad (2c)$$

$$130\ E_{61}^{0.054} \phi^{+0.11} \quad \text{km/s}$$

$$M_{cool,11} = 3 \times 10^5 (1+z)^{-3.6} E_{61}^{0.6} \Omega_0^{0.4} h^{0.8} \quad , \qquad (2d)$$

$$42\ E_{61}^{0.88} \phi^{-0.21}$$

where for the sake of brevity we have defined the repeated function $\phi \equiv h^2 \Omega_0 (1+z)^3$. The temperature at cooling is given by $86 v_{cool}^2$ (km/s) °K. Two other quantities of interest are the dimensionless amplification factor $\eta \equiv (M_{cool}/M_{seed})$ and the Mach number $\mathcal{M}_{cool}$ defined to be the velocity of the shock divided by the isothermal sound speed in the cool shell:

$$\eta = 1.0 \times 10^5 (1+z)^{-3.6} \varepsilon_{-4}^{0.6} M_{seed,11}^{-0.4} \Omega_0^{+0.4} h^{0.8} \quad , \qquad (3)$$

$$730\ \varepsilon_{-4}^{0.88} M_{seed,11} \phi^{-0.21}$$

$$\mathcal{M} = 35(1+z)^{1.3} E_{61}^{-0.4} \Omega_0^{-0.2} (T_c/T_{BB})^{-1/2} h^{-0.4} \quad , \qquad (4)$$

$$16\ E_{61}^{0.05} \phi^{0.11} T_{c4}^{-1/2}$$

where T_{c4} is the temperature of the cool shell in units of 10^4 °K in the radiative era, (T_c/T_{BB}) is the temperature in units of the black body temperature in the inverse Compton era and $\varepsilon_{-4} \equiv \varepsilon/10^{-4}$. These factors are expected to be close in unity.

As expected, the amplification factor is large and, in the radiative era, rather insensitive to epoch and seed mass. However, arbitrarily large explosions cannot cool in a Hubble time. The maximum energy explosion that can cool will produce a radiative shell with mass

$$M_{cool,max} = 4.2 \times 10^{13} h^{-0.35} \Omega_0^{1.68} (1 + z)^{1.34} \varepsilon_{-4}^{-0.1} M_\odot \qquad (5)$$

It is attractive that this is larger than the mass of most galaxies and comparable to the mass of groups within which galaxies are typically formed. The corresponding seed mass is $M_{s,max} = 0.56 \times 10^{11} h^{0.092} \varepsilon_{-4}^{-1} \Omega^{2.4} (1 + z)^{2.23}$. The cooling time for an explosion from a seed smaller than $M_{seed,max}$ is correspondingly shorter than the Hubble time [as $(M/M_{s,max})^{1/4}$] and the amplification large so that in one Hubble time all seeds regardless of initial size should have grown to the maximum possible size producing finally shells of mass $\approx 10^{13.5} M_\odot$. In addition to checking that there is time for an explosion to cool we must also check that the neglect of the Hubble expansion in the treatment of blast waves was justified. It is easy to show that if this is not the case ordinary cooling shells will not form (because $E_{Hubble} > E_{explosion}$), but that if the criterion $t_{cool} < t_{Hubble}$ is satisfied then it also follows that $v_{cool} > v_{Hubble}(R_{cool})$ and neglect of the cosmological expansion is justified.

Next let us look in greater detail at the inverse Compton cooled era. We see that if $(1 + z) > 6.2\ h^{1/3}$, the Compton cooling time is shorter than the Hubble time and this will be the dominant cooling process for large explosions. The earliest epoch at which amplification can occur is determined by the condition that the amplification factor must be greater than unity. The shortest energy injection time we can consider is that appropriate for the lifetime of massive stars or $\sim 10^7$ years, which corresponds to an epoch $z_{max} = 20$. Had we, more conservatively, taken $t_{evol} = 10^8$ years we would have found $z_{max} = 10$. The other limit ($\eta > 1$) can be expressed as $M_{mass} = 3 \times 10^{23} (1 + z)^{-9} \varepsilon_{-4}^{-1.5} \Omega_0 h^2$ which for $z = 20$ is $6 \times 10^{11} M_\odot$. It is interesting to note that the number of Compton cooling times available within a local Hubble time increases rapidly with z so that the number of cycles possible increases with z. The result of this is that, if a disturbance forms at $z > 10$ and remains isolated, it will always reach a characteristic shell mass at $z = 5$ of $M_{fin} \approx 10^{15} \varepsilon_{-4}^{-0.6} \Omega_0^{0.4} M_\odot$ for any initial seed mass.

In order to have amplification of the original seed, the shell must fragment so we now consider the gravitational instability of a dense cool shell (radius R, velocity v_s). We assume for definiteness in this section, since seeds must still be forming at the epoch in question, that $\Omega(z_f) \approx 1$. Specifically, we shall consider that $z_f \approx 5$. We will base our arguments on rough, but probably fairly accurate energy considerations. A detailed discussion is given elsewhere (Ostriker and Cowie 1980). Here we note that there is normally a range of unstable wavelengths, all of which are large compared to the shell thickness but small compared to its radius. The masses of the most rapidly growing fragments are

$$M_m = 24(1+z)^{3.2} E_{61}^{-1} \Omega_0^{1/2} h^{-3.2} \quad , \qquad (6)$$

$$2.5 \times 10^8 E_{61}^{0.18} h^{-1.22} (1+z)^{0.18} \Omega_0^{2.78}$$

for the two cases of interest. This gives fragments the size of dwarf ellipticals or Magellanic irregulars ($\sim 10^9$ $M_\odot$) in the radiatively cooled era but objects of order $10^{3.5}$ to $10^{5.5}$ $M_\odot$ in the inverse Compton era.

Amplification will always proceed rapidly unitl $(t/T) \gtrsim 1$ when the largest (and slowest cooling) explosions occur. In the radiative era this will always lead to bound groups with sizes of order 1 Mpc, velocity dispersion of order 200 km/s and with the largest bound objects of order 10^{12} $M_\odot$ made of fragments (some of which will remain as satellites) of order 10^9 $M_\odot$. Explosions occurring in the inverse Compton era will produce very massive stars which, according to existing theories will burn some fuel and then implode to black holes, contaminating their surroundings with some debris which should be characteristic of that processed in "little big bangs" (cf. Wheeler 1977, Wallace and Woosley 1980).

OVERLAP AND THE GENERATION OF A HOT INTERGALACTIC MEDIUM

There are two possible routes which the subsequent evolution may take. If the seed formation ceases owing to the decrease in IGM density prior to the overlap of the hot cavities, the individual galaxies will be enclosed in hot bubbles embedded in a cool IGM. This possibility, while elegant in predicting the relation between the epoch of galaxy formation and Ω_0, does not appear consistent with the observations. In particular, the present number density of galaxies is too high for this to be the case. However, it may be relevant to local regions of low galaxy number density where quiescent cool IGM could continue to exist.

As many galaxies form, overlap between the hot bubbles generated in the final noncooling explosions will take place. This will result in the formation of a hot intergalactic medium with cool material remaining as embedded clouds. This process is analogous to that now believed to occur in the formation of hot gas in the interstellar medium [Cox and Smith 1974, McKee and Ostriker 1977].

We may now estimate the fraction of gas mass processed into galaxies during this stage. If we consider hot regions at overlap, their age will be uniformly distributed between the age at which the last outburst occurs (t_1) to that at which the next will occur ($t_2 \gg t_1$). Thus on average the ratio of the mass in the hot bubble to that in the generating explosion is $M_{gal}/M_{hot} \approx \eta^{-1}$ or about .2%.

A consequence of the above result is that at overlap most regions are within a factor 2 of being radiative and hence that the temperature of the IGM at overlap will be a few times 10^5 °K, irrespective of the intergalactic density. The medium is thermally unstable with timescale only slightly greater than the current age of the Universe. Since the remnants at overlap are near cooling, we may expect a spectrum of cold gas regions to form in the hot intergalactic gas. Of order 1/2 of the IGM would end up in this phase as dense clouds in a tenuous coronal medium after the overlap process.

There is now an additional source of galaxies available in the cold material residual to the overlap process. Geometric considerations suggest that the cold remaining regions will have a mass spectrum stretching to the mass of the overlapping bubbles and dominated by the high mass end of the spectrum. As the pressure in the IGM rises these will themselves become gravitationally unstable, collapse and release yet more energy into the IGM. It is easy to conclude that this is a fast process on the cosmological timescale. Whether this or the prior amplification process is the dominant galaxy producing machine is not of great significance, since the mass spectrum will be similar in the two cases, and in particular the upper bound on the cloud masses should be less than, but of order of, the mass initially in the hot cavities.

The two stages together allow us to set bounds on the fraction of mass formed into galaxies to that in gas. We conclude that $0.002 < \Omega_{gal}/\Omega < 0.5$. It is probable that the fraction lies towards the upper end of this range (i.e., that most galaxy formation occurs in the second stage and that this is efficient). For $\Omega_0 = 0.2$ this provides good agreement with observation. We shall subsequently adopt a value of $\Omega_{gal} = 0.05$ F in our numerical estimates. In this case the density of matter in galaxies at the epoch of formation is $\rho_{gal}(z_f) = 9 \times 10^{-31}(1 + z_f)^3 Fh^2$ gm cm^{-3}. The pressure generated in the hot IGM is then

$$P_{IGM}(z_f) = \frac{2}{3}\, \varepsilon \rho_{gal}(z_f) c^2$$

$$= 5 \times 10^{-14}(1 + z_f)^3 \varepsilon_{-4} Fh^2 \text{ ergs cm}^{-3} \qquad (7)$$

$$P_{IGM}(z) = P_{IGM}(z_f) \left(\frac{1 + z}{1 + z_f}\right)^5$$

The temperature of the intergalactic medium is

$$T(z_f) = 1.3 \times 10^8 \varepsilon_{-4} F \left[\frac{.15}{\Omega_{gas}}\right] \tag{8}$$

$$T(z) = T(z_f) \left(\frac{1 + z}{1 + z_f}\right)^2$$

The spectrum of the clouds themselves is in turn determined by this rise in pressure and temperature in the intergalactic medium. In order to survive clouds must satisfy the basic physical constraints that they do not gravitationally collapse and that they are not destroyed by the hot intergalactic medium (cf. Sargent et al. 1980). The constraints are most stringent at the epoch of formation when the pressure and temperature of the IGM is highest, so that the cloud mass spectrum is essentially frozen after this point.

An upper bound to the cloud mass is given by the condition of stability to gravitational collapse (Mestel 1965) $M_{crit} = 1.3\ c^4/G^{3/2}P^{1/2}$ where c is the isothermal sound speed in the cloud and P is the external pressure. This constrains the cloud masses to be less than

$$M_{crit}(z) = 3 \times 10^8 (1 + z)^{5/2} (1 + z_f)^{-4} \varepsilon_{-4}^{-1/2} F^{-1/2} L^{-1}\ M_0 \tag{9}$$

or about $10^7\ M_\odot$ for $z_f = 5$.

If clouds are too small they may be thermally evaporated. Evaporation is highly saturated (Cowie and McKee 1977) and we estimate an evaporation time,

$$t_{evap} = 2 \times 10^9 \left(\frac{\Omega}{.15}\right)^{3/2} M_7^{1/3} (F\varepsilon_{-4})^{1.83} (1 + z_f)^{-1}\ , \tag{10}$$

where M_7 is the mass of the cloud in units of $10^7\ M_\odot$. The evaporation rate and the various parameters are sufficiently uncertain in this regime that we probably cannot rule out the existence of $10^7\ M_\odot$ clouds, however much smaller clouds would be rapidly evaporated.

The properties of the clouds are ($T_{cl} = 10^4\,°K$)

$$n_{cl} = 1.5 \times 10^{-2}\ F\varepsilon_{-4} h^2 (1 + z_f)^{-2} (1 + z)^5\ \mathrm{cm}^{-3} \tag{11a}$$

$$R_{cl} = 2000\ M_7^{1/3}(1 + z_f)^{+2/3}(1 + z)^{-5/3} \times h^{-2/3}(F\varepsilon_{-4})^{-1/3}\ \mathrm{Pc} \qquad (11b)$$

$$N_{cl} = 10^{20}\ M_7^{1/3}(1 + z_f)^{-4/3}(1 + z)^{10/3} \times h^{4/3}(F\varepsilon_{-4})^{5/3}\ \mathrm{cm}^{-2} \qquad (11c)$$

These properties are close to those deduced by Sargent et al. (1980) for their intergalactic clouds on the basis of the statistics of Lyman α absorption lines in quasars. However, the temperature of the intercloud gas is a factor of 30 times higher than that of Sargent et al.

DISCUSSION

The Intergalactic Gas

Grain Opacity: The forming protogalaxies eject metals along with energy into the intergalactic medium. If dust survives this expulsion process it may blanket observations at larger redshifts and render some of the remaining observations discussed in this and the following subsection impossible.

We first note that a fraction of the grains probably do survive the expulsion from the galaxy. We may estimate the effective exposure time to the hot gas as (nt), where n is the density of the hot gas in the galactic wind and t the flow time, to be $\overline{nt} = \Delta M(R)/4\pi R v^2\ \bar{m}$ where $\Delta M(R)$ is the mass loss within radius (R), v the outflow velocity, τ the mass loss time and $\bar{m}$ the average mass per hydrogen atom. Bookbinder et al. (1980) estimate for an L_* galaxy that $v = 10{,}000\ \mathrm{km\ s^{-1}}$ and that at 10 kpc, $\tau = 10^7$ years and $\Delta M = 1.6 \times 10^{11}\ M_\odot$. Here we obtain $nt = 4 \times 10^4$ years $\mathrm{cm^{-3}}$ comparable to the critical value of $2 \times 10^4\ \mathrm{yr\ cm^{-3}}$ estimated by Draine and Salpeter (1979) for survival of 0.01 μ refractory in a hot gas. Since comparable amounts of gas are lost at larger radii where grains more easily survive, it seems certain that a substantial fraction of dust will survive and in addition a minimum of 10% of ejected grain material will be present at any time.

How significant is this opacity? If we assume one L_* galaxy density of .01 $\mathrm{Mpc^{-3}}$ and that each galaxy ejects $2 \times 10^{11}\ M_\odot$ of gas of twice cosmic metallicity, the equivalent hydrogen density at cosmic value of the metals in the IGM is approximately $10^{-7}\ \mathrm{cm^{-3}}$ at the present time. We adopt the relation (Mathis et al. 1977) $\tau(\lambda)/N_H = 1.6 \times 10^{-22}\lambda_\mu^{-1}$ for the dust optical depth (absorption and extinction) at wavelength λ_μ in microns for hydrogen column density N_H, valid for interstellar grains except in the neighborhood of the 2200 Å bump. We then find

$$\tau(z,\lambda) = 10^{-28.5}\left\{\frac{5000\ \text{A}}{\lambda_0}\right\}\phi\int_0^z (1+z)^4\,\frac{cdt}{dz}\,dz$$
$$\approx 0.1\left\{\frac{5000\ \text{A}}{\lambda_0}\right\}\phi[(1+z)^3 - 1] \tag{12}$$

for the optical depth to an object at redshift z at the observed wavelength λ_0. Here ϕ is the fraction of surviving grains. An equivalent normalization is obtained if one uses as calibration the metals seen in gas in the Coma cluster divided by the visual light of the cluster.

This optical depth is interesting since it assumes a significant value for high redshift quasars. McKee and Petrosian (1974) have measured positive curvature effects in the spectra of a sample of low redshift quasars ($z < 2$) to obtain an upper limit to τ. However, the predicted effect is much larger in large redshift quasars ($z \sim 3$) and we have estimated τ by a method analogous to that of McKee and Petrosian in the sample of Osmer (1979) finding good support for the hypothesis in the observed curvature of the spectra. Conservatively all the τ's could be considered upper limits following McKee and Petrosian since positive curvature may be intrinsic to the quasar spectra rather than owing to extinction and Lyman α absorption, although primarily confined to a small region near the rest wavelength of Lα, will produce some curvature. However the presence of this curvature at the theoretical level together with an indication of the expected increase of τ with z suggests that the IGM is indeed dusty, and distant objects significantly obscured. We estimate ϕ = 0.1 to 0.3.

The apparent absence of large redshift quasars ($z > 3.5$) could be understood quite simply as due to dust obscuration. They should still be seen in X-rays since, for any reasonable intergalactic conditions, the IGM is transparent for $E(z = 1) > 1$ keV (Shapiro and Bahcall (1980). But if indeed our estimate of intergalactic absorption is correct (ϕ = 0.1-0.3) then at z = 5, $\tau(\lambda_0 = 5000\ \text{A})$ = 1.2 to 3.7 and all optical observations discussed subsequently would not be possible.

Optical Observations: Photons emitted at optical wavelengths from the cooling shells are unimportant but harder emitted photons may be a powerful contributor to the diffuse background light at op-ical and UV wavelengths. Emitted at 10 eV, these are redshifted into bands where observational constraints are most severe. If a fraction x of the injected energy is radiated (this is an upper bound to the fraction of IGM which cools), then the photon energy density at the present time is

$\varepsilon = 1.4 \times 10^{-14} xFh^2\{6/(1 + z_f)\}$ ergs/cm^3; Dube et al. (1977) give an upper limit of 2.9×10^{-15} ergs cm^{-3} at 5.5×10^{14} Hz giving $x < 1/6$. This fraction is just consistent with our estimate of the fraction of IGM formed into galaxies.

Numbers and Luminosities of Protogalaxies

The integrated luminosity of the protogalaxy is around 5×10^{61} ergs on the basis of the metallicity arguments used to estimate the energy release into the IGM. Denoting the period of maximum luminosity as 10^7 years $\times\ \tau_7$ we obtain $L_{max} = 2 \times 10^{46}$ ergs/s τ_7^{-1} for a galaxy of present luminosity L_*. The total number of such galaxies on the sky is $N = [4\pi d_L^2/(1 + z_f)]n_0 c$, where d_L is the usual luminosity distance. At $z_f = 5$, $d_L = 5 \times 10^5$ Mpc the number of sources per square degree is $4000\ \tau_7$ at a bolometric apparent magnitude of $22 + 2.5 \log \tau_7$; for $z_f = 2$, the corresponding numbers are $d_L = 1.2 \times 10^4$ Mpc, # = $500\ \tau_7$ per sq deg at a magnitude of $19 + 2.5 \log \tau_7$. The images will be semi-stellar irrespective of z_f. In the absence of intergalactic dust these are comparable to present observations (cf. Kron 1979, Davis and Wilkinson 1974). As we have quoted in the previous subsection, extinction may substantially decrease the expected number if $z_f = 5$ but not if $z_f = 2$.

Spatial Distribution of Galaxies

It is clear from the geometrical arguments presented in section 2 that unvirialized groups of galaxies (those with $t_{crossing} > t_{Hubble}$) should tend to lie on two-dimensional surfaces and have velocity dispersions of order 100–200 km/s.

The Dark Matter

A possible outcome of explosions in the Compton Cooled Epoch is production of a significant number of black holes in the mass range $10^2 - 10^4\ M_\odot$. These may be observable as gravitational lenses in front of distant objects (cf. Press and Gunn, 1973, Gott 1980) or as nearby accreting X-ray or XUV sources (cf. Carr 1978). A more definitive discussion of model predictions and test awaits the result of detailed numerical simulations now in progress.

ACKNOWLEDGEMENTS

The author would like to thank L. L. Cowie who collaborated in this work and P. J. E. Peebles, R. D. Blandford and R. McCray for giving the benefit of their thoughts and especially C. McKee for significant contributions to this work. Thanks also are due to the National Science Foundation for grant AST79-22074.

FOOTNOTE

1. This was pointed out to me by Dr. Roger Blandford.

REFERENCES

Bahcall, J. N. and Shapiro, P. R.: 1980, Preprint.
Binney, J. J.: 1977, Astrophys. J. 215, 483.
Bookbinder, J., Cowie, L. L., Krolik, J. H., Ostriker, J. P. and Rees, M.: 1980, Astrophys. J. 237, 647.
Carr, B. J.: 1978, Comments on Astrophys. VII, 161.
Carr, B. J. and Rees, M. J.: 1979, Nature, 278, 605.
Chevalier, R. A.: 1974, Astrophys. J. 188, 501.
Cowie, L. L. and McKee, C.: 1977, Astrophys. J., 211, 135.
Cox, D. P. and Smith, B. W.: 1974, Astrophys. J. Lett. 189, L105.
Davis, M. and Wilkinson, D.: 1974, Astrophys. J. 192, 251.
Doroskevich, A. G., Sunyaev, R. A. and Zeldovich, Y. B.: 1979, Proceedings IAU Symposium #79, ed. M. S. Longair (Reidel: Dordrecht), p. 213.
Draine, B. T. and Salpeter, E. E.: 1979, Astrophys. J. 231, 438.
Dube, R. R., Wickes, W. C. and Wilkinson, D. T.: 1977, Astrophys. J. Lett. 215, L51.
Gott, J. R.: 1977, Ann. Rev. Astron. and Astrophys. 15, 235.
Gott, J. R.: 1980, Preprint.
Kron, R. G.: 1980, Physica Scripta 21, 652.
Mathis, J. S., Rumpl, W. and Nordsieck, K. H.: 1977, Astrophys. J. 217, 425.
McKee, C. F. and Ostriker, J. P.: 1977, Astrophys. J. 218, 148.
McKee, C. F. and Petrosian, V.: 1974, Astrophys. J. 189, 17.
Mestel, L.: 1965, Quart. J. Roy. Astron. Soc. 6, 161.
Osmer, P. S.: 1979, Astrophys. J. 227, 18.
Ostriker, J. P. and Cowie, L. L.: 1980, submitted to Astrophys. J.
Ostriker, J. P. and Thuan, T. X.: 1975, Astrophys. J. 202, 353.
Peebles, P. J. E.: 1971, Physical Cosmology (Princeton Univ. Press: Princeton, NJ).
Peebles, P. J. E.: 1980, The Large Scale Structure of the Universe (Princeton Univ. Press: Princeton, NJ).
Press, W. H. and Gunn, J. E.: 1973, Astrophys. J. 185, 397.
Rees, M. and Ostriker, J. P.: 1977, Monthly Notices Roy. Astron. Soc. 179, 541.
Sargent, W. L. W., Young, P. J., Boksenberg, A. and Tytler, D.: 1980, Astrophys. J. Suppl. 42,41.
Schwarz, J., Ostriker, J. P. and Yahil, A.: 1975, Astrophys. J., 202, 1.
Shapiro, P. and Bahcall, J. N.: 1980, Preprint.
Silk, J.: 1977, Astrophys. J., 211, 638.
Wallace, R. K. and Woosley, S. E.: 1980, Lick Observatory Bull. #801.
Wheeler, J. C.: 1977, Astrophys. and Sp. Science 50, 125.

SUBJECT INDEX

GPSR Compliance

The European Union's (EU) General Product Safety Regulation (GPSR) is a set of rules that requires consumer products to be safe and our obligations to ensure this.

If you have any concerns about our products, you can contact us on ProductSafety@springernature.com

In case Publisher is established outside the EU, the EU authorized representative is:

Springer Nature Customer Service Center GmbH
Europaplatz 3
69115 Heidelberg, Germany

Batch number: 10607222

Printed by Printforce, the Netherlands